D0850683

STUDIES ON THE DEVELOPMENT OF BEHAVIOR AND THE NERVOUS SYSTEM

Volume 1

BEHAVIORAL EMBRYOLOGY

CONSULTING EDITORS

STUDIES ON THE DEVELOPMENT OF BEHAVIOR AND THE NERVOUS SYSTEM

Volume 1

BEHAVIORAL EMBRYOLOGY

Edited by

GILBERT GOTTLIEB

Psychology Laboratory
Division of Research
North Carolina Department of Mental Health
Raleigh, North Carolina

ACADEMIC PRESS New York and London 1973

A Subsidiary of Harcourt Brace Jovanovich, Publishers

ACADEMIC PRESS, INC.
111 Fifth Avenue, New York, New York 10003

United Kingdom Edition published by
ACADEMIC PRESS, INC. (LONDON) LTD.
24/28 Oval Road, London NW1

LIBRARY OF CONGRESS CATALOG CARD NUMBER: 72-12194

PRINTED IN THE UNITED STATES OF AMERICA

CONTENTS

Section 3 HATCHING: HORMONAL, PHYSIOLOGICAL, AND BEHAVIORAL ASPECTS

Section 4 SENSORY PROCESSES: EMBRYONIC BEHAVIOR IN BIRDS

Prenatal Origins of Parent–Young Interactions in Birds: A Naturalistic Approach

Monica Impekoven and Peter S. Gold

LIST OF CONTRIBUTORS

Numbers in parentheses indicate the pages on which the authors' contributions begin.

WALTER L. BAKHUIS (245), Netherlands Central Institute for Brain Research, Amsterdam, The Netherlands

MICHAEL BERRILL (141), Biology Department, Trent University, Peterborough, Ontario, Canada

MICHAEL A. CORNER (245), Netherlands Central Institute for Brain Research, Amsterdam, The Netherlands

RAINER F. FOELIX (103), Neuroembryology Laboratory, Division of Research, North Carolina Department of Mental Health, Raleigh, North Carolina

PETER S. GOLD (325), Department of Biology, State University of New York at Buffalo, Buffalo, New York

GILBERT GOTTLIEB (3), Psychology Laboratory, Division of Research, North Carolina Department of Mental Health, Raleigh, North Carolina

VIKTOR HAMBURGER (51), Department of Biology, Washington University, St. Louis, Missouri

MONICA IMPEKOVEN (325), Institute of Animal Behavior, Rutgers University, Newark, New Jersey

RONALD W. OPPENHEIM (103, 163), Neuroembryology Laboratory, Division of Research, North Carolina Department of Mental Health, Raleigh, North Carolina

ROBERT R. PROVINE (77), Department of Psychology, Washington University, Saint Louis, Missouri

MARGARET A. VINCE (285), Cambridge Psychological Laboratory, University of Cambridge, Cambridge, England

CORA VAN WINGERDEN (245), Netherlands Central Institute for Brain Research, Amsterdam, The Netherlands

PREFACE

The plans for the present works on the topic of prenatal neural and behavioral development got underway when it became apparent that none of the numerous books on neurobehavioral development deal with prenatal ontogeny except in a highly selective or rather superficial way. One of the main reasons for this state of affairs is the fact that the field of behavioral embryology is peopled by representatives of various disciplines who do not have a professional background in common and who tend to publish their findings in their own specialty journals. Thus, a significant amount of the literature in this field is spread over more journals, published in more countries and in more languages than many other areas of study, and it is not until one begins to try to gather all the ends together that one realizes there is a "field" and something approximating a coherent body of knowledge which is being contributed to rather regularly by many active researchers.

The first two (companion) volumes in this serial publication are meant to present a fairly basic orientation to some of the more persistent and most important "philosophical," theoretical, and empirical problems of behavioral embryology. Also included are a few studies on new and interesting animal groups, and some new and different approaches, to give some sample of the less conventional, more ground-breaking aspects in the field.

There is a regrettable omission in this introductory volume. An article on "Human Fetal Behavior and It's Relation to Prenatal and Postnatal Development" remained undone as a terminal illness took the life of our revered colleague, Professor Tryphena Humphrey. She was the single most active worker on problems of human prenatal neurobehavioral development for a period of over 30 years, and the majority of what we know and understand about that topic is due to her efforts and that of her long-time senior collaborator, the late Davenport Hooker. We miss her personally as well as professionally, and the introductory volumes are the poorer for the lack of her contribution and editorial advice. In succeeding volumes we hope to treat Dr. Humphrey's topic and others such as the genetic aspects of neuroembryology, the prenatal "organizing" effect of gonadal hormones on the brain and later behavior, and so on.

The problem of defining the field or scope of behavioral embryology is grappled with in the introductory article. Suffice it to say here that the findings of neuroembryology, neurogenesis, or development neurobiology are candidates for inclusion, especially as their relation to sensory, motor, or central neural function, overt embryonic or fetal sensitivity, overt motility, and actual behavior becomes clear. Behavioral embryology is neither exclusively reductionistic nor constructionistic in strategy, and it is fairly widely recognized that both tactical approaches are necessary. The limitations are, as is usual, partly personal. For example, it is to be hoped that more neuroembryologists may gradually become convinced that behavioral study need not necessarily be "soft," and that more behaviorists may eventually see neuroembryological study as something other than "irrelevant" or frighteningly esoteric.

The Editor is grateful for the generous assistance and expert counsel of John I. Johnson, Jr., Ronald W. Oppenheim, Viktor Hamburger, Stanley M. Crain, Marcus Jacobson, and Joseph Altman in the selection and editing of contributions for this and the next volume in the serial publication. Mary Catharine Vick provided cheerful and competent assistance in copy editing.

Gratitude is also expressed to the Animal Behavior Society and the American Society of Zoologists for sponsoring a symposium on the Prenatal Ontogeny of Behavior and the Nervous System at the annual meeting of the American Association for the Advancement of Science in Philadelphia, Pennsylvania, from December 28 through December 30, 1971. That meeting gave the contributors to Volumes 1 and 2, and other interested persons, an opportunity for considerable fruitful interchange in formal and informal discussions. The National Science Foundation and the American Society of Zoologists provided financial assistance toward the travel of the foreign participants and that support is gratefully acknowledged.

GILBERT GOTTLIEB

W. Preyer

DEDICATION TO W. PREYER (1841–1897)

This volume is dedicated to the memory of the British-born, psychological physiologist William Preyer who, with the publication of his monumental tome, *Specielle Physiologie des Embryo*, in 1885, launched the field of study which we now call behavioral embryology. Preyer's book is broadly comparative and it combines physiological, neurological, behavioral, and psychological approaches to the study of fetuses and embryos of various species, traditions which are still evident today but rarely in the practice of any single investigator. While Preyer is celebrated here for his prenatal ontogenetic studies, it is his self-designated companion book, *Die Seele des Kindes*, published in 1882 in Leipzig, for which he is perhaps best known. That work went through numerous editions in the original German language version and a number of editions in the English translation as well—it was translated by H. W. Brown and published in New York by D. Appleton and Company in two volumes in 1888 and 1889, under the title *The Mind of the Child*.

While the embryo book is remarkable for its diverse coverage, the child book is remarkable for its singularity. *The Mind of the Child* is based on a diary recounting the daily activities of Preyer's only child (a son) from birth to the end of his third year. In Preyer's words, the unifying theme of both these books is the problem of "psychogenesis" (the genesis of mind) as studied from the natural scientific point of view, and, as such, both monographs show the extraordinary power and reach of the developmental method of study, especially in the hands of a broadly trained, widely read, and acute observer such as their author. Preyer's books are notable for their thorough and critical integration of the *world* literature on the topic under discussion. His *Specielle Physiologie des Embryo*, for example, contains over 500 references by more than 400 authors, extending back to Aristotle.

It is of especially great interest in the present context that Preyer's *Die Seele des Kindes*, when compared to other works, is regarded by some authorities as providing ". . . the greatest stimulation for development of modern ontogenetic psychology" (Norman L. Munn, page 4, in the second edition of his text *The Evolution and Growth of Human Behavior*). Edwin G. Boring, in his classic *A History of Experimental Psychology*, notes that it was Preyer, along with a few other notable psychologically minded physiologists such as Helmholtz, Hering, and von Kries, who helped to establish the new natural science outlook of psychology in 1890 in Europe by joining the editorial board of the *Zeitschrift für Psychologie und Physiologie der Sinnesorgane*, a new journal founded by the German psychologist Hermann Ebbinghaus and his physiologist colleague Arthur König in an effort to isolate the scientific study of psychology from the metaphysical impingements of philosophy. It is somewhat ironic, to say the least, that it was *Die Seele des Kindes* and not the *Specielle Physiologie des Embryo* which appears to have kindled the great embryologist Hans Spemann's interest in biology, according to a passing remark in his autobiography *Forschung und Leben*.

It is most unfortunate from a historical standpoint that only the barest details of Preyer's life have survived. He was born near Manchester, England on July 4, 1841, and named William Terry Preyer. (He spent most of his adult life in Germany where he became known as Wilhelm Thierry.) Preyer acquired his D. Phil. in 1862 and for several years thereafter worked in the laboratory of the noted physiologist Claude Bernard in Paris. In 1865, he did advanced study in Zoophysics and Zoochemistry while on the Philosophical Faculty at the University of Bonn, and he received his medical degree in 1866. In 1869, at the relatively young age of 28, Preyer was promoted to professor of physiology at Jena in Germany, where he remained until 1888. At that time he resigned his chair to move to Berlin which, according to Boring, provided a more attractive intellectual atmosphere for him. Around 1893 Preyer became chronically ill and moved to Wiesbaden, Germany, where he died on July 15, 1897, at the age of 56.

Preyer's physiological interests were unusually varied and he brought an excellent background to the study of prenatal development. His early published research was on the biochemistry of hemoglobin, other gaseous aspects of blood, and muscle physiology (1868–1871), as well as color vision (1868). In 1876 and 1879, he determined the lower limit of hearing in a comparative survey which included human subjects, and he also described the response of the pinna to sound in certain mammalian species (sometimes called the Preyer reflex in the modern literature). In 1877, Preyer published on the causes of sleep, and in the next year published the first of at least four works on hypnotism (both animal and human), one of these being a German translation (1882) of Braid's original English text on the subject. It was also in 1882, as already noted, that Preyer published what has been far and away his most popular book, *Die Seele des Kindes*, which went through eight editions (1912) in the German language and a number of editions in English.

(The photograph of Preyer in the present book appeared as the frontispiece to the German edition of 1905.) In 1883, he published a treatise on general physiology; two years later he published his second book on development, *Specielle Physiologie des Embryo. Untersuchungen über die Lebenserscheinungen vor der Geburt* (Special Physiology of the Embryo: Investigations on the Phenomena of Life before Birth), the tome which has such historical and conceptual importance for the field of behavioral embryology.

Shortly after his move to Berlin in 1889, Preyer published a monograph on the general topic of (what would be called in English) *Contemporary Questions in Biology*, followed one year later by another treatise titled, simply, *Der Hypnotismus*, which was published simultaneously in Vienna and Leipzig. In 1893, he produced *The Genetic System of Chemical Elements* and his third book on development, *Die geistige Entwicklung der ersten Kindheit*, published simultaneously in Stuttgart and New York, where it was titled *Mental Development in the Child*, with H. W. Brown again serving as translator and Appleton as publisher. In 1895, Preyer wrote *On the Psychology of Writing*, and, in 1896, he produced a sympathetic, popular work on *Darwin: His Life and Works*. This incomplete list of Preyer's own works gives a fair idea of his breadth as well as his literary fecundity.

Although it is virtually sheer guesswork, it is of some interest to speculate on the external intellectual influences, if any, which may have prompted Preyer to turn his attention to the problem of development, especially prenatal development. Jane Oppenheimer, in her fascinating and highly readable *Essays in the History of Embryology and Biology*, suggests that Preyer was influenced by the recapitulation theory of Ernst Haeckel ("ontogeny recapitulates phylogeny"), who was a colleague of Preyer's at Jena. Preyer's 1882 and 1885 books on development, however, give little evidence of such an influence, neither in their intent nor in their theoretical summation. Whereas, as Professor Oppenheimer makes clear, Haeckel saw embryonic developmental study as solving *other riddles* (e.g., confirming or establishing evolutionary relationships among different animal groups), Preyer saw such study as necessary to solving the riddles of development itself (e.g., "psychogenesis"). Although Preyer was certainly congenial to Darwin's writings (as already noted, he wrote a popular book on Darwin's life and works), it does not seem that he was influenced by Darwin in any very direct way. One inclines to the necessarily tentative conclusion that Preyer merely saw and adopted the developmental approach for what it is: the only way to understand the origin of any function or structure in the life of the individual. He went to the study of embryos and fetuses, it would seem, because in his own view psychogenesis was dependent on physiogenesis, and he knew at the outset that physiogenesis begins before birth. Preyer was certainly not the first person to make behavioral observations on embryos and fetuses, but he was the most comprehensive and possibly the most original (he devised an oöscope whereby he could view the chick embryo's activity inside the egg, for example).

As mentioned before, Preyer regarded his 1882 and 1885 books on the child and embryo, respectively, as companion works, the division being made for the sake of convenience of exposition. The theme which unites these two works is, in Preyer's words, the problem of "Psychogenesis" or the genesis of mind. Based on his investigations of sensory function and overt motor behavior in embryos and fetuses, Preyer came up with his motor primacy theory of behavioral development which he enunciated in both the child and the embryo books. Before describing that view, it is pertinent to note that Preyer aptly summed up the guiding premise of his investigations (and ours) in one sentence in the preface to his 1882 child book: "Above all, we must be clear on this point, that the fundamental activities of mind, which are manifested only after birth, do not originate after birth." Another one of our modern premises from the same pen: "Heredity is just as important as individual activity in the genesis of the mind. No man is in this matter a mere upstart, who is to achieve the development of his mind through his individual experience alone; rather must each one, by means of his experience, fill out and animate anew his inherited endowments . . ."

Preyer's motor primacy theory arose from his observations of the chick embryo and other forms which indicated that all of these embryonic organisms are active before they are reactive—that is, the earliest movements stem exclusively from ". . . the organic processes going on in the central organs of the nervous system, especially the spinal cord, and take place without any peripheral excitement of any of the sensory nerves. To these belong the remarkable, aimless, ill-adapted movements of the arms and legs of children just born, and their grimaces [from page 334 of Part I of *The Mind of the Child.*]" The autogenous character of early movement received significant verification and extension in the exhaustive behavioral studies of H. C. Tracy on fish embryos in the 1920s—at least some species appear to develop even swimming movements before they are capable of responding to externally applied sensory stimulation. In more recent years the motor primacy viewpoint has been at last unequivocally confirmed by the elegant experimental work of the neuroembryologist Viktor Hamburger and his young psychologist colleagues Martin Balaban and Ronald Oppenheim, much of which is reviewed in the present volume. And, in the chick embryo at least, Preyer's hypothesis about the importance of the spinal cord in autogenous motility has received splendid verification in the recent virtuosic work of the psychologist Robert Provine, which is also reported in the present volume.

Preyer's *Specielle Physiologie des Embryo*, in which the empirical aspects of motor primacy and other views are considered in great detail, has rarely received its due acclaim. The outstanding neuroembryologist George E. Coghill believed that this state of affairs was in part the consequence of a lack of an English translation, so he and a German-language professor, W. K. Legner, undertook to translate those portions (about 150 pages) of Preyer's book which deal with embryonic movement and sensory function in particular. This translation was duly published in 1937 in the *Monographs of the Society for Research in Child Development*, Vol. 2, no. 6, pp. 1–115, under the title "Embryonic Motility and Sensitivity"—a truly scholarly and appreciative gesture of one outstanding scientist for the work of another. It is clear that Coghill recognized Preyer as his forerunner in the establishment of behavioral embryology as a field of study—in the preface to the translation of the embryo book Coghill wrote: "*Specielle Physiologie des Embryo* (Leipzig 1885) by W. Preyer is an epochal work in its field." That book may yet take its place next to—or perhaps even above—its companion piece *Die Seele des Kindes* in historical importance for modern studies of behavioral development. Toward this end, it seems particularly fitting to note that Preyer's *Spezielle Physiologie des Embryo* is being reprinted by the Zentralantiquariat der Deutschen Demokratischen Republik. It is scheduled for publication in Leipzig next year, some 89 years after it first appeared in that city.

Section 1

BEHAVIORAL EMBRYOLOGY

INTRODUCTION TO BEHAVIORAL EMBRYOLOGY

GILBERT GOTTLIEB

Psychology Laboratory
Division of Research
North Carolina Department of Mental Health
Raleigh, North Carolina

I. Aims

Behavioral embryology, which incorporates neurogenesis and developmental neurobiology, involves the study of the very early development of the nervous system and behavior with a view toward understanding how the formative periods of neural and behavioral development affect later stages of neurobehavioral ontogeny. The guiding philosophy is that neural and behavioral development at any given point in time can only be comprehended fully in light of the immediate and remote developmental history of the organism. For a truly comprehensive picture, the "forward reference" of development must also be considered. A most important and pervasive aspect of embryonic behavior is its "anticipatory" or "preparatory" nature —crucial adaptive functions always develop well in advance of their necessity for the survival of the newborn, and several writers have emphasized that aspect of development in particular (e.g., Anokhin, 1964; Carmichael, 1970; Coghill, 1929).

A subsidiary aim of behavioral embryology involves the establishment of detailed and intimate relationships between neuroanatomy, neurophysiology, neurochemistry, and behavior. It is felt that these relationships can be established most readily and most meaningfully during the formative stages of embryonic development, at which time the investigator is in a position to actually observe the increasingly complex changes in organization manifest themselves. A "naturalistic" theme pervades behavioral embryology in that most studies involve living specimens in their ordinary surroundings and, as far as is possible, there is an attempt to relate the results of *in vitro* studies to the *in vivo* and *in situ* conditions.

To paraphrase the words of Pearl (1904), the study of the ontogenetic history of an organism is regarded of prime importance in elucidating the adult condition. This method of study can gain the complete explanation of many structures and functions which are inexplicable when only the adult condition is considered. Thus, in many quarters, embryological study has come to be regarded as a necessary part of almost any anatomical, physiological, or behavioral investigation which aims at completeness, including human psychology. [See, for example, the recent review of behavioral embryology by Trevarthen (1973) for *The Handbook of Perception.* Carmichael's classical review of the older literature has been a standard feature of handbooks of child psychology for many years (Carmichael, 1933, 1970).]

In sum, the developmental method is basic to all disciplines which deal with organisms, whether from the genetic, biochemical, anatomical, physiological, behavioral, or psychological points of view, and behavioral embryology pushes this method of study to its logical extreme. The developmental method is an analytic tool *par excellence.*

II. Tradition

Since its inception in the modern era by W. Preyer (1885), to whose memory this book is dedicated, the field of behavioral embryology has remained broadly comparative in its interests and has aligned itself closely with the study of neurogenesis. These two trends are manifest in the work of the contributors to the first two volumes in this serial publication: the species under study include crustaceans, amphibians, birds, and various placental mammals, and many of the behavioral observations on these forms speak at least indirectly to the neural mechanisms which may underlie them (and vice versa).

While, in a very real sense the study of prenatal behavioral development entails problems (methodological and theoretical) peculiar to itself, and the same can be said of neurogenesis, there is an attempt to try to unite the relevant aspects of these two disciplines in the study of behavioral embryology.

This attempt at unified study has obvious advantages, but it also entails "communication" problems which are attributable to the background of the research workers in the field. To wit, most of the contributors in behavioral embryology have been trained almost exclusively either in biology or in psychology. The former bring great expertise to the study of neurogenesis, while the latter bring sophistication to the study of behavior. The twain do meet, of course, but when they do the most "natural" and flagrant biases or blind spots frequently surface. This sort of problem, which has important implications for both theory and method, can not be entirely rectified until a new generation appears on the scene, all of whom have a broad-based education in neuroembryology as well as behavior (developmental psychobiology).

III. Procedures

Regardless of the particular aims and aspirations of individual researchers in behavioral embryology, work proceeds almost inevitably through two stages, the first of which is necessarily descriptive-correlational and the second of which is causal-analytic or deductive-manipulative.

A. Descriptive-Correlational Stage

In this preliminary stage, cross-sectional relationships are established between behavior and age (or stage), or nervous system and age (stage), eventually yielding a fairly complete longitudinal picture (in ideal instances) of behavior–stage–nervous system correlations. The behavior in question can be a highly specific activity (e.g., the manifestation of tactile or other sensitivity, sucking reflex, locomotion) or rather general movements (such as the cyclic motility of avian and other embryos). The investigator usually tries to establish the onset and subsequent course of the activity in question and, if he or she is in a position to do so, goes on to designate its neuroanatomical, neurophysiological, and/or neurochemical counterparts. In the not too distant future, it will probably become common to push such correlations to the molecular domain, which will be quite exciting and most rewarding for those who are competent to do it.

Another approach employed during the descriptive stage of inquiry is peculiar to developmental study and provides an exceedingly fertile basis for the derivation of hypothetical "mechanisms" which can later be put to experimental test. Namely, after a number of cross sections have been compiled into a longitudinal picture of embryonic neural and/or behavioral development, the investigator begins to see possible functional relationships between earlier and later states of the system such that a certain earlier state (or event) comes to be regarded tentatively as the "precursor" to a particular later state. Such precursors come in two varieties: facilitative and determinative. *Facilitative precursors* exercise a direct and specific quantitative or temporal influence on the later state, such that quantitative or temporal alterations in the earlier state produce quantitative or temporal changes in the later state. That is, these precursors affect the *rate* of behavioral development or neural maturation, the *number* or *size* of neural elements, changes in behavioral or neural *threshold*, the *degree* of behavioral or neural differentiation, etc. [Some examples: the rate of "clicking" within and between quail embryos regulates their time of hatching (Vince, this volume); increased postnatal stimulation leads to an increase in the density of dendritic spines in neonatal rats (Schapiro and Vukovich, 1970); reduction of auditory input in duck embryos leads to an elevated behavioral threshold of response to species-specific auditory stimulation after hatching (Gottlieb, 1971a).] *Determinative precursors* differ from facilitative ones in that they force or channel neurobehavioral development in one direction rather than another, and the particular direction taken is a consequence of the action of the precursor, such that if the particular precursor had not been active, the particular direction would not have been taken. Genes, hormones, and the inductors of classical embryology are prime examples of such precursors, of course, but there is as yet little firm evidence for determinative precursors (motor and sensory stimulative) at the behavioral level in prenatal neurobehavioral ontogeny. Possible examples of determinative precursors in the sensory-perceptual realm are noted later (in Section V,D on Sensory and Perceptual Processes), especially in premetamorphic insects and amphibians. In the motor sphere, Kuo (1967, p. 47) hypothesizes that it is the yolk sac which forces the legs and feet of the chick embryo into a position which is essential for the normality of later postnatal postures such as sitting, standing, and walking. Further, Kuo (1967) goes on to say that the positional relationship of the yolk sac to the legs influences the kind of alternating movements which the embryo's limbs make and ". . . that this process of embryonic development is precursory to the posthatch walking pattern with its alternating and coordinated movements of the legs [p. 48]."

The descriptive-correlational approach is a necessary first stage in the mode of inquiry in behavioral embryology just as it is in other disciplines. This approach provides the baseline information which is essential to the development of meaningful hypotheses and a further delineation or explanation of the phenomenon in question. Thus, although the descriptive-correlational approach often provides highly suggestive and otherwise significant insight into the wherewithal of organismic function at any given stage of development, this approach alone can not conclusively demonstrate the precursors or developmental mechanisms involved. For that we must employ what has been traditionally called the causal-analytic method in experimental embryology, or what is here called the deductive-manipulative method (merely to signify that it usually entails the controlled experimental testing of hypotheses which have been derived from the initial stage of inquiry).

B. Deductive-Manipulative Stage

In this stage, the investigator sets about the testing of hypotheses (or hunches) derived from the descriptive-correlational stage, and this is done by experimentally altering earlier neural, chemical, mechanical, sensory stimulative, or behavioral states and predicting (or merely examining) the effect of such alterations on the later states of larval, embryonic, fetal, postmetamorphic, or even postnatal development. This approach can conclusively demonstrate mechanisms of development, whereas the descriptive-correlational approach can only suggest them. Thus, it is in the deductive-manipulative stage that the investigator determines the precise nature or function of the tentative precursors or mechanism derived from the descriptive-correlational stage of inquiry.

Up to now there have been relatively few deductive-manipulative experiments in behavioral embryology. The chief reasons for the paucity of experimental studies are (*1*) the primarily empirical (descriptive-correlational) orientation of many researchers and (*2*) the lack of profusion of hypotheses and theories of behavioral embryology. Another reason is that the experimental manipulation of embryos and fetuses of placental mammals is virtually impossible at the moment; thus, research with these specimens must be necessarily of a descriptive-correlational nature. Embryos of birds and reptiles, and the larvae of amphibians, fish, and certain invertebrates, are much more accessible to experimental study, so it is with these forms that the greatest number of manipulative studies are performed.

IV. Conceptions of Development

Before turning to the very few lively theoretical issues in neurobehavioral development, it is most appropriate to delineate contemporary conceptions of behavioral embryological development. These conceptions are the assumptions behind the theoretical issues and as such are rarely stated explicitly. Though they are to a large extent implicit, such conceptions (assumptions) play a most important role in scientific inquiry of any kind in that they guide experimentation via the construction of hypotheses.

A. Epigenesis

All modern workers in behavioral embryology share the assumption, now well documented, that embryonic neural and behavioral development is epigenetic in character. This means that the embryo's nervous system and behavior arise through a successive series of transformations whereby new structures and functions and new patterns of organization come into existence. As far as the nervous system is concerned, epigenesis involves the recognition that neural cells and their axons and dendrites [by means of which the cells come into contact (synapse) with each other, including sensory receptors and motor effectors] arise via maturational processes such as proliferation, migration, differentiation, and growth. In the era before the invention of high-resolution microscopes, many scientist-philosophers held the view that the embryo is completely formed at the time of conception, and that development therefore involves merely (solely) the *growth* of preexisting structures. This view, called preformation, has now been discarded in favor of epigenesis (which incorporates the other three processes of maturation: proliferation, migration, and differentiation). When applied to behavioral development, the term epigenetic denotes the fact that all patterns of activity and sensitivity are not immediately evident in the initial stages of development and that the various behavioral capabilities of the organism become manifest only during the course of development.

While there is unanimous agreement on the epigenetic character of embryonic neural and behavioral development, there is disagreement on the character of neurobehavioral epigenesis itself: some theories assume that the epigenesis of the nervous system and/or behavior is *predetermined*, while others hold that it is *probabilistic.*

B. Predetermined Epigenesis

To put it in the clearest possible way, the concept of predetermined epigenesis, as it applies to behavior, means that the development of behavior in larvae, embryos, and fetuses can be explained entirely in terms of neuromotor and neurosensory maturation (i.e., in terms of the proliferation, migration, differentiation, and growth of neurons and their axonal and dendritic processes). In this view, factors such as the use or exercise of muscles, sensory stimulation, mechanical agitation, environmental heat, gravity, etc., play only a passive role in the development of the nervous system. Thus, according to predetermined epigenesis, the nervous system matures in an encapsulated fashion so that a sufficiently comprehensive account of the maturation of the nervous system will suffice for an explanation of embryonic behavior, the key idea being that structural maturation determines function, and not vice versa. The notion that function is required to *maintain* the integrity of fully mature systems is very well documented. The possibly formative or facilitative role of function prior to complete structural maturation has not been widely appreciated, however, so very few experiments have been designed in such a way that this possibility could even be evaluated. This state of affairs should be borne in mind when we hear or read critical remarks to the effect that we have little or no evidence for the role of function on the structural maturation process itself. For example, ". . . the bulk of the nervous system must be patterned without the aid of functional adjustment [Sperry, 1951, p. 271]." Or, "Development in many instances . . . is remarkably independent of function, even in . . . [the] . . . sense . . . [of] . . . function as a general condition necessary to healthy growth [Sperry, 1951, p. 271]." More recently, Sperry (1971) has said: "In general outline at least, one could now see how it would be entirely possible for behavioral nerve circuits of extreme intricacy and precision to be inherited and organized prefunctionally solely by the mechanisms of embryonic growth and differentiation [p. 32]."

Perhaps the most dramatic and explicit statement of the classical neuroembryological bias concerning the nonparticipation of behavior or neural function in the maturation of the nervous system is the following:

> The architecture of the nervous system, and the concomitant behavior patterns result from self-generating growth and maturation processes that are determined entirely by inherited, intrinsic factors, to the exclusion of functional adjustment, exercise, or anything else akin to learning [Hamburger, 1957, p. 56, reiterated *in toto* in Hamburger, 1964, p. 21].

(In the next article in this volume, Dr. Hamburger considers these and other matters in light of the recent developments in the field, to which he himself has made an enormous contribution. It is clear that the above quotation does not reflect his present position.)

Rather extreme statements have been made on both sides of the theoretical fence, and the above statements may be necessary "correctives" for the outmoded point of view expressed by an influential psychologist some 40 years ago: "... John Locke's doctrine of the *tabula rasa* rests on solid embryological as well as psychological ground [Holt, 1931, p. 35]." Or, more explicitly: "It [the nervous system] allows afferent impulses to diffuse or spread so widely that the possibility is open for any sense-organ to acquire functional connection with any muscle [Holt, 1931, p. 33]."

Statements such as the above, coming as they do from leading practitioners in their respective fields, are apt to be taken as empirically justified by young students inside each field, and that can only lead to the persistence of long-standing antithetical attitudes which are not in fact justified by the evidence.

While the concept of predetermined epigenesis may be correct for the very earliest stages of embryonic and larval development, when, in fact, there is no embryonic behavior or motility as such, or when the nervous system is not yet functioning, it seems unduly conservative to apply this conception to later stages when the embryo is capable of motility or behavior, or when neurons have become capable of function (spontaneous or evoked). Whereas the view of predetermined epigenesis rightly focuses on the enormous self-differentiating prowess of the nervous system and embryonic organismic development in general, it tends to become unduly narrow in conception when, in advance of supporting evidence, even a facilitative role is denied to functional factors.

Although it may at first seem paradoxical, those who adhere to the predetermined conception of embryonic behavior may not hold a similar conviction about the intrinsic maturational properties of the nervous system itself. For example, G. E. Coghill (1929), who can be regarded as the father of behavioral embryology if Preyer is regarded as the grandfather, held a strongly predeterministic view of the development of behavior while favoring rather tenuous and radically probabilistic theories of neural development (the "laws" of neurobiotaxis and stimulogenous fibrillation to be described in Section V). This apparent paradox resolves itself when one realizes that Coghill was trained exclusively in the neuroanatomical tradition. As I have pointed out elsewhere in some considerable detail (Gottlieb, 1970), many developmental neuroanatomists tend to view behavior as a mere epiphenomenon of the nervous system. Thus, while the intricacies of intrinsic neural development may be well appreciated through such training, it hardly prepares one for the vicissitudes of behavioral development as such. This may be unfair to Coghill, especially in view of the fact that his favorite specimens for study (salamander larvae) are almost inert unless provoked by experimental stimulation and thus do not "behave" very much. But it must be recalled that Coghill felt that his principle of behavioral development (partial motor patterns differentiating out of total patterns) had widespread application in the vertebrate subphylum (Coghill, 1929, Lecture Three). In any event, the conceptual inclination of certain researchers toward predeterminism as regards embryonic behavior and probabilism with respect to neural development is not restricted to Coghill, as a reading of some of the ensuing articles will reveal.

C. Probabilistic Epigenesis

While the adherents to predetermined epigenesis cast a dubious if not downright jaundiced eye on the possible effects of stimulation, musculoskeletal use, and neural function on the development of embryonic behavior, some theorists of the probabilistic persuasion seem almost overexuberant or utterly naïve in their assumptions of the immediate and subsequent effects of such factors. While, by definition, the predeterministic conception is loathe to grant even a facilitating role to functional factors in the progressive expansion and development of the nervous system and behavior, the probabilistic conception holds that function not only facilitates neural maturation and behavioral development but that it is capable of exerting a determinative influence as well. Further, an analysis (Gottlieb, 1970) of the assumptions underlying the main probabilistic theories of prenatal behavior development (Holt, 1931; Kuo, 1967) reveals that these theories seem to demand (implicitly) that functional factors be operative during the actual maturation of the various sensory and motor systems of the embryo and fetus—that is, functional and environmental factors are assumed to contribute in both a facilitative and a determinative way to neural maturation and not merely to maintain and/or channel behavior only after neural maturation is completed.

Thus, in summary, whereas the key assumption of predetermined epigenesis holds that there is a unidirectional relationship between structure and function whereby structural maturation determines function (structural maturation ⟶ function) but not the reverse, probabilistic epigenesis assumes a bidirectional or reciprocal relationship between structural maturation and function, whereby structural maturation determines function and function alters structural maturation (structural maturation ⟷function).

Before turning to the specific theoretical issues of behavioral embryology, it is perhaps important to mention briefly two other more or less subsidiary features which distinguish predetermined and probabilistic conceptions of behavioral development, both of which are implicit in the bidirectional structure ⟷function concept. Namely, probabilistic theories (e.g., Holt,

1931; Kuo, 1967) hold that (*1*) behavioral development is a gradual and continuous process, wherein (*2*) certain features of late embryonic, fetal or early neonatal behavior can be traced to, and are in some sense (facilitative or determinative) dependent upon, earlier behaviors or stimulative events. Certain other viewpoints (e.g., Hamburger, 1968), on the other hand, are explicit in their emphasis of discontinuities in embryonic behavior such that, for example, later movements are envisaged as arising *de novo*, in which case these movements could bear nothing other than a permissive relationship to earlier movements. (Dr. Hamburger has enlarged on this topic considerably in the present volume.)

Another distinction betwen predetermined and probabilistic conceptions —one which must at the moment be considered tentative because it only infrequently and inconsistently comes to the fore—concerns the question of variability in neural and behavioral *outcomes* and neural and behavioral *processes*. Probabilistic perspectives sometimes tend to emphasize (or suggest) variable means to variable and reversible end states, whereas predeterministic viewpoints sometimes emphasize (or imply) invariant means to invariable and irreversible terminal events. (For an example of the latter see Dr. Jacobson's views on the maturation of the so-called class I neurons, expressed in Volume 2.)

In summary, with regard to behavioral development, the predetermined conception stresses the unidirectionality of the structure-function relationship and the discontinuities between earlier and later stages. The probabilistic conception emphasizes the bidirectionality of the structure-function relationship (whether the function is regarded as spontaneous or whether it is evoked by stimulation), the facilitative and determinative nature of precursors, and thus the continuities between earlier and later behavior. While adherents to either approach recognize and appreciate the stability of behavioral development which is manifested within (and sometimes even between) species, they of course differ in their conceptualization of the events which are responsible for these highly predictable and recurrent regularities. The specific ways in which these conceptualizations differ on particular topics are presented below.

V. Theoretical Issues

In the introductory portion of this essay, it was noted that the study of neurogenesis entails problems peculiar to itself and that the study of behavior entails problems peculiar to *itself*. Behavioral embryology begins to realize its potential when the relevant aspects of the two areas are brought to bear on each other. From the standpoint of certain investigators with particular backgrounds, the relevant aspect for study is how structure determines function, while, for others, due to their particular background and interests, the relevant aspect for study is how function affects structure (or how contemporary function affects later function or structure). In ideal instances, all too infrequent at the present time, one investigator may be able and willing to study (or at least to entertain) both of these aspects. Despite the conceptual controversy described in the previous section, it is correct to say that the unidirectional structure-function relationship (S $\longrightarrow$ F) and the bidirectional structure-function (S $\longleftrightarrow$ F) relationship are *the* relevant aspects of behavioral embryology, because there is virtually unanimous agreement at the moment (based on the currently available evidence) that the unidirectional structure-function relationship seems to hold for at least the earliest stages of neurobehavioral ontogeny, and that the bidirectional structure-function relationship seems to hold for at least the very latest stages of neurobehavioral ontogeny. The conceptual dispute concerns the generality of either conception (i.e., their applicability to early, intermediate, as well as late stages; different neural or behavioral systems; different species; and so on). Proponents of predetermined epigenesis would hardly agree, nor is there any compelling reason for them to agree, that their conception of development holds *only* for the earliest stages, or *only* for certain systems, or *only* for certain behaviors, or *only* for certain species. Adherents to the probabilistic conception, on the other hand, seek to determine how early their conception of development operates, and whether it is not in fact the dominant theme in development, neural and behavioral, regardless of the species or the system under consideration.

A. Motor Primacy: Autogenous Motility

In 1885, Preyer observed that the chick embryo is capable of overt movement prior to the time when an overt response to exteroceptive sensory stimulation can be detected (Preyer, 1885). The significance of this very important observation resides in the implication that the early motility of embryos (of certain species at least) is *autogenous*, which means that it is generated exclusively in the interneurons and/or the motor neurons of the spinal cord, so that such movements are not necessarily instigated by exteroceptive, proprioceptive, or interoceptive sensory stimulation.

An interesting feature of these movements is their periodicity. That is, as first indicated by Clark and Clark (1914), Visintini and Levi-Montalcini (1939), Rhines (1943), and expanded upon in recent years by Hamburger (this volume) and his associates (Provine, and Foelix and Oppenheim, in this volume), motility in chick embryos is characterized by alternating periods of activity and inactivity, the extent of which are more or less age- or stage-specific. Subsequently, Tracy (1926) observed the same phenomenon (periodic or cyclic motility) in certain teleost fish embryos, again before the embryos showed an overt response to exteroceptive sensory stimulation. The Clarks hypothesized that the rhythmic activity of chick embryos might be controlled by the periodic pulsations of the embryonic lymph heart (a pair of organs located at the base of the spinal cord which are part of the embryonic lymphatic system and which disappear around hatching), but they were unable to determine whether the lymph heart pulsations cause the rhythmic activity of the chick's neuromusculature, or vice versa, or whether both rhythms are under the control of some other part of the nervous or circulatory systems. Tracy, on the other hand, attempted to show that the periodicity of movements of his fish embryos might be related to the periodic build-up of CO_2 in the circulatory system of the embryos, a hypothesis later reiterated by Visintini and Levi-Montalcini (1939) in connection with their observations of cyclic activity in chick embryos, and confirmed in part by Hamburger and Balaban (Hamburger, 1963). Specifically, Hamburger and Balaban found that the administration of O_2 to chick embryos served to increase the length of their activity phases, and decrease the length of their inactivity phases, while it did not alter the number of activity–inactivity cycles as such. Hamburger's (1963) interpretation is that the motility is triggered in the motor system, and that O_2 affects merely the length of the phases of activity and inactivity within each cycle. While O_2 may affect only the length of the activity–inactivity cycles, Tracy's (1926) evidence that the accumulation of CO_2 may trigger embryonic or larval movement (at least in toadfish) still remains.

In recent years, Hamburger and his colleagues have performed a number of experiments designed to assess the influence of sensory stimulation on the level of embryonic motility. The crux of the idea that embryonic motility is autogenous resides in the notion that such motility is not typically instigated by sensory stimulation, at least for the first two-thirds of the embryonic period in chicks. Despite the fact that in some studies (Bogdanov, 1963; Bursian, 1965; Sviderskaya, 1967) an effect (inhibitory or excitatory) has been found in one or more of the various behavioral measures on one or more days of chick embryonic development, the interpretation of the evidence has favored the autogenous nature of motility under normal conditions of development, because the stimulative effects have been demonstrated only in the presence of what seems to be overly intense sensory stimulation, and such effects have not consistently been inhibitory or excitatory, nor have they occurred on each day of development or in each behavioral measure (e.g., Hamburger & Narayanan, 1969; Helfenstein & Narayanan, 1970; Oppenheim, 1972b). In dealing with a very rapidly developing organism such as the chick embryo, it may of course be incorrect to assume that the effects of sensory stimulation (if any) are consistent from day-to-day, etc. In light of the methodological problems encountered in the study of sensory influences on motility (Oppenheim, 1972b), it would seem that the issue of the specific nature of the influences of sensory stimulation on motility at different ages will remain ambiguous until such time that mechanical, electronic, or other objective means of recording the motility of the embryo are instituted. Outside of the Russian work, all of the studies have involved sheerly visual observation and it has not been possible to resolve differences between investigators, investigations, and behavioral measures. [See, for example, the recent study by Oppenheim (1972a) in which he reviews the various discrepancies between his own findings and those of Kuo on the chick embryo.]

Apart from the mundane but all too important methodological problems associated with the study of embryonic motility and prenatal behavior in general, certain of the aspects of autogenous motility, as previously elaborated by Hamburger (see his article in this volume for changes in his position), made it the epitome of the viewpoint of predetermined epigenesis of behavior. Specifically, quite independent of the question of whether or not sensory stimulation plays any role whatsoever in the early movements of avian and other embryos, there remains the question of the influence of these early movements on later movements. On the basis of sheerly observational data, Hamburger (1963, 1968) has tentatively concluded that the early movements of chick embryos are of a random and uncoordinated character, and thus bear no relationship to the coordinated movements which are observed later in development during the prehatching period. This tentative conclusion of course does not take into account the fact that at least *certain* of the movements in young avian embryos are neither random nor uncoordinated. For example, the earliest movements of both chick and duck embryos are highly stereotyped and coordinated head and pelvic movements (Oppenheim, 1970); and during the latter half of the embryonic period, the embryos show a clearly nonrandom pattern in their lateral head movements, namely, they begin turning their head more often to their right than to their left (Gottlieb & Kuo, 1965; Hamburger & Oppenheim, 1967). Although it has not been determined experimentally, the latter may be a facilitative or determinative precursor to the avian embryo's tucking its head under its right wing during the preliminary stages of hatching. (The various movements and postures related to hatching in various species are described in detail by Dr. Oppenheim in this volume.)

In any event, nonrandom and coordinated movements *do* occur early in avian embryonic development, and it is entirely an unanswered question whether these movements (or random and uncoordinated movements) contribute in either a quantitative or determinative way to later coordinated or uncoordinated behavior. It is most pertinent to the continuity-discontinuity issue whether (*1*) specific patterns of early movement (spontaneous or reflexive) are or are not related to later behavior patterns, and (*2*) whether early motility (spontaneous, random, or otherwise) does or does not contribute quantitatively to later levels of motility. In the service of clarity, these questions should be held distinct from the question of the influence or non-influence of sensory stimulation on embryonic motility which, in itself, is not a developmental question per se. With regard to the influence of sensory stimulation on motility, I would like to reiterate once again (see, e.g., Gottlieb, 1968, p. 148, and 1970, p. 117) that all of the evidence taken together indicates (*a*) that embryonic motility is in a very basic and primary sense self-generated, as Dr. Hamburger and his colleagues have shown, but (*b*) that it appears also to be susceptible to at least transient inhibitory and excitatory modulation by sensory and mechanical factors. (See the articles by Miss Vince and by Drs. Impekoven and Gold, this volume, for further evidence on the last point.)

Although the above discussion has dwelled largely on the avian embryo, the same problems and issues apply to the study of the (apparently) spontaneous and cyclic motility of invertebrate forms (Dr. Berrill, this volume) and lower vertebrate species (the amphibians discussed by Dr. Hughes in the next volume). The peculiarities of mammalian fetal activity with respect to the notion of autogenous motility are discussed by Dr. Hamburger in the present volume and by Dr. Bergström in Volume 2.

B. *Patterned Movement*

There have been three approaches to the understanding of the development of patterned movement. The earliest of these approaches (Coghill, 1929) postulated as the general principle of motor development a progression from total to local (or partial) patterns of activity. According to this view, during the early stages of movement the embryo moves as a whole, and it is only later on that the capacity for discrete movements develops. The alternative view (Swenson, 1929; Windle, 1944) held that total patterns of movement arise secondarily out of an integration of local reflexes; in other words, that the local reflex is the primary building block in the development of total patterns of behavior. The generality of both of these views has been challenged so repeatedly by contrary evidence (Carmichael, 1934; Faber, 1956; Kuo, 1939) that it has become obvious that these formulations can have only a very restricted utility. For a lack of adherents, this particular controversy has tended to die out since the early 1940s. According to the evidence, the best approximation at the present time is that both local and total patterns of activity coexist during the development of most or many species, and that this particular way of approaching the understanding of the development of motor behavior is no longer fruitful (i.e., with respect to the establishment of general principles of behavioral development). Perhaps the best evidence of the difficulty of applying the total pattern–partial pattern dichotomy to the study of behavioral embryology came from Coghill's own laboratory: his students, Swenson (1929) and Angulo y González (1932), reached opposite conclusions on the development of behavior in the same species of rodent.

From the standpoint of the history of behavioral embryology, Coghill's principle, however ambiguous and uncertain it eventually became in practice, stimulated more interdisciplinary interest and more comparative research in embryonic, larval, and fetal behavior than any other formulation to date, and the field owes much to the clarity of his thinking and writing on its basic problems, as well as to the vigor with which he promulgated his own particular theoretical views and his research [under the very trying and depressing circumstances described by Herrick (1949) in his biography of Coghill].

A different theoretical approach to the study of the ontogeny of patterned movement has been espoused by Kuo (1967). In this view, many of the important movement patterns evident in the neonate are said to represent a repetition and extension of the various components of embryonic motor activity, now *reorganized* and incorporated into a new motor pattern in response to the changed environmental context (birth or hatching introducing the new context), as well as changes in physiology and morphology. The central idea here is that the emergent neonatal motor behavior pattern is in some sense (facilitative or determinative) dependent upon certain of its various components having been active during the embryonic state, although these components may not have entered into the same relationship prior to birth or hatching. The most complicated, best known, and widely misunderstood example of Kuo's view on this subject has to do with the act of pecking in neonatal chicks. Though the act of pecking can be construed rather broadly (to include postural mechanisms, visual perceptual selectivity, motivation, etc.), at a bare minimum the motor components must include extension of the neck (head lunging in the neonate) and opening and closing of the beak (corresponding to seizure of the object with the beak in the neonate). Although the act of pecking never occurs as such during embryonic development, each of its minimum motor components has been activated numerous times during embryonic development. (These same embryonic movements—neck stretching, opening and closing of the beak or bill—occur in species which do not peck in the neonatal stage. Gaping and other food-getting or food-soliciting motor patterns in various avian species all involve head, neck, and beak activity akin to that observed in the embryo. Whether the embryonic movements are facilitative or determinative precursors to the neonatal food-getting or food-soliciting behavior is not known.)

Kuo's theory assumes that throughout embryonic and neonatal development, new patterns of movement manifest themselves (due to changes in morphology, embryonic position, etc.), and that these almost always include some components which were active earlier in development. The extent to which the onset, threshold, rate of development, and/or coordination of the later movement patterns are dependent upon the prior activity or exercise of certain of their components has not been tested in an experimental fashion, so the validity of Kuo's viewpoint on this issue has not been tested.

The third major view on the ontogeny of patterned movement assumes that sequences of certain movements which occur in late embryonic or fetal life (or in early neonatal or perinatal behavior) are rather straightforward repetitions of the sequences of previous embryonic or fetal motor activity. Humphrey (1969, 1971) has adopted this point of view in attempting to establish relationships between certain patterns of movement observed in human fetuses and those observed in the (later) perinatal period. She stresses (Humphrey, 1971, p. 35) that it is the repetition of fetal motor *sequences*, not of identical movements, which occurs when, for example, the neonate makes a certain train of oral and facial movements. While it is virtually certain that the central "control" mechanisms change during development, Humphrey (1969, p. 80) points out that the motor neurons involved are most likely the same in each case, so, in this sense, the earlier activity is the precursor to the later one.

> Neurophysiologically, the neurons that have functioned the longest require the least stimulation to fire them, and less oxygen to react, so they are discharged more readily. . . . These factors will contribute also to a repetition of the same order (or sequence) of motor events as the higher nervous system levels become active. However, there will be a variation in the movements themselves, even though the sequence is retained [Humphrey, 1971, pp. 35–36].

It is not yet clear, however, whether the earlier movements of human fetuses should be regarded as facilitative or determinative precursors to the later movements. While placental mammalian fetuses do not lend themselves readily to experimentation on this point, a number of experiments has been carried out on this question with respect to the ontogeny of swimming behavior in amphibian forms. Beginning early in this century, numerous workers subjected amphibian embryos to motor paralysis prior to the free-swimming stage in order to assess the role of prior neuromuscular function on later swimming behavior. All of these early studies indicated that the embryos, once released from the paralyzing anesthetic, could swim (Carmichael, 1926; Harrison, 1904; Matthews & Detwiler, 1926). Building on these results, Fromme (1941) conducted a somewhat finer grain analysis of the later swimming behavior of the previously paralyzed embryos and found heretofore undetected quantitative deficiencies in the manner in which they propelled themselves through the water. The swimming of the previously paralyzed embryos was less sustained than that of the normal embryos and, consequently, they took considerably more time to get from one place to another under considerable prodding. Thus, according to the definition of precursors offered earlier, the swimming movements exhibited by amphibian embryos prior to the free-swimming stage may be a facilitative precursor to the later behavior. Whether these results apply to the repetitional relationship between prenatal and postnatal movement patterns *generally* is, of course, not known, but it does not seem unreasonable that later movements which are solely or primarily repetitions of earlier movements are facilitated by their earlier exercise, as implied in the quotation from Dr. Humphrey.

The early studies of the ontogeny of swimming behavior in amphibians have been mimicked in a very clever way by Crain, Bornstein, and Peterson (1968, and Volume 2 in this serial publication), using an *in vitro* preparation. Fetal rodent cortical and spinal cord cells were explanted and exposed to blocking agents so they could not fire spontaneously. When the blocking agent was removed from the culture medium, the very first electrical stimulation often produced complex, long-lasting discharges, indicating that functional synaptic networks had formed in the culture in the absence of prior functional activity. It has not yet been possible to determine whether the function of such cells shows quantitative deficiencies (e.g., higher threshold, less sustained firing, different temporal pattern, etc.) compared to cells not exposed to blocking agents during the formative period of synaptogenesis. There can be no doubt that synapses can be formed and that amphibians can swim in the absence of prior functional activity (see Foelix and Oppenheim, this volume, for further suggestive evidence on this point), but the question remains entirely open whether the function (neural or behavioral) which typically accompanies or precedes such activities plays a facilitative role in the process. By way of breaking out of the strait jacket of a purely dichotomous nature–nurture approach to the study of ontogeny, it would now appear useful to inquire not only if, but how well, later functions are performed in animals or tissue cultures which have been subjected to a curtailment of their usual (typical) functional activity. The very likely possibility that many deficiencies may be momentary or transient and can be rectified by subsequent activity would be no cause for the exclusive celebration of either nature or nurture, but would merely indicate that function *normally* plays a role in the maturational process. The equally likely possibility that certain processes are merely slowed down in their rate of development by withholding the opportunity for function, and that in the continued absence of function these processes (neural or behavioral) eventually (i.e., belatedly) achieve a level of attainment or end point indistinguishable from that of nondeprived neural systems or animals again would attest to the *facilitative* participation of function in the normal or usual course of ontogenetic events and, as a corollary, the susceptibility of maturational processes to the influence of function during the usual course of events.

C. *Nerve Fiber Growth and Synapse Formation*

As implied earlier, one of the central problems of neural ontogeny as it relates to behavior concerns the question, so aptly put by Coghill (1929): "How do conduction paths of the central nervous system come to be where they are, or how do they acquire their definitive function [p. 76]?"

As mentioned before, in the process of getting themselves mature, nerve cells multiply (proliferate), travel from their site of origin to another place in the nervous system (migrate), sprout axons and dendrites (differentiate), and become larger in size (grow). Coghill's question, then, is why do the cells migrate here rather than there and, more particularly, why do they enter into synaptic relations with these cells rather than those cells? While the answer to this question is still unknown, three hypotheses of nerve fiber outgrowth and synaptic formation have been proposed: electrical field hypothesis, mechanical or contact guidance hypothesis, and chemoaffinity hypothesis. [Although it is usual to make a distinction between neural connections which occur inside the central nervous system (intracentral connections) and central-peripheral connections (central connections with sensory receptors or muscles), I am going to momentarily ignore these distinctions for the sake of a simplified and hopefully more understandable presentation of each hypothesis.]

1. Electrical Field Hypothesis

In its most general form, this hypothesis holds that the bioelectric potentials present in maturing nerve tissue exert an orienting influence on the direction of growth of nerve fibers. Neural growth and synaptic connections are said to occur between fibers that are stimulated (spontaneously or otherwise) at the same time or in close temporal contiguity. In its more specific form, proponents of the electrical field hypothesis differ in their predictions concerning the effects of electrical activity on axons and dendrites. Ariëns Kappers (1936), incorporating Bok's (1915) idea of stimulogenous fibrillation, formulated the "law" of neurobiotaxis which asserts that axons of maturing nerve cells grow in the same direction taken by the electric current that radiates from the growing nerve bundle (i.e., axons grow with the current), whereas dendrites grow *toward* nerve bundles from which they receive the greatest amount of electrical stimulation (i.e., dendrites grow against the current). In addition, Ariëns Kappers held that the nerve cell itself migrates in the direction taken by the dendrites.

Electrical activity is intimately connected with (or reflective of) metabolic, oxidative, and other chemical activities of neural cells, and other theorists (Child, 1921) have suggested that axons tend to grow into regions of low metabolic activity ("down the gradient," so to speak); still others (Coghill, 1929) suggested that axons grow into regions of relatively high metabolic activity (up the gradient), and dendrites into regions of relatively low metabolic activity (down the gradient). In fact, Coghill (1929), in line with his interpretation of the then available facts, went so far as to suggest ". . . that those processes of neuroepithelial cells which grow into regions of higher rate of a metabolic gradient *become* axones while those which grow into regions of lower rate of the gradient *become* dendrites [p. 59]." (Emphasis added.)

Thus, there are two separate questions associated with the electrical field hypothesis. First, do electrical fields orient the direction of growth of immature nerve fibers and, second, if so, in what way is the direction of growth of axons and dendrites different under such circumstances?

The most direct way to test the electrical field hypothesis is to explant some immature neural tissue in a culture, subject the culture to an electrical field, and observe the direction taken by the outgrowing nerve fibers with reference to the direction of current flow in the field. If the first part of the electrical field hypothesis is correct, the flow of current should orient or reorient outgrowing nerve fibers and, if the second part of the hypothesis is correct, axons and dendrites should grow in different directions under the influence of the electrical field.

The first investigator to use the tissue culture approach to test the electrical field hypothesis was Ingvar (1920). In a regrettably brief report which omitted important procedural details and, through error, may have severely underestimated the current employed [see the comments of Marsh and Beams (1946) and the hypothetical recalculation of Ingvar's current by Crain (1965, p. 416)], Ingvar observed that the outgrowing fibers of chick nerve cells oriented themselves almost entirely along the lines of current flow and, further, that there were morphological differences between those processes which grew toward the anode (+ pole) and those which grew toward the cathode (– pole). In 1934, Weiss repeated the experiment and failed to replicate Ingvar's results (Weiss, 1934). He used current strengths in the region *reported* by Ingvar. (Weiss does not mention the duration of exposure.) In the same year, Karssen and Sager (1934), apparently using a current as weak or weaker than that reported by Ingvar and exposing the cultures to an electrical field for a period of 6 days, observed the majority of the fibers to grow along the lines of current flow and a tendency toward greater fiber outgrowth on the cathode side of the field.

In 1946, Marsh and Beams, in the most careful and sophisticated study of this problem up to that time, exposed nerve tissue cultures composed of chick embryo medulla to various current densities for periods ranging from 3 to 28.5 hours (Marsh & Beams, 1946). At the highest densities, fibers did not emerge from the anode side of the culture and, at all but the lowest current densities, the growth pattern of the fibers was deflected toward the cathode side of the field (see Fig. 1). When the axis of the current was rotated 90°, a majority of the fibers bent toward the new cathode direction, and the growth rate of the fibers on the cathode side surpassed that of the fibers on

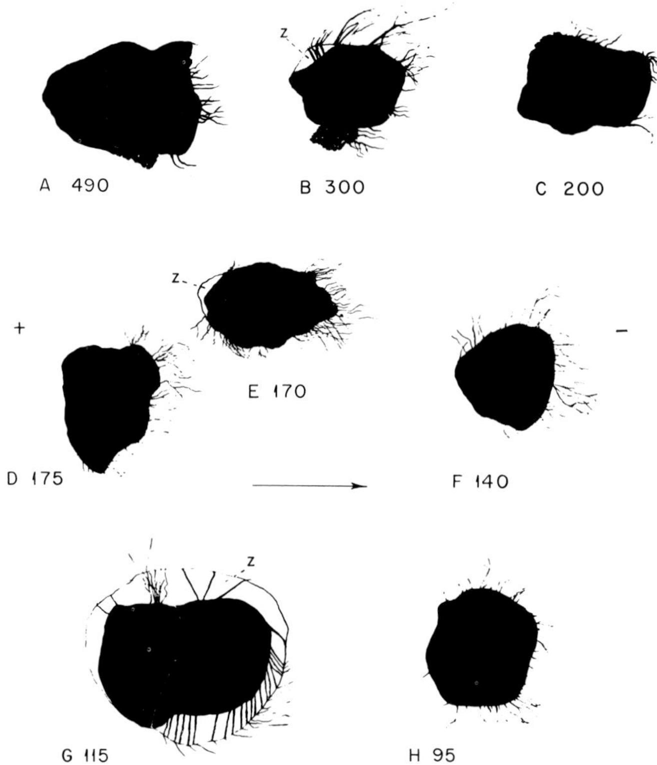

FIG. 1. Nerve fiber growth pattern in chick embryo medulla explants subjected to electrical field for 22 hours. Direction of current flow designated by arrow. Current densities (A–H) shown beneath each explant. At the highest current density (A) (490) there is complete suppression of fiber outgrowth on the anode side (+) of the field. Except in zones of liquefaction (*z*), the majority of the fibers are reflected toward the cathode (−) side of the field. From Marsh & Beams (1946).

the anode side of the field (shown in Fig. 2). The effective orienting current densities were 50,000 times greater than that reported by Ingvar. In this connection, it is important to note that chick embryo spinal ganglion explants growing in culture do exhibit resting potentials as well as evoked action potentials (Crain, 1956), and spinal cord muscle cell preparations also show spontaneous as well as evoked neural potentials (Fischbach, 1970; Kano & Shimada, 1971). Due to differences in the culture mediums, it has not yet been possible to equilibrate the strength of artificial electric fields with those produced by neural tissue in culture. An even more formidable problem is encountered in relating the effective field strength in culture to those *in situ*.

In recent years, some workers (Grosse, Lindner, & Schneider, 1969a, 1969b), using light microscopy, have determined certain quantitative

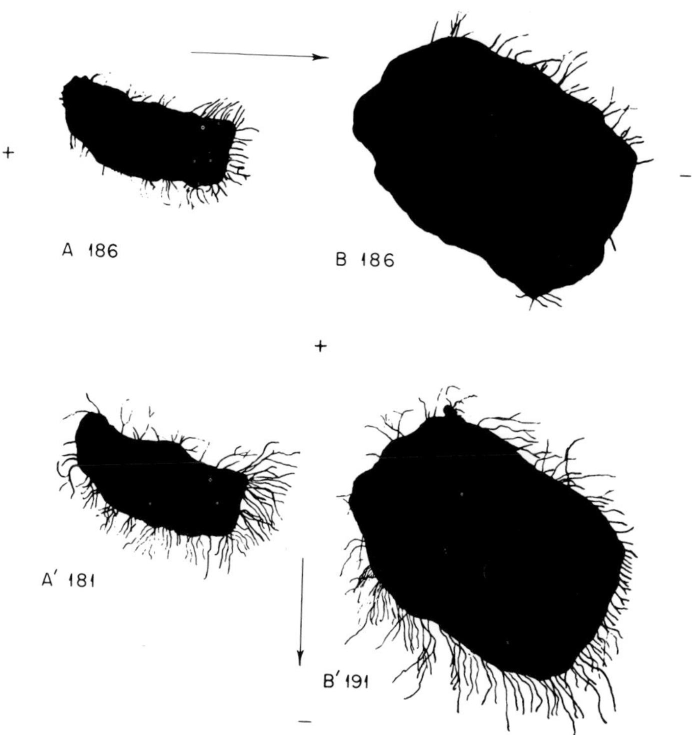

FIG. 2. At top, neural explants A and B after 8.5 hours of exposure to current flowing in direction of arrow. At bottom, A′ and B′ after 11.5 hours of additional exposure to current rotated through 90°. Note that the majority of the nerve fibers now bend toward the new cathode (−) direction, and the growth rate of the fibers on the cathode side surpasses those on the anode side. From Marsh & Beams (1946).

effects of electrical fields on the growth of nerve terminals and supporting cells in tissue culture, but nothing definite is known about the supposedly different *directional* effects which such treatment should have on the growth of axons and dendrites. With the use of the electron microscope and microelectrode recording from synapses formed in culture, it should be possible to answer that question.

In summary, the present status of the electrical field hypothesis is as follows. Experimentally applied electrical fields of sufficient current density, which may or may not be comparable to that generated by embryonic neural tissue, influence outgrowing nerve fibers to orient toward the cathode (i.e. in the direction taken by the flow of current). Whether axons and dendrites respond differently to the flow of current is not known. The extent to which the influence of electrical fields on neural tissue in culture mimics the actual situation *in vivo* or *in situ* is purely conjectural, but the recently discovered phenomena of neuroid (i.e. preneural or nonneural) conduction (Mackie, 1970) and the electrical coupling of cells (Sheridan, 1968) in early embryonic stages suggests that the electrical field hypothesis, *or some variant of it*, is worthy of further investigation. (It should be borne in mind that electrical activity is intimately associated with chemical activity, so perhaps the hypothesis is best regarded as an electrochemical one.) For example, merely to illustrate one possible variant of the electrical field hypothesis, the electrical coupling of cells very early in development involves the exchange of ions, and this early exchange could somehow (cytochemically?) determine the later highly specific functional interconnection of these cells in the mature neural network.

2. Contact Guidance Hypothesis

Early in the history of tissue culture work, it was noted that cells did not grow or migrate unless they made contact with a solid body, such as the coverslip, the fibrin net, fibers immersed in fluid, or the surface film of the culture medium (Harrison, 1914). On the basis of these early observations, Weiss (1934) put forth the hypothesis that the outgrowth of neural fibers is oriented by the lines laid down by tensions in the ultrastructure of the medium. These tensions represent the physical fields of force present in the medium, and they establish directional vectors for growth. Weiss and others have adduced considerable evidence for the channeling of growth of nerve fibers by the mechanical substrate, and this feature of directional nerve growth is widely accepted, even by authors of competing theories (e.g., Sperry, 1951, 1965). The "hypothetical" part of the fact of contact guidance comes in the determination of how far-reaching mechanical factors can be in the *selective* growth of nerve fibers over any but the shortest of distances. In other words, it is generally agreed that the advancing tip of the nerve fiber needs something to grow on or along, and the *inital* outgrowth of fibers has been shown to follow the path taken by the advance of connecting tisue and sheath cells which have preceded them (see the photographs in Harrison, 1969, pp. 128–129). After this initial outgrowth many channels of advance are possible, and the question remains why the fiber tip would take this direction rather than that one. In this connection, perhaps it is appropriate to mention that Weiss (1934) believed that the orienting effect of the electrical field could be achieved by vectors it might produce in the substrate, and, in this sense, perhaps the two theories could be harmonized.

Since the leading proponent of the chemoaffinity hypothesis (Sperry, 1951) holds "... that mechanical factors are ubiquitous in nervous development ... [p. 240]" and that "... the universal importance of mechanical factors has been clearly demonstrated [p. 239]" and, further, that electrical current has "... a slight directive effect ... [p. 239]," it remains to inquire what special role chemoaffinity plays in the drama of nerve growth and synaptic formation.

3. Chemoaffinity Hypothesis

Up to this point, we have been speaking about the nervous system as if it were a rather "general system" in its developmental characteristics, not necessarily totally equipotential or unlimited in its growth possibilities, but also not following very strict rules of connection between its individual units (neurons). In the first view, for example, one envisages that cells in the retina somehow manage to get themselves connected in a general way with cells in the visual cortex (in the case of mammals) or the optic tectum (in the case of birds, amphibians, and fishes). Suppose, however, that not only do retinal cells get connected with cells in the optic tectum in a sort of *regionally* specific way, but that particular retinal cells in certain very specific, highly local parts of the retina get connected to particular other cells in very specific, highly local places in the optic tectum. Since it is difficult to envision that electrical activity (whether spontaneous or evoked by visual stimulation) could be invariant between individuals of the same species, or that the physical substrate along which individual fibers travel (or are guided) are likewise so highly invariant, if a point-to-point specificity holds between retinotectal connections in individuals of the same species, it would seem virtually impossible that such highly specific connections could be formed on the basis of the electrical field hypothesis or the contact guidance hypothesis. So goes the reasoning of Sperry (1965), who, emphasizing the specificity of connections between neurons within the motor systems as well as within the sensory systems, has revived and greatly extended Ramón y Cajal's and Forssman's idea of chemical affinities mediating selective growth between similar types of fibers in the nervous system (reviewed in Ramón y Cajal, 1928). The central idea here, in the hands of Sperry, is that nerve cells (and their fiber tips) have a distinctive biochemical (cytochemical) composition which is matched only by cells and fibers in some other part of the system and, because of these cytochemical identities, synapses are formed between these particular cells and not other cells. While in his earlier articles on the subject, Sperry (1951, for example) took a somewhat more cautious and tentative stand on the ability of the chemoaffinity hypothesis to account for the selective formation of synapses in embryos, as the evidence for apparent anatomical specificity in the development of the nervous system mounted, he has taken a much bolder stance, one which discounts electrical activity or function of any kind, on the one hand, and embraces early behavioral development, on the other.

> This chemoaffinity interpretation not only offered an explanation of the perplexing 18-year-old problem of homologous response and related phenomena, obviating the concept of impulse specificity, but it also provided the basis for an entire(ly) new and attractive hypothesis for the normal developmental patterning of behavioral nerve nets. . . . This proved to be much more satisfactory, detailed, and explanatory than any of the ideas previously available, such as neurobiotaxis, disuse atrophy, contact guidance, bioelectric fields, functional trial and error, conditioning, and so on [Sperry, 1965, p. 164].

The chemoaffinity hypothesis, attractive as it is, has had a curious intellectual history. Namely, it has waxed and waned in direct relationship with the waxing and waning of evidence indicating a high degree of *functional* (i.e., electrophysiological and behavioral) specificity in the developing nervous system. Toward the end of a rather confusing trail, it comes as a disappointment to finally realize there is no direct (or unequivocally direct) evidence for the chemoaffinity hypothesis itself, nor is there any direct evidence for the anatomical specificity upon which the chemoaffinity hypothesis rests. Neural tissue grown in mediums which contain neurochemicals as well as other chemicals in some cases does not show any selective chemoaffinities in its growth pattern (e.g., Weiss & Taylor, 1944), while in others it does (Forssman, 1900), and it has not been possible to trace the anatomical development of individual fiber tracts in the cases where functional evidence of specificity has been found (reviewed in detail by Gaze, 1970). The last author faces these difficult problems in the clearest possible way, and his book is a good source of substance, if not solace. These difficulties notwithstanding, the chemoaffinity hypothesis lays a strong claim to a field which is not without its challenges and challengers. Among the latter are G. Székely and M. Jacobson, and they deal with alternative interpretations of the evidence for specificity in separate articles in Volume 2 of this serial publication.

Again, it should be pointed out that the aforementioned electrical coupling of cells early in embryonic development apparently involves the exchange of chemical substances, and this kind of exchange could be involved in some way in the cytochemical attractions posited by the chemoaffinity hypothesis later in development. At present, however, any connection between the electrical coupling of cells early in development and the chemoaffinity hypothesis of fiber outgrowth and synaptic development later in development is purely conjectural.

Finally, by way of introduction to other viewpoints on the topic of specificity itself, it should be mentioned that some writers hold that specificity may be a secondary outcome and not a primary phenomenon of neural development. The idea here is that the initial cellular connections are diffuse and widespread, most of which atrophy as a consequence of a biochemical (trophic) mismatch or the absence of function (Jacobson, 1970 and Volume 2 of this serial publication; Ramón y Cajal, 1928). According to this view, the initial connections are very general and only subsequently become highly specific as a consequence of function (Jacobson's notion of functional validation). That is, the connections which are "validated" through function are retained and those that are not so validated disintegrate, the outcome being a highly specific set of functional pathways in the central nervous system. Whereas the integrity of such neural connections would rest, in some ultimate sense, on chemoaffinity, in this view chemoaffinity serves as a survival mechanism rather than a guidance mechanism.

4. Summary

By way of attempting a synthetic summary of the various points of view on nerve fiber outgrowth and synapse formation, it seems reasonable to tentatively conclude that (*1*) nerve fibers must grow along or on some other mechanical structures which precede them, that (*2*) diffuse bioelectric phenomena may determine the direction of growth in a general way over relatively long distances, and that (*3*) rather specific chemical affinities determine which fibers terminate where (and survive) in the final stage of growth. Since these same conclusions appear to have been reached by Tello in 1923 (cited in Ramón y Cajal, 1928, p. 395), definitive empirical progress on the mechanisms involved in this area awaits further experimentation. [More detailed and expert accounts of these matters can be found in Dr. Berry's article in Volume 2 of this serial publication and in Dr. Jacobson's (1970) textbook.]

D. *Sensory and Perceptual Processes: Physiology and Anatomy*

The onset and subsequent development (or maturation) of each sensory system provides a fertile ground for testing predetermined and probabilistic conceptions of epigenesis, and it is surprising that relatively few investigators have taken advantage of this very basic approach to the study of ontogenesis. As far as we know now, from evidence in a wide variety of species, parts of each sensory system become capable of function before the entire system has completed maturation (Gottlieb, 1971b).* Thus, the sensory systems lend themselves to relatively simple analytic study of behavioral, electrophysiological, and histological procedures during their formative period of development, and the role of function in the maturation process can be precisely determined by such studies. The basic approach, of course, is to withhold, augment, replace, or precociously administer normally occurring sensory stimulation during the formative period of development and determine the effects of such manipulations on the development of the system.

Perhaps it may heighten the significance of what is being discussed here to mention that, when investigators refer to "critical periods," "sensitive periods," or "periods of maximum susceptibility," they are very likely referring to a period of development when a partially functional system is in the process of becoming completely functional. Since the development of perception is based on the development of the sensory systems, this idea applies to sensory as well as perceptual processes, so, in order to establish the validity of the problem, it is necessary at the outset to document the functional capability of immature sensory systems. [The material immediately below is summarized from a recent review (Gottlieb, 1971b) containing 141 references. For the sake of space, those references are not repeated here.]

1. Function Prior To Complete Maturation

Somesthetic or cutaneous (skin) sensitivity is the first sensory system to develop functional capability in a wide variety of vertebrate species. All vertebrate animals that have been examined show such sensitivity early in embryonic or fetal development, and the first region to become sensitive to tactile stimulation is usually the snout or oral region, especially that innervated by the trigeminal nerve. At the time of the first overt response to tactile stimulation, the trigeminal nerve endings possess no discernible specialized receptors, and they are located well below the skin. So, upon the first manifestation of cutaneous sensitivity, the trigeminal nerve itself is not fully mature, and it is only later on in development that the remainder of the surface areas of the body (innervated by other nerves) become capable of responding to tactile stimulation. In amphibians, the peculiar neuroid (non-

*The same holds true for the motor systems (e.g., in the chick and duck), so the effects of function on the maturation of the motor systems can also be studied by experimentally diminishing or increasing, or perhaps even changing (by mechanical or electrical means), the usual motor activity of the embryos at given stages of development. The excellent technical breakthrough of Dr. Provine in his electrophysiological work on chick embryos makes such manipulative experiments feasible (see his chapter in this volume).

neural) conducting system is present in the skin in advance of conventional neural innervation.

The vestibular system develops functional capability after the somesthetic system, again usually in the embryonic or fetal stage of development. The earliest response to a rapid change in bodily position is usually a generalized one; later in development the capability for head or body righting manifests itself and, finally, head or ocular nystagmus is exhibited in response to systematic rotation.

The auditory system becomes functional after the first signs of function in the vestibular system. When the auditory system first becomes functional, it is responsive to sounds in the low and mid-frequency regions, and only later on does it become sensitive or responsive to sounds of higher frequency. [This generalization appears to hold for bird species (Gottlieb, 1968) as well as mammalian species (Gottlieb, 1971b).]

With the possible exception of the circuit mediating the pupillary reflex (Heaton, 1971), the visual system becomes capable of function after the auditory system begins functioning. [The gradual development of the retinal, cortical, and visually mediated behavioral functions has a relatively large *postnatal* literature, part of which has been reviewed elsewhere (Gottlieb, 1971b) and part of which is covered in the next section.]

Thus, all the evidence at hand (behavioral, biochemical, anatomical, and physiological) indicates that the four sensory systems mentioned above begin to function prior to complete maturity in a variety of species. The other sensory systems are omitted from consideration because there is little or no information on their development.

2. Role of Function in Immature Sensory Systems

The investigative question of great practical and theoretical significance is whether the spontaneous or evoked firing of cells in each system at its functional inception plays a facilitative or determinative role in the subsequent maturation of that system and, secondly, whether function in one system can affect the maturation of another system. In seeking an answer to these questions, we must distinguish sharply between studies that allow a clear statement about the positive constructive role of function during the maturation process itself from studies which deal primarily with the maintenance or preservation of an already mature system. Many postnatal developmental studies of the effects of sensory deprivation are of such a long-term nature that they extend well beyond the formative period of maturation of a particular system and cannot give a clear answer to the first question. The degenerative changes observed in these overly long-term studies are most readily interpreted as support for Ramón y Cajal's (1928) law of disuse atrophy, the most recent version of which appears to be Jacobson's (1970 and Volume 2) notion of functional validation. The basic (and undoubtedly correct) idea is that aspects of recently mature systems disintegrate in the absence of stimulation, use, or function. Conversely, studies which expose young animals to a single invariant pattern of stimulation for weeks or months on end, extending well beyond what might be thought of as the formative period for the particular modality in question, are most useful and necessary for a first approximation to the problem. They cannot, however, be directly applied to either the spatial or temporal aspects of events taking place in the usual course of ontogeny without subsequent experimental refinement in light of knowledge of the usual period of maturation of the system in question, as well as the varying parameters of stimulation ordinarily encountered by the embryo, fetus, or neonate. In light of these restrictions, and because we wish to answer the question of the effects of function on the maturation process itself, the results of very few currently available studies can be used, and most of these are postnatal studies.

The visual system is the last one to develop, it is a very important sensory system for many animals, and it is a relatively simple one to augment or deprive, so it is the one most frequently used in developmental studies of augmentation and deprivation. To take up the question of facilitation first, the few studies available indicate that deprivation leads to a slower rate of functional maturation in the peripheral part of the visual system (kitten: Zetterström, 1956), and the precociously administered or augmented photic stimulation leads to an acceleration of functional maturation in the peripheral and central parts of the visual system (duck embryo: Paulson, 1965; chick embryo: Peters, Vonderahe, & Powers, 1956). On the anatomical side, light deprivation decelerates the rate of myelin deposition (e.g., Gyllensten, Malmfors, & Norrlin, 1966b), as well as the rate of maturation of the retina, lateral geniculate body, and visual cortex in young rodents (Gyllensten & Malmfors, 1963; Gyllensten, Malmfors, & Norrlin, 1965, 1966a). As evidence of an interconnection between sensory systems during development, at least at the cortical level, Gyllensten *et al.* (1966a) observed an initial hypotrophy in the *auditory* cortex as a consequence of light deprivation. Later on, the auditory cortex showed hypertrophy, which suggests that prolonged visual deprivation does not merely affect the visual system directly, but also indirectly, through the increased reliance the deprived animal may come to place on other sensory modalities.

In order for the retina to be able to transmit impulses to the higher levels of the visual system, synapses must of course be present. In a study of the neuroanatomical correlates of the *a*- and *b*-waves of the electroretinogram (ERG), the findings of Nilsson and Crescitelli (1969, 1970) indicate the *a*-

wave (a negative potential produced in response to light) is a response limited to the receptor itself, whereas the *b*-wave (a positive potential which occurs later in development) corresponds to the establishment of the synaptic mechanism. In light of these findings, it is interesting that the *b*-wave of the ERG of visually deprived kittens has a reduced magnitude (Ganz, Fitch, & Satterberg, 1968), suggesting some quantitative deficiency in the synaptic mechanism as a consequence of deprivation. Since the kittens in question were deprived of light for up to 14 weeks after birth, it is not possible to say whether the deficiency resulted from atrophy or whether the synaptic mechanism never achieved its usual functional efficiency in the absence of visual stimulation.

This is perhaps an appropriate time to raise a rather knotty issue, one which hinders a clear-cut interpretation of the results of several recent studies of postnatal augmentation and deprivation of the developing kitten's visual system (Blakemore & Cooper, 1970; Ganz *et al.*, 1968; Hirsch & Spinelli, 1970, 1971; Shlaer, 1971). To be specific, it is quite different to maintain that highly specialized cells in the visual cortex of kittens establish their "proper" functionally specific connections previsually (Hubel & Wiesel, 1963), and that such connections are merely "vulnerable" during the first months after birth (Hubel & Wiesel, 1970), than to say, "It may be that the critical period coincides with the time when a particular system is being formed and developed and that use during this time is essential for its *full maturation* and sustenance [Hubel & Wiesel, 1970, p. 435; emphasis added]." These statements can be reconciled only if it is conceded that the initial connections (i.e. functional specificities) in the visual system, although certainly indicating some degree of organization in advance of visual stimulation, are not in fact fully formed (specified) at that time and only become so under the influence of subsequent visual stimulation *and* the further maturation of the system. [As a matter of fact, according to a recent but very brief report by Barlow and Pettigrew (1971), the previsual functional organization of the kitten's visual cortex is not highly specific, as was once believed, and it only becomes so with normal visual experience. This finding is consonant with a *constructive* effect of visual stimulation.] For a correct understanding of the role of sensory stimulation in the usual (normal) course of maturation of the nervous system, we must be able to distinguish phenomena such as atrophy (Hubel & Wiesel, 1970) or dystrophy (Ganz *et al.*, 1968) and adaptive modification (Hirsch & Spinelli, 1970)—all of which seem to assume that connections have been made and then *changed* as a consequence of visual stimulation, or the lack of it—from the possible constructive (facilitative or determinative) role of visual stimulation in the *initial* and subsequent stages of formation of such connections. With the currently available experimental data, we are not yet in a position to distinguish clearly between these not mutually exclusive alternatives.*

If kittens are exposed to an invariant pattern of visual stimulation during the first 10–12 weeks of life, the cells in their visual cortex show the influence of this exposure by responding in a highly specific way to the visual patterns to which they have been previously exposed, a way in which they would not respond in the absence of such exposure (Blakemore & Cooper, 1970; Hirsch & Spinelli, 1970; Shlaer, 1971). If kittens are deprived of patterned visual stimulation for only a few days during the fourth through twelfth weeks of life, the cells of their visual cortex respond to visual stimulation differently than if they had not been subjected to this limited deprivation (Hubel & Wiesel, 1970). Thus, it appears that visual stimulation could play both a facilitative and a determinative role in the usual course of maturational events in the developing kitten's visual system. It is hard to say for sure, but it does seem to be a strong possibility, especially in view of the recent evidence of Barlow and Pettigrew cited above.

Much the same problem plagues the interpretation of the electrophysiological mapping results on the development of specific synaptic connections in the amphibian visual system. The specificity of connections between the retina and the ipsilateral optic tectum (right eye-right tectum, left eye-left tectum) is the same if both eyes are deprived of stimulation of if one eye is deprived of stimulation (Jacobson, 1971), but the specificity is altered (i.e., the pattern of functional connections is different) if both eyes are exposed to stimulation (Gaze, Keating, Székely, & Beazley, 1970). Around the time of metamorphosis, when the tadpole is becoming a frog or a toad, the eyes move from the side of the head to a more central position such that the visual fields overlap and stimulation from both eyes can affect both tecta (Gaze, 1970). Thus, binocular interaction affects the retinotectal connection in the case where both eyes are stimulated (Gaze *et al.*, 1970), but, on the other hand, such stimulation is not a prerequisite for the establishment of "the" pattern (Jacobson, 1971).† A different but related situation

*In a more recent paper, which appeared in a journal that afforded a more generous opportunity to discuss the various possible interpretations of their results, Hirsch and Spinelli (1971) note that visual stimulation could play a constructive role (as well as a modifying role) in the maturation of the cells in the kitten's visual cortex and that, at present, it is not possible to distinguish between these alternatives.

† In addition to procedural differences in the recent studies of Gaze and Jacobson, there may be a very important species differences: the former has used the African toad (*Xenopus*) and the latter bullfrog (*Rana*). The importance or use of visually guided behavior in the two species (e.g., in feeding) seems quite different, so their mode of development may also be different. *Rana* would appear to be much more visually dependent in its behavior than *Xenopus*, but whether such a difference implies a different developmental mechanism for the establishment of retinotectal connections in anyone's guess.

holds for binocular interaction in the kitten. Namely, when kittens are reared with one eye open and one eye closed, only a small proportion of their visual cortical cells are later capable of responding to binocular stimulation. If, however, they are reared with both eyes open or both eyes closed, a very high percentage (approx. 80%) of their cortical cells can be activated with both eyes (Ganz *et al.*, 1968). In terms of the receptive fields in the kitten's cortex, the deprived eye shows a much less discrete field than does the eye which has been open (Ganz *et al.*, 1868). Again, whether this result is attributable solely to atrophy or dystrophy is not clear.

3. Summary

With regard to sensory development, some writers hold the view that function merely serves to maintain or preserve neuronal connections after they have fully matured, while others hold the view that function may also participate, at least facilitatively, in the actual maturation process itself (e.g., in the establishment of such connections). Function can of course serve in both ways. They are not mutually exclusive. While there is considerable evidence that function (behavioral and neural) helps to maintain anatomical, physiological, and behavioral activities, there is *some* evidence that function can play a facilitative role in the formative stages of sensory maturation (evidence reviewed in Gottlieb, 1971b). Thus, most modern workers would acknowledge that function can play a facilitative as well as a maintenance role in sensory maturation. There is also some evidence, pro (Gaze, 1970) and con (Jacobson, 1971), that function can play a determinative role in the maturation of the nervous system. Whereas there is no disagreement on the maintenance role attributed to function in preserving sensory processes, the extent to which function plays a facilitative role in the establishment of such processes has only recently been realized or appreciated, and the determinative role of function has received too little investigative attention to warrant any general conclusions. The fact that the sensory systems begin to function before they reach maturity prompts us to emphasize once again the possibility that function (spontaneous or evoked) may play a contributory role in the maturation of the sensory systems in the usual course of events. Naturally, since motor systems become functional while they are still immature (see footnote on page 28), these same general considerations hold for the maturation of the motor systems.

E. Perceptual Processes: Behavior

Although electrophysiology has much to tell us about the actual neuronal activities involved in the development of sensory and perceptual processes, behavior can often supply more incisive (and for some, more relevant) information about these processes. There is rarely a clear-cut relationship between electrophysiological (single-unit) studies of perceptual processes and behavioral studies of such processes (e.g., see Dews & Wiesel, 1970; Hirsch, 1970). For this reason, some of the most sophisticated workers in the field of postnatal development combine electrophysiological and behavioral methods of study. With behavioral methods, it is possible to get at the facilitative and determinative stimulative factors involved in the development of perception in an unequivocal way, if indeed prior sensory stimulative factors are at all involved. In the prenatal area, our information is as meagre here as elsewhere, and the material below is more illustrative than it is definitive on the prenatal or embryonic roots of perception.

1. Embryonic Conditioning

Learning depends upon perception and the ability of various vertebrate and invertebrate forms to form conditioned responses in embryonic or larval stages is testimony to the early onset of perceptual processes. For example, several species of parasitic insects show evidence of olfactory or chemical conditioning, effected in the larval or pupal stage, which may in some cases determine the choice of the host on which they deposit their eggs in adulthood (Thorpe, 1939; Thorpe & Jones, 1937).

Mealworms conditioned in the larval stage retain the conditioned response through metamorphosis and show evidence of its retention in adulthood (avoidance learning: Somberg, Happ, & Schneider, 1970; maze learning: Somberg, Happ, & Schneider, 1970; maze learning: Borsellino, Pierantoni, & Schieti-Cavazza, 1970; Von Borell du Vernay, 1942).

Larval salamanders have likewise been shown to learn mazes and to avoid noxious stimulation (Fankhouser, Vernon, Frank, & Slack, 1955; Kuntz, 1923; Schneider, 1968).

In an exposure learning situation (which is unlike a conditioning paradigm), larval frogs show evidence of a selective learning of substrate pattern *which persists through metamorphosis* (Wiens, 1970). In this instance, the tadpoles selectively acquired a preference for a striped substrate rather than a square substrate. The results of this kind of study are more readily related to the ecology of the species than are most conditioning studies (especially those involving unusually noxious stimulation like electric shock), so this kind of paradigm seems more fruitful for demonstrating ecologically meaningful relationships, on the one hand, and more durable (less transient) influences on subsequent behavior, on the other.

Bird embryos are capable of tactile (Gos, 1935) as well as auditory conditioning (Gos, 1935; Hunt, 1949; Sedláček, 1964), and human fetuses show some evidence of conditioning *in utero* (Spelt, 1948).

While it is always somewhat difficult to demonstrate the relevance of prenatal conditioning studies to events which occur during the usual course of embryonic development in a given species, the results of such studies certainly indicate (*1*) the capacity of the embryo's nervous system to react in an adaptive fashion to changes or contingencies in its immediate environment, (*2*) the early functional development of perception, and, in some cases, (*3*) the influence of embryonic experience on behavior in adulthood.

2. Species-Typical Perceptual Preferences

Many newly born or newly hatched animals show distinctive perceptual preferences which are characteristic of their species, and the prenatal ontogeny of these kinds of preferences presents an ideal opportunity for behavioral embryonic analysis. Again it is surprising that very few investigators have taken advantage of such opportunities.

Newly hatched chicks, for example, tend to approach intermittent auditory or visual stimulation which has a rate of change of 2–4 pulses per second in preference to lower and higher pulse rates (Gottlieb & Simner, 1969). Simner (1966) has suggested that the embryonic heart rate (3–4 beats per second) may be instrumental in establishing such a rate preference but, as yet, there have been no manipulative studies on this question. There are other rhythmic occurrences (behavioral and physiological) during embryonic development besides the beating of the heart, so the cardiac self-stimulation hypothesis is not the only possible embryonic precursor to the postnatal rate preferences exhibited by chicks. Certain rates of clicking (in the region of 3 clicks per second) alter the hatching time of quail more effectively than other rates (Vince, this volume), and peking duck embryos give their first overt response to auditory stimulation which is pulsed at 4 per second in contrast to 1 or 6 pulses per second (Heaton, 1970), but the prior ontogeny of these preferences has not been investigated.

In a manipulative study, Grier, Counter, and Shearer (1967) have shown that exposing chick embryos to a 200-Hz tone increases their preference for it after hatching, and Tschanz (1968) has found that young guillemots can learn certain calls of either of their parents during a fairly circumscribed period just before hatching. My own work (reviewed below) indicates that auditory self-stimulation and/or sib stimulation is necessary for the usual development of the Peking duck embryo's response to the maternal call of its species, both before and after hatching. Dimond (1968) found that chicks exposed to light both before and after hatching avoid a moving object more often than chicks exposed to light only after hatching.

Other than rate preferences, chicks and ducklings of various species show very strong color preferences (Hess, 1956; Kear, 1964; Oppenheim, 1968).

These kinds of visual preferences may reflect the region of greatest sensitivity of the eye (analogous to the region of greatest sensitivity of the ear). All waterfowl species tested to date (41) give their maximal behavioral response (pecking) to green, despite the different feeding habits and ecology of these species (Kear, 1964). Rails and pheasants also prefer green, whereas gulls, moorhens, and coots prefer red (Kear, 1964). The bill of the parent gull and moorhen is red, and the young birds peck at the parent's bill to obtain food. In the case of the coot, however, which also prefers red and also pecks at the parent's bill to obtain food, the color of the parent's bill is white. Insofar as the color preferences of the various species reflect the region of greatest sensitivity of the eye, it may be possible to enhance or depress particular preferences (responsiveness) within that region but not to induce a preference for another color by exposure of the embryo to that other color. In fact, Oppenheim (1968) failed to induce a postnatal preference for yellow by exposing duck embryos to that color late in incubation; the postnatal consequences of earlier exposure, or exposure to green itself, have not yet been examined.

In my own work (Gottlieb, 1971a), I have examined the prenatal ontogeny of the domestic mallard (Peking) duckling's selective response to the maternal assembly call of its species. The selectivity of the response in the duckling is not dependent upon prior exposure to the maternal call itself, but it turns out to be dependent upon exposure of the embryo to its own vocalization and/or those of sibs (these birds begin vocalizing 3–4 days before hatching). The embryo's response to the maternal call goes through two developmental phases. In the first (earliest) phase, the embryos show an inhibition of overt activity in response to the call, whereas in the second phase they show an excitation of overt activity in response to the call. If the embryos are precociously exposed to sib vocalizations, they show the more mature response (excitation) to the maternal call 1 day earlier than usual. If the embryos are isolated from sibs, they do not show the more mature response to the maternal call at any time prior to hatching. Thus, the embryonic auditory system, although functional, is still undergoing maturation prior to hatching, and normally occurring stimulation regulates the rate and further development of the auditory perceptual mechanism for species identification during the embryonic period. Most importantly, when embryos which have been deprived of hearing their own vocalizations as well as those of sibs are tested after hatching, they show an imperfect development of their usual auditory discrimination abilities, indicating the dependence of the duckling's auditory system on normally occurring auditory stimulation in order to reach its full maturity. These experiments are described in full detail in a recent monograph (Gottlieb, 1971a).

Aside from general theoretical considerations concerning the extent to which embryonic sensory stimulation plays a facilitative and/or determinative role in the usual ontogeny of species-typical perception, this problem assumes further importance because at least one theory of the development of perception in neonates rather specifically holds that early stimulative events occurring *in embryo* cause the neonate to approach certain stimulative configurations and withdraw from others after hatching. This is T. C. Schneirla's (1965) approach/withdrawal theory of neonatal behavior.

Schneirla's theory rests importantly on a most interesting and as yet untested assumption, namely, that *quantitative* features of tactile and proprioceptive stimulation (rhythms of activity, etc.) to which the young bird is recurrently exposed as an embryo affect its response to previously unencountered auditory and visual stimuli after hatching. This intermodality generalization mechanism, for which there is no direct developmental evidence at present, is akin to Klüver's (1933) principle of stimulus equivalence derived from perceptual experiments with adult monkeys. The essence of Klüver's principle is that if an animal gives the same response to physically dissimilar stimuli, these stimuli are psychologically (perceptually) equivalent to the animal. In his theory, Schneirla applies Klüver's principle to the development of perception, particularly the very early behavior of avian neonates, in order to account for the responses young animals address to previously unencountered forms of stimulation (e.g. auditory or visual patterns). Based on certain evidence that young birds tend to approach gradually changing, regular, low, or low-medium intensity forms of auditory and visual stimulation and tend to withdraw from abruptly changing, irregular, high-intensity forms of auditory and visual stimulation, Schneirla links the former (approach) tendency to the kinds of tactile and proprioceptive stimulation encountered by the embryo and the latter (withdrawal tendency) to the relative absence of such stimulation during embryonic development.

Thus, Schneirla's theory represents an effort to account for postnatal species-typical perceptual preferences on the basis of embryonic experience. It not only assumes (*1*) the *determinative* role of embryonic sensory stimulation on the development of perception, it also assumes (*2*) that the quantitative aspects of experience in one sensory modality can influence the features to which the embryo or neonate will respond in other (later developing) sensory modalities. The second assumption is potentially very significant for our understanding of the development of species-typical perception, so it warrants a very careful experimental examination.

3. Summary

The investigation of the embryonic roots of perception by behavioral means offers promise for the future. At the present time, the meagre amount of evidence available indicates that the embryos of various species (invertebrate and vertebrate) are capable of making discriminative responses within various modes of stimulation (chemical, tactile, auditory), and there is some evidence of a carry-over effect (facilitative and determinative) into later life. The various species-typical perceptual preferences which neonates manifest shortly after hatching or birth provide ideal phenomena for behavioral embryonic analysis. This opportunity has yet to be fully exploited.

VI. Summary and Conclusions

The developmental method is basic to all disciplines which deal with organisms, whether from the genetic, biochemical, anatomical, physiological, behavioral, or psychological points of view, and behavioral embryology pushes this method of study to its logical extreme. The specific aim of behavioral embryology is to determine how the earliest stages of neural and behavioral development affect later stages of neurobehavioral ontogeny. From its inception in the modern era by Preyer, the tradition of behavioral embryology has remained broadly comparative. After a relatively long period of primarily descriptive-correlational studies, behavioral embryology seems now to be moving into an era of increasingly manipulative studies, aimed at testing particular hypotheses.

The overriding background theme in behavioral embryology has to do with the role of functional factors—sensory stimulation, the use or exercise of sensory and motor organs, and the spontaneous functional activity of neural or behavioral systems—in the neural maturation process and behavioral development. (Maturation refers to the proliferation, migration, differentiation, and growth of neural cells, including their axonal and dendritic processes.) One conception of development (predetermined epigenesis) holds that the development of behavior can be explained entirely in terms of neurosensory and neuromotor maturation without the participation of any of the functional factors mentioned above; in other words, that structural maturation determines function (S ⟶ F) and not vice versa. In this view, sensory stimulation, the use of motor and sensory organs, and spontaneous neural or behavioral activity merely serve to maintain or preserve the system when it becomes completely mature. Another conception (probabilistic epigenesis) holds that the above functional factors play an active role in the structural maturation processes themselves, as well as in the development of behavior; in other words, that during the maturation process itself, the structure-function relationship is bidirectional (S ⟷ F). There is very little clear evidence for or against this latter point of view, largely because very few experiments have been designed in such a way that it could be tested. The role of function in maintaining already mature organs or systems is widely appreciated. In any case, the various possible forms which the structure-function relationship may take during development is the unifying investigative theme and core problem of behavioral embryology. In light of extremely misleading and dogmatic assertions which have been made in the past by leading exponents of all views, a strong plea has been entered in this introductory article for an experimentally oriented open-mindedness rather than a premature conceptual closure on the role of function in neural maturation and behavioral development. It seems most reasonable, and in keeping with the meagre evidence at hand, to take as a preliminary working assumption that neural maturation and behavioral development involve both unidirectional and bidirectional structure-function relationships, with the former relationship predominating at the early stages and the latter at the later stages of development.

At the behavioral level, it is possible to study the relationships between early function and later function (function ⟶ function relationships) in a very fruitful way, without necessarily studying the neuroanatomical or neurophysiological correlates of the activity in question, although such studies always gain in significance when the latter are included. There are likewise no strictures on the neuroanatomical and neurophysiological levels of analysis, except that such analyses do tend to hold more obvious promise when overt organismic motility, behavior, or sensitivity is kept in view.

With these considerations in mind, the most salient and concrete problems of behavioral embryology involve:

1. The ontogeny of overt motor activity, specifically the possible relations between early movements and later ones.

2. Nerve fiber growth and the formation of synapses, especially the problem of the mechanism of the functional specificity of interneuronal connections.

3. The physiological and anatomical aspects of sensory development, particularly the possible influence of function in *immature* systems.

4. The behavioral aspects of perceptual development, such as the prenatal roots or origins of postnatal perceptual preferences.

A review of the main findings and hypotheses in these four areas reveals several viewpoints, many promising leads, and much unexplored territory.

References

Angulo y González, A. W. The prenatal development of behavior in the albino rat. *Journal of Comparative Neurology*, 1932, **55**, 395–442.

Anokhin, P. K. Systemogenesis as a general regulator of brain development. *Progress in Brain Research*, 1964, **9**, 54–86.

Ariëns Kappers, C. U. *The comparative anatomy of the nervous system of vertebrates including man.* Vol. 1. New York: Macmillan, 1936.

Barlow, H. B., & Pettigrew, J. D. Lack of specificity of neurones in the visual cortex of young kittens. *Journal of Physiology (London)*, 1971, **218**, 98P–100P.

Blakemore, C., & Cooper, G. Development of the brain depends on the visual environment. *Nature (London)*, 1970, **228**, 477–478.

Bogdanov, O. V. The significance of proprioceptive input for functional maturation of the central nervous system in the chick embryo. *Sechenov Journal of Physiology (USSR)*, 1963, **49**, 701–705.

Bok, S. T. Stimulogenous fibrillation as the cause of the structure of the nervous system. *Psychiatrie en Neurologie*, Vol. 19. Amsterdam: Bladen, 1915. Pp. 393–408.

Borsellino, A., Pierantoni, R., & Schieti-Cavazza, B. Survival in adult mealworm beetles (*Tenebrio molitor*) of learning acquired at the larval stage. *Nature (London)*,1970, **225**, 963–964.

Bursian, A. V. Primitive forms of photosensitivity at early stages of embryogenesis in the chick. *Journal of Evolutionary Biochemistry and Physiology*, 1965, **1**, 435–441 (translated from Russian).

Carmichael, L. The development of behavior in vertebrates experimentally removed from the influence of external stimulation. *Psychological Review*, 1926, **33**, 51–58.

Carmichael, L. Origin and prenatal growth of behavior. In C. Murchison (Ed.), *A handbook of child psychology.* (2nd ed.) Worcester, Mass.: Clark University Press, 1933. Pp. 31–159.

Carmichael, L. An experimental study in the prenatal guinea-pig of the origin and development of reflexes and patterns of behavior in relation to the stimulation of specific receptor areas during the period of active fetal life. *Genetic Psychology Monographs*, 1934, **16**, 338–491.

Carmichael, L. The onset and early development of behavior. In P. H. Mussen (Ed.), *Carmichael's manual of child psychology.* Vol. 1. New York: Wiley, 1970. Pp. 447–563.

Child, C. M. *The origin and development of the nervous system from a physiological viewpoint.* Chicago: University of Chicago Press, 1921.

Clark, E. L., & Clark, E. R. On the early pulsations of the posterior lymph hearts in chick embryos: their relation to the body movements. *Journal of Experimental Zoology*, 1914, **17**, 373–394.

Coghill, G. E. *Anatomy and the problem of behaviour.* Cambridge, Eng.: Cambridge University Press, 1929.

Crain, S. M. Resting and action potentials of cultured chick embryo spinal ganglion. *Journal of Comparative Neurology*, 1956, **104**, 285–330.

Crain, S. M. In M. R. Murray. Nervous tissues *in vitro*. In E. N. Willmer (Ed.), *Cells and tissues in culture.* Vol. 2. New York: Academic Press, 1965. Pp. 373–455.

Crain, S. M. Bornstein, M. B., & Peterson, E. R. Development of functional organization in cultured embryonic CNS tissues during chronic exposure to agents which prevent bioelectric activity. In L. Jílek & S. Trojan (Eds.), *Ontogenesis of the brain.* Prague: Charles University Press, 1968. Pp. 19–25.

Dews, P. B., & Wiesel, T. N. Consequences of monocular deprivation on visual behavior in kittens. *Journal of Physiology (London)*, 1970, **206**, 437–455.

Dimond, S. J. Effects of photic stimulation before hatching on the development of fear in chicks. *Journal of Comparative and Physiological Psychology*, 1968, **65**, 320–324.

Faber, J. The development and coordination of larval limb movements in *Triturus taeniatus* and *Ambystoma mexicanum* (with some notes on adult locomotion in *Triturus*). *Archives Néerlandaises de Zoologie*, 1956, **11**, 498–517.

Fankhauser, G., Vernon, J. A., Frank, W. H., & Slack, W. V. Effect of size and number of brain cells on learning in the larvae of salamander, *Triturus viridescens*. *Science*, 1955, **122**, 692–693.

Fischbach, G. D. Synaptic potentials recorded in cell cultures of nerve and muscle. *Science*, 1970, **169**, 1331–1333.

Forssman, J. Zur Kenntnis des Neurotropismus. *Beiträge zur Pathologischen Anatomie und zur Allgemeinen Pathologie*, 1900, **27**, 407–430.

Fromme, A. An experimental study of the factors of maturation and practice in the behavioral development of the embryo of the frog. *Genetic Psychology Monographs*, 1941, **24**, 219–256.

Ganz, L., Fitch, M., & Satterberg, J. A. The selective effect of visual deprivation on receptive field shape determined neurophysiologically. *Experimental Neurology*, 1968, **22**, 614–637.

Gaze, R. M. *The formation of nerve connections*. New York: Academic Press, 1970.

Gaze, R. M., Keating, M. J., Székely, G., & Beazley, L. Binocular interaction in the formation of specific intertectal neuronal connections. *Proceedings of the Royal Society, Series B*, 1970, **175**, 107–147.

Gos, M. Les réflexes conditionnels chez l'embryon d'oiseau. *Societé Royale des Sciences de Liége, Bulletin*, 1935, **4**, 194–199, 246–250.

Gottlieb, G. Prenatal behavior of birds. *Quarterly Review of Biology*, 1968, **43**, 148–174.

Gottlieb, G. Conceptions of prenatal behavior. In L. R. Aronson, E. Tobach, D. S. Lehrman, & J. S. Rosenblatt (Eds.), *Development and evolution of behavior. Essays in memory of T. C. Schneirla*. San Francisco: Freeman, 1970. Pp. 111–137.

Gottlieb, G. *Development of species identification. An inquiry into the prenatal determinants of perception*. Chicago: University of Chicago Press, 1971. (a)

Gottlieb, G. Ontogenesis of sensory function in birds and mammals. In E. Tobach, L. R. Aronson, & E. Shaw (Eds.), *The biopsychology of development*. New York: Academic Press, 1971. Pp. 67–128. (b).

Gottlieb, G., & Kuo, Z.-Y. Development of behavior in the duck embryo. *Journal of Comparative and Physiological Psychology*, 1965, **59**, 183–188.

Gottlieb, G., & Simner, M. L. Auditory versus visual flicker in directing the approach response of domestic chicks. *Journal of Comparative and Physiological Psychology*, 1969, **67**, 58–63.

Grier, J. B., Counter, S. A., & Shearer, W. M. Prenatal auditory imprinting in chickens. *Science*, 1967, **155**, 1692–1693.

Grosse, G., Linder, G., & Schneider, P. Der Einfluss des elektrischen Feldes auf gliale und mesenchymale Sellen in vitro. *Zeitschrift für Mikroskopisch-Anatomische Forschung*, 1969, **80**, 532–540. (a)

Grosse, G., Lindner, G., & Schneider, P. Der Einfluss des elektrischen Feldes auf in vitro kultiviertes Nervengewebe. *Zeitschrift für Mikroskopisch-Anatomische Forschung*, 1969, **80**, 260–268. (b)

Gyllensten, L., & Malmfors, T. Myelinization of the optic nerve and its dependence on visual function—a quantitative investigation in mice. *Journal of Embryology and Experimental Morphology*, 1963, **11**, 255–266.

Gyllensten, L., Malmfors, T., & Norrlin, M.-L. Effect of visual deprivation on the optic centers of growing and adult mice. *Journal of Comparative Neurology*, 1965, **124**, 149–160.

Gyllensten, L., Malmfors, T., & Norrlin, M.-L. Developmental and functional alterations in the fiber composition of the optic nerve in visually deprived mice. *Journal of Comparative Neurology*, 1966, **128**, 413–418. (a)

Gyllensten, L., Malmfors, T., & Norrlin, M.-L. Growth alteration in the auditory cortex of visually deprived mice. *Journal of Comparative Neurology*, 1966, **126**, 463–470. (b)

Hamburger, V. The concept of "development" in biology. In D. H. Harris (Ed.), *The concept of development*. Minneapolis: University of Minnesota Press, 1957. Pp. 48–58.

Hamburger, V. Some aspects of the embryology of behavior. *Quarterly Review of Biology*, 1963, **38**, 342–365.

Hamburger, V. Ontogeny of behaviour and its structural basis. In D. Richter (Ed.), *Comparative neurochemistry*. Oxford: Pergamon, 1964. Pp. 21–34.

Hamburger, V. Emergence of nervous coordination. Origins of integrated behavior. In M. Locke (Ed.), *The emergence of order in developing systems*. New York: Academic Press, 1968. Pp. 251–271.

Hamburger, V., & Narayanan, C. H. Effects of the deafferentation of the trigeminal area on the motility of the chick embryo. *Journal of Experimental Zoology*, 1969, **170**, 411–426.

Hamburger, V., & Oppenheim, R. Prehatching motility and hatching behavior in the chick. *Journal of Experimental Zoology*, 1967, **166**, 171–204.

Harrison, R. G. An experimental study of the relation of the nervous system to the developing musculature of the frog. *American Journal of Anatomy* 1904, **3**, 197–220.

Harrison, R. G. The reaction of embryonic cells to solid structures. *Journal of Experimental Zoology*, 1914, **17**, 521–544.

Harrison, R. G. In S. Wilens (Ed.), *Organization and development of the embryo*. New Haven: Yale University Press, 1969.

Heaton, M. B. Stimulus coding in the species-specific perception of peking ducklings. Unpublished doctoral dissertation, North Carolina State University, 1970.

Heaton, M. B. Ontogeny of vision in the peking duck (*Anas platyrhynchos*): the pupillary light reflex as a means for investigating visual onset and development in avian embryos. *Developmental Psychobiology*, 1971, **4**, 313–332.

Helfenstein, M., & Narayanan, C. H. Effects of bilateral limb bud extirpation on motility and prehatching behavior in chicks. *Journal of Experimental Zoology*, 1970, **172**, 233–244.

Herrick, C. J. *George Ellett Coghill, naturalist and philosopher*. Chicago: University of Chicago Press, 1949.

Hess, E. H. Natural preferences of chicks and ducklings for objects of different colors. *Psychological Reports*, 1956, **4**, 477–483.

Hirsch, H. V. B. The modification of receptive field orientation and visual discrimination by selective exposure during development. Unpublished doctoral dissertation, Stanford University, 1970.

Hirsch, H. V. B., & Spinelli, D. N. Visual experience modifies distribution of horizontally and vertically oriented receptive fields in cats. *Science*, 1970, **168**, 869–871.

Hirsch, H. V. B., & Spinelli, D. N. Modification of the distribution of receptive field orientation in cats by selective visual exposure during development. *Experimental Brain Research*, 1971, **13**, 509–527.

Holt, E. B. *Animal drive and the learning process*. Vol. 1. New York: Holt, 1931.

Hubel, D. H., & Wiesel, T. N. Receptive fields of cells in striate cortex of very young, visually inexperienced kittens. *Journal of Neurophysiology*, 1963, **26**, 994–1002.

Hubel, D. H., & Wiesel, T. N. The period of susceptibility to the physiological effects of unilateral eye closure in kittens. *Journal of Neurophysiology*, 1970, **206**, 419–436.

Humphrey, T. Postnatal repetition of human prenatal activity sequences with some suggestions of their neuroanatomical basis. In R. J. Robinson (Ed.), *Brain and early behaviour*: *Development in the fetus and infant*. New York: Academic Press, 1969. Pp. 43–84.

Humphrey, T. Human prenatal activity sequences in the facial region and their relationship to postnatal development. *American Speech and Hearing Association* (ASHA) *Report*, 1971, No. 6, 19–37.

Hunt, E. L. Establishment of conditioned responses in chick embryos. *Journal of Comparative Psychology*, 1949, **42**, 107–117.

Ingvar, S. Reaction of cells to galvanic current in tissue culture. *Proceedings of the Society for Experimental Biology and Medicine*, 1920, **17**, 198–199.

Jacobson, M. *Developmental neurobiology*. Hew York: Holt, 1970.

Jacobson, M. Absence of adaptive modification in developing retinotectal connections in frogs after visual deprivation or disparate stimulation of the eyes. *Proceedings of the National Academy of Sciences, U.S.*, 1971, **68**, 528–532.

Kano, M., & Shimada, Y. Innervation of skeletal muscle cells differentiated *in vitro* from chick embryo. *Brain Research*, 1971, **27**, 402–405.

Karssen, A., & Sager, B. Sur l'influence du courant electrique sur la croissance des neuroblastes *in vivo*. *Archiv für Experimentelle Zellforschung Besonders Gewebezüchtung*, 1934, **16**, 255–259.

Kear, J. Colour preference in young Anatidae. *Ibis*, 1964, **106**, 361–369.

Klüver, H. *Behavior mechanisms in monkeys*. Chicago: University of Chicago Press, 1933.

Kuntz, A. The learning of a simple maze by the larva of *Amblystoma tigrinum*. *University of Iowa Studies in Natural History*, 1923, **10**, 27–35.

Kuo, Z.-Y. Studies in the physiology of the embryonic nervous system: II. Experimental evidence on the controversy over the reflex theory in development. *Journal of Comparative Neurology*, 1939, **70**, 437–459.

Kuo, Z.-Y. *The dynamics of behavior development*. New York: Random House, 1967.

Mackie, G. O. Neuroid conduction and the evolution of conducting tissues. *Quarterly Review of Biology*, 1970, **45**, 319–332.

Marsh, G., & Beams, H. W. In vitro control of growing chick nerve fibers by applied electric currents. *Journal of Cellular and Comparative Physiology*, 1946, **27**, 139–157.

Matthews, S. A., & Detwiler, S. R. The reactions of Amblystoma embryos following prolonged treatment with chloretone. *Journal of Experimental Zoology*, 1926, **45**, 279–292.

Nilsson, S. E. G., & Crescitelli, F. Changes in ultrastructure and electroretinogram of bullfrog retina during development. *Journal of Ultrastructure Research*, 1969, **27**, 45–62.

Nilsson, S. E. G., & Crescitelli, F. A correlation of ultrastructure and function in the developing retina of the frog tadpole. *Journal of Ultrastructure Research*, 1970, **30**, 87–102.

Oppenheim, R. W. Color preferences in the pecking response of newly hatched ducks (*Anas platyrhynchos*). *Journal of Comparative and Physiological Psychology*, 1968, **66**, (Monogr. Suppl.), 1–17.

Oppenheim, R. W. Some aspects of embryonic behavior in the duck (*Anas platyrhynchos*). *Animal Behaviour*, 1970, **18**, 335–352.

Oppenheim, R. W. The embryology of behavior in birds: A critical review of the role of sensory stimulation in embryonic movement. In K. H. Voous (Ed.), *Proceedings of the XVth international congress of ornithology*. Leiden: Brill, 1972. (a)

Oppenheim, R. W. An experimental investigation of the possible role of tactile and proprioceptive stimulation in certain aspects of embryonic behavior in the chick. *Developmental Psychobiology*, 1972, **5**, 71–91. (b)

Paulson, G. W. Maturation of evoked responses in the duckling. *Experimental Neurology*, 1965, **11**, 324–333.

Pearl, R. On the behavior and reactions of Limulus in early stages of its development. *Journal of Comparative Neurology and Psychology*, 1904, **14**, 138–164.

Peters, J. J., Vonderahe, A. R., & Powers, T. H. The functional chronology in developing chick nervous system. *Journal of Experimental Zoology*, 1956, **133**, 505–518.

Preyer, W. *Specielle Physiologie des Embryo. Untersuchungen über die Lebenserscheinungen vor der Geburt*. Leipzig: Grieben, 1885.

Ramón y Cajal, S. *Degeneration and regeneration of the nervous system*. (R. M. May, Trans.) Vol. 1. 1928. (Reprinted: New York, Hafner, 1959.)

Rhines, R. An experimental study of the development of the medial longitudinal fasciculus in the chick. *Journal of Comparative Neurology*, 1943, **79**, 107–126.

Schapiro, S., & Vukovich, K. R. Early experience effects upon cortical dendrites. *Science*, 1970, **167**, 292–294.

Schneider, C. W. Avoidance learning and the response tendencies of the larval salamander *Ambystoma punctatum* to photic stimulation. *Animal Behaviour*, 1968, **16**, 492–495.

Schneirla, T. C. Aspects of stimulation and organization in approach/withdrawal processes underlying vertebrate behavioral development. In D. S. Lehrman, R. A. Hinde, & E. Shaw (Eds.), *Advances in the study of behavior*. Vol. 1. New York: Academic Press, 1965. Pp. 1–74.

Sedláček, J. Further findings on the conditions of formation of the temporary connection in chick embryos. *Physiologia Bohemoslovenica*, 1964, **13**, 411–420.

Sheridan, J. D. Electrophysiological evidence for low-resistance intercellular junctions in the early chick embryo. *Journal of Cell Biology*, 1968, **37**, 650–659.

Shlaer, R. Shift in binocular disparity causes compensatory change in the cortical structure of kittens. *Science*, 1971, **173**, 638–641.

Simner, M. L. Cardiac self-stimulation hypothesis and the response to visual flicker in newly hatched chicks: preliminary findings. *Proceedings, 74th Annual Convention, American Psychological Association*, 1966, **1**, 141–142.

Somberg, J. C., Happ, G. M., & Schneider, A. M. Retention of a conditioned avoidance response after metamorphosis in mealworms. *Nature* (*London*), 1970, **228**, 87–88.

Spelt, D. K. The conditioning of the human fetus *in utero*. *Journal of Experimental Psychology*, 1948, **38**, 338–346.

Sperry, R. W. Mechanisms of neural maturation. In S. S. Stevens (Ed.), *Handbook of experimental psychology*. New York: Wiley, 1951. Pp. 236–280.

Sperry, R. W. Embryogenesis of behavioral nerve nets. In R. L. DeHaan & H. Ursprung (Eds.), *Organogenesis*. New York: Holt, 1965. Pp. 161–186.

Sperry, R. W. How a developing brain gets itself properly wired for adaptive function. In E. Tobach, L. R. Aronson, & E. Shaw (Eds.), *The biopsychology of development*. New York: Academic Press, 1971. Pp. 27–44.

Sviderskaya, G. E. Effect of sound on the motor activity of chick embryos. *Bulletin of Experimental Biology and Medicine* (*USSR*), 1967, **63**, 24–28.

Swenson, E. A. The active simple movements of the albino rat fetus: the order of their appearance, their qualities, and their significance. *Anatomical Record*, 1929, **42**, 40. (Abstract)

Thorpe, W. H. Further experiments on pre-imaginal conditioning in insects. *Proceedings of the Royal Society Series, B*, 1939, **127**, 424–433.

Thorpe, W. H., & Jones, F. G. W. Olfactory conditioning and its relation to the problem of host selection. *Proceedings of the Royal Society, Series B*, 1937, **124**, 56–81.

Tracy, H. C. The development of motility and behavior reactions in the toadfish (*Opsanus tau*). *Journal of Comparative Neurology*, 1926, **40**, 253–369.

Trevarthen, C. Behavioral embryology. In E. C. Carterette & M. P. Friedman (Eds.), *The handbook of perception*. New York: Academic Press, 1973.

Tschanz, B. Trottellummen, Die Entstehung der persönlichen Beziehungen zwischen Jungvogel und Eltern. *Zeitschrift für Tierpsychologie, Supplement*, 1968, **4**, 1–103.

Visintini, F., & Levi-Montalcini, R. Relazione tra differenciazone strutturale e funzionale dei centri e delle vie nervose nell'embrione di pollo. *Schweizer Archiv für Neurologie und Psychiatrie*, 1939, **43**, 1–45.

von Borell du Vernay, W. Assoziationsbildung und sensibilisierung bei *Tenebrio. Zeitschrift für Vergleichende Physiologie*, 1942, **30**, 84–116.

Weiss, P. In vitro experiments on the factors determining the course of the outgrowing nerve fiber. *Journal of Experimental Zoology*, 1934, **68**, 393–448.

Weiss, P., & Taylor, A. C. Further experimental evidence against "neurotropism" in nerve regeneration. *Journal of Experimental Zoology*, 1944, **95**, 233–257.

Wiens, J. A. Effects of early experience on substrate pattern selection in *Rana aurora* tadpoles. *Copeia*, 1970, No. 3, 543–548.

Windle, W. F. Genesis of somatic motor function in mammalian embryos: a synthesizing article. *Physiological Zoology*, 1944, **17**, 247–260.

Zetterström, B. The effect of light on the appearance and development of the electroretinogram in newborn kittens. *Acta Physiologica Scandanavica*, 1956, **35**, 272–279.

Section 2

EMBRYONIC MOTILITY AND ITS NEURAL CORRELATES

INTRODUCTION

A first principle of behavioral embryology holds that, in some submammalian species at least, the motor system becomes active in advance of its synaptic connection to any of the sensory systems, so that the earliest overt movements in these species can truly be called spontaneous or autogenous. The principle of motor primacy was first enunciated by W. Preyer, in 1885, with reference to the early motility of chick embryos. In 1926, H. C. Tracy made the same discovery in his behavioral studies of toadfish. The conclusion of Preyer and Tracy hinges on the fact that both the chick embryo and the toadfish larva are active before they are capable of showing a response to exteroceptive sensory stimulation. Thus, to the extent that the movements are not instigated by proprioceptive or interoceptive sensory stimulation, they might very well be an intrinsic derivative of the neuromuscular system.

In the 1930's, several neuroanatomical investigations of the chick embryo supported Preyer's hypothesis in that no reflex pathways were found between the dorsal (sensory) root fibers and the internuncial neurons which connect to the motor neurons until 6 or 7 days of development, whereas the chick embryo becomes overtly active around day 4. It is to the lasting credit of Viktor Hamburger and his colleagues, Eleanor Wenger and Ronald Oppenheim, that any possible doubts about the chick embryo's capability for autogenous motility were entirely removed when, in 1966, they published a study which showed that leg movements occurred in chick embryos which had been surgically deprived of the sensory fibers at the leg level of the spinal cord. The leg movements of these chick embryos were clearly nonreflexogenic.

In the following section, Dr. Hamburger discusses the possible occurrence of spontaneous motility in a variety of vertebrate forms, perhaps even in mammalian embryos where there is no nonreflexogenic period prior to the onset of movement. (The onset of movement and the overt response to exteroceptive sensory stimulation appear to coincide in most, if not all, mammalian species.) His novel suggestion that, in the chick embryo, there may be an incongruity between neurogenesis and the development of behavior, if correct, has rather pessimistic implications for establishing correlations between the developing nervous system and later embryonic behavior, already a most difficult task. Dr. Hamburger also discusses, in a way he has not done previously, the possible influence of sensory stimulation, the autogenous movements themselves, and other factors on the development of behavior and the maturation of the nervous system.

In Robert Provine's chapter we are introduced to one of the neural correlates of early embryonic motility in the chick: the electrical burst discharges of the spinal cord that correlate very well with the overt motility of the embryo throughout the period of incubation. Dr. Provine's finding is a testimony to the prescience of Preyer, who emphasized the probable importance of the spinal cord in particular as the source of early embryonic motility in the chick. We know from other studies that there are influences on motility from the brain, but these would appear to be more of a contributory nature rather than the primary source of motility, at the early ages at least.

Rainer Foelix and Ronald Oppenheim broaden the neural basis of our understanding of chick embryonic motility as they relate the outcome of their pioneering search for synapses in the chick embryo's spinal cord to the onset of motility and later inhibitory changes. Although the leap from the ultrastructural level of the electron microscope to the overt behavior of the embryo is huge indeed, the authors' attempt to bridge the gap makes their findings all the more interesting and meaningful.

Very little behavioral research has been performed on invertebrate embryos, so Michael Berrill introduces us to the behavior of the only invertebrates (six crustacean species) included in the present volume. Since very refined neuroanatomical and neurophysiological work is proceeding apace on the *adults* of other aquatic forms, the apparently spontaneous embryonic motility, as well as the embryonic behavioral differences, in the various crustacean forms studied by Dr. Berrill may very well excite some *developmental* neuroanatomical and neurophysiological interest in these forms. One can not refrain from noting that, depending on the temperature at which the "berried" female is kept, the eggs of the lobster (*Homarus americanus*) remain in their embryonic state anywhere from 5 to 12 months, thereby providing ample time for leisurely experimental analysis and, with continued patience, eventually, perhaps even a delicious repast.

ANATOMICAL AND PHYSIOLOGICAL BASIS OF EMBRYONIC MOTILITY IN BIRDS AND MAMMALS

VIKTOR HAMBURGER

Department of Biology
Washington University
St. Louis, Missouri

I. Introduction

In previous communications, I have paid tribute to Professor Wilhelm Preyer, to whom this volume is dedicated. The importance of his theoretical and observational contributions to the embryology of behavior has been acknowledged particularly in a recent essay (Hamburger, 1971). If his name does not appear in the following pages, this means that his influence has reached the stage of anonymity which is the fate of all great innovators.

Observations and experiments on embryonic motility in amniotes have been reviewed repeatedly in recent years (e.g., Gottlieb, 1971b; Hamburger, 1968, 1970, 1971), and I shall not go once more over familiar grounds. The task which I have undertaken is to scrutinize the relationships of processes going on at three different levels of organization: the level of motility and behavior, the level of bioelectrical activity, and the level of structural and ultrastructural organization. The discussion is limited to birds and mammals, but it includes spontaneous motility and evoked responses. I have selected a few key issues, but in these I have tried to go beyond generalities and to understand relatioships in depth and in specific detail.

The basic motility type of the chick embryo has three characteristics: it is spontaneous (nonreflexogenic); it is intermittent; and the parts of the embryo, such as head, trunk, limbs, beak, move in a seemingly uncoordinated fashion. This type of motility has been designated as Type I. Type II are startles, and Type III, coordinated hatching movements (Hamburger & Oppenheim, 1967). In a previous review (Hamburger, 1971) I have tried to establish spontaneity, based on autonomous discharges of neurons, as the constitutive element in embryonic motility, and I have made the point that neither periodicity nor the forms or patterns of overt motility are of equally universal character. In many of our investigations, durations of activity phases or number of movements per time unit had been used preferentially, for the simple reason that this parameter of Type I motility can be quantified. However, periodicity cannot be considered as a necessary prerequisite of spontaneity. Clear periodicity is obvious in chick embryos only up to about 13 days of incubation. Thereafter, motility and its correlate, burst activity, remain discontinuous, but the inactivity phases become very short and ill-defined.

The form, or pattern, of motility likewise is not universally uniform. Spontaneous motility can be either uncoordinated, as in reptiles, birds, and mammals, or coordinated, as in teleosts. In the latter, Tracy (1926) has found that embryonic and larval movements, including swimming, are spontaneous (nonreflexogenic) for about 2 weeks. The situation in the salamander is not quite clear. Spontaneity remains the one basic characteristic of overt, unelicited motility in vertebrate embryos.

II. Some Remarks on Lack of Coordination in Amniote Embryos

The unstructured performance of the chick embryo up to 17 days, and similar performances of reptilian and mammalian embryos, hardly deserve the designation of "behavior." I have usually referred to Type I and II ("startless") as motility, and only to the integrated prehatching and hatching performance (Type III) as behavior, since this term usually implies an integrated activity which subserves the whole organism (see also Gottlieb, 1971a, p. 6). In contrast, although Type I also involves the movements of several or all parts, the resulting motility appears to be merely the summation or combination of movements of individual parts, such as head, trunk, limbs, beak, eyelids. Such performance hardly occurs in postnatal behavior except perhaps in infants of immaturely born forms, including the human. The jerky uncontrolled movements of the chick have more resemblance to neuropathological convulsive conditions. The phenomenon of uncoordinated motility which, moreover, does not involve sensory input to any extent, may not seem to be of much interest to developmental psychobiologists. Yet, there remains the inescapable truth that this type of motility with all its monotony seems to be the basic type that amniote embryos and fetuses perform, notwithstanding occasional responses to stimulation. Hence, this activity is obviously of considerable importance for the embryo; it cannot be ignored in generalizations and theories concerning prenatal behavior and its relation to postnatal action patterns.

It has been objected that our claim that the combinations of movements in Type I motility are uncoordinated has not been subjected to a rigorous statistical analysis. It would be desirable, indeed, to make the recording method more objective and to do a correlation analysis of sequence of movements, perhaps using slow-motion moving pictures. However, even if such analysis would reveal a low degree of correlation between those particular types of movements that become integrated later, this would not necessitate a revision of my basic position, as I hope to be able to show when we analyze the electrophysiological patterns underlying Type I motility, and evoked responses.

In his Introduction, Dr. Gottlieb has stated that we have failed to recognize two particular instances of coordinated patterns in bird embryos. The one instance is nonrandom movement of the head, or, more specifically, head turning which is performed in the following context. Up to several days before hatching the beak is buried in the yolk sac and the neck is bent downward. The head is then lifted out of the yolk by head movements which are preferentially to the right. Eventually, the twist of the neck to the right becomes permanent and the head, which is now near the right wing, is subsequently tucked under the right wing. Oppenheim and I (Hamburger & Oppenheim, 1967) have described the preponderance of head movements to the right and considered this as the beginning of coordinated prehatching motility (Type III) and as preparatory to tucking. We have called this the pretucking phase. This interpretation is based (1) on a rotatory component that is involved and which is not observed in Type I, and (2) on the continuity of pretucking and tucking movements. We think that, in the chick, the head turning to the right is part of Type III and has nothing to do with Type I. Gottlieb and Kuo (1965) have apparently traced the preferential head turning to the right in the duck embryo to earlier stages than we did. The point deserves further consideration. Perhaps in the duck this nonrandom performance has to be considered as an aspect of Type I motility and as an exception to the rule.

The second instance are sinusoid waves performed by early chick embryos. They have been described by us and by others before us. In fact, S-waves are characteristic for very early motility stages in many vertebrate embryos, and they are, indeed, a type of coordinated movements. In fishes and amphibians, they are the direct precursors of swimming movements (Coghill, 1929; Tracy, 1926). But in the chick, this pattern breaks down soon after its inception at 4 days. At 5 days, only the first waves in a sequence usually go down all the way. Subsequent waves may end in the trunk or begin in the trunk and spread in both directions. At the same time, local contractions of trunk somites make their appearance (Hamburger & Balaban, 1963). In subsequent stages, the exceptions become the rule, and the S-waves disappear altogether.

Perhaps the neural mechanism available to the early embryo permits only the coordinated S-waves. A longitudinal fiber tract, which sends collaterals to the segmentally arranged motoneurons, and a ventral commissure are formed very early (Visintini & Levi-Montalcini, 1939), and perhaps only the rostrally located neurons of this fiber tract are at first sufficiently mature to initiate spontaneous discharges. The pattern breaks up, when sufficient numbers of neurons in the spinal cord become interconnected.

One might ask, why the chick embryo, and the amniote embryo in general, does not continue on its original path, following Coghill's scheme of continuously integrated behavior from beginning to end. Several answers to this question can be given; I shall come back to it later. One answer can be dismissed right away. The occurrence of S-waves in early embryos of higher forms could be considered as the recapitulation of an ancestral pattern, characteristic of aquatic forms, where it leads to swimming. While such a viewpoint is legitimate, it cannot be accepted as an adequate "explanation" of currently displayed behavior.

III. Spontaneous Motility in Rat Fetuses

Before we continue the discussion, I propose to broaden the basis by including mammals. Our recent studies of spontaneous motility in rat fetuses (Narayanan, Fox, & Hamburger, 1971) have shown basic similarities with the chick, and some differences. A considerable difference is in the stage of initiation of motility. Mammalian embryos are much more advanced in overall body development at the time when motility begins; their limbs are in possession of movable joints, whereas chick embryos have small limb buds, when the first neck flexions begin. In mammals reflex arcs are completed at the onset of motility, whereas chick embryos have a prereflexogenic period of 4 days. Motility in the rat embryo begins at $15\frac{1}{2}$ days; our observations cover the period of 16–20 days.

The females were immobilized by thoracic spinal transection to avoid anesthesia. Their lower body parts were submerged in a water bath at 37° C. Fetal movements were observed through the uterine walls, or after exposure of the fetus, with placenta and amnion intact. Such preparations show no changes in motility for several hours. Recordings of activity phases were made as in the chick, during 15-minute observation periods, both *in utero* and *ex utero*.

We have distinguished between total, regional, and local movements. By regional we mean, for instance, a combination of head and forelimb movements, with other parts at rest, or hindlimb, tail, and pelvis combinations. The movements are in general smoother than in the chick, though local movements of individual parts may be jerky. As in the chick, total activity builds up to a peak which is reached at 18 days and then declines. As in reptiles and birds, motility is intermittent; the activity phases occur at irregular intervals. The movements of parts are not correlated with each other, much like in Type I motility of birds. This is particularly striking in the lack of coordination of the movements of right and left forelimbs, which are the most active parts throughout.

Altogether, the similarities in the three major characteristics, namely spontaneity, periodicity, and lack of coordination, are so great that we have no doubt that we are dealing with the same basic type in birds and mammals. Although other investigators of motility in mammals have paid little attention to spontaneous motility, their casual observations on guinea pig, cat, sheep, and other fetuses, including the human, are in line with ours. What we call uncoordinated Type I motility is usually referred to as generalized or mass or total activity.

The nonreflexogenic nature of the fetal movements in the rat has not been established experimentally. We have excluded one possible source of extraneous stimulation: contractions of the uterus. Fetal movements are five times as frequent as uterine contractions, and their periodicities are unrelated. The amnion is not contractile in the rat.

IV. The Continuity–Discontinuity Problem on the Behavioral Level

The shift from the coordinated S-waves to an uncoordinated type of motility, and then to the coordinated hatching behavior in bird embryos, raises a larger and more general issue: the continuity-discontinuity problem. In the following discussion, we shall keep in mind that we are dealing with phenomena on several levels, the behavioral, the electrophysiological, and the structural-ultrastructural. Much confusion can be avoided if this point is clearly recognized.

On the behavioral level, Oppenheim and I (Hamburger & Oppenheim, 1967) have asserted a discontinuity between Type I and the integrated prehatching and hatching behavior, Type III. We have given repeatedly the arguments for our stand that it is not possible to assume a smooth transformation of Type I into Type III. Briefly, they are the following (*1*) The two types are very different in their form, apart from the highly integrated character of Type III; a rotatory component is observed in most of the phases of prehatching and hatching motility; this is entirely missing in Type I. (*2*) The Type III movements are performed in episodes that are separated by intervals of different durations. The Type I movements do not disappear altogether in the prehatching period between 17 and 20 days. They are merely suspended during Type III episodes, but resumed during the intervals, though at reduced frequency.

Oppenheim (1970) has examined the question of gradual transition versus rather sudden appearance of a behavioral act. He has analyzed the origin of one important component of Type III; namely, the vigorous back thrusts of head and beak which are instrumental in pipping and hatching. By recording head movements directed backward, lateral, or in other directions during prehatching stages in the duck, he has found that the frequency of back movements does not build up gradually; they increase rather suddenly beginning about 16 hours before initiation of hatching, compared to lateral and other movements. Oppenheim (1970) points out that individual back movements occur sporadically in earlier stages. In this sense, there is "not a de novo appearance of these component movements [p. 348]." The case is well stated when he points out that the novel aspect, apart from increase in frequency, is the incorporation of back thrusts in an integrated activity pattern, whereby they become effective in pipping and hatching. I think this kind of continuity whereby behaviorally uncommitted individual movements become integrated in a more complex behavioral act, at a higher organizational level, is paradigmatic for much of embryonic and fetal behavior. *In this way, aspects of continuity and discontinuity become compatible.* In a similar way, complex postnatal patterns, like walking or pecking, can be said to arise "*de novo*"; the newly hatched chick stands up and walks within $\frac{1}{2}$ hour; so does the newborn sheep or colt. Most of the component movements have been performed before, but not in the context of an integrated activity. To give another example: Embryonic swallowing is incorporated in postnatal feeding and drinking. Kuo (1932) has expressed a similar notion, except that in his view the component movements result from stimulation or self-stimulation.

We have observed another form of discontinuity within the framework of hatching behavior. I refer to the sudden onset of the climax, that is, the act of hatching, which involves a change in the temporal pattern of Type III movements. Kovach (1970) has also studied this and preceding processes in more detail, using a movement transducer for recording from intact eggs and eggs with windows. Between days 17 and 20, Type III rotatory movements are performed sporadically or in short repetitive sequences, at varying and sometimes hour-long intervals. Then, suddenly, at an unpredictable moment, at day 20+, they are enacted in regular cycles with short intervals of 10–30 seconds, for periods lasting $\frac{1}{2}$ to $1\frac{1}{2}$ hours, until the chick is hatched. In this kind of discontinuity, only the temporal pattern is changed, but not the form of behavior. The causation of this shift is not known; other contributors also discuss this question (Oppenheim article and Vince article, this volume). But it is clear that we are dealing with a systemic factor. Balaban and Hill (1969) have found that vocalization and raising of the upper eyelid, two parameters which are irrelevant for hatching, also increase suddenly and manifold exactly synchronously with the other climax symptoms.

There is yet another facet to hatching behavior, this time along the line of continuity. Kovach (1970) finds a great similarity between the rotatory Type III movements and the postnatal righting reflexes. In fact, he considers the latter as a resumption of Type III movements in postnatal life. Even if one recognizes differences in the release of the two acts and in their adaptive functions, the idea of a continuity of a specific embryonic behavior into postnatal life, with a change of function, is appealing. Time does not permit to pursue this interesting theme of prenatal antecedents to postnatal action patterns, though it is very much part of the continuity-discontinuity problem.

Before we turn the discussion to lower levels of organization, I shall try to clarify another point. In the Introduction, Dr. Gottlieb connects the continuity-discontinuity problem with the important question of *causal* or determinative relations between different behavioral stages. He states that "probabilistic theories. . . .hold (*1*) that behavior development is a gradual and continuous process wherein (*2*) certain features of late embryonic, fetal or early neonatal behavior can be traced to, and are in some sense (facilitative or determinative) dependent upon, earlier behaviors or stimulative events." (p. 17). He continues, "Certain other viewpoints (e.g., Hamburger, 1968), on the other hand, are explicit in their emphasis of discontinuities in embryonic behavior such that, for example, later movements are envisaged as arising *de novo*, in which case these movements could bear nothing other than a permissive relationship to earlier movements." The last part of this sentence imputes a *causal* relationship, whereas Oppenheim and I have claimed discontinuity only with respect to the *forms* of the motility patterns. We have not claimed that Type III could not be modified by experimental manipulation of Type I; there are no experimental data that shed light on this question. Nor have I claimed that our deafferentation experiments touch upon the more specific question of whether sensory input in earlier stages might have an effect (facilitative, determinative, or otherwise) on hatching behavior. We could not tackle this question because the deafferented embryos did not hatch. This problem is also unresolved. Our only claim is that *ongoing Type I motility up to day 17 is not dependent on sensory input* (e.g., Hamburger, 1970; Hamburger & Narayanan, 1969; Hamburger, Wenger, & Oppenheim, 1966). An earlier statement which went beyond this claim (Hamburger, 1963, 1964) stands corrected.

In summary, as far as the probabilistic claim of gradualness and continuity is concerned, it is fully realized in the behavior of anamniote embryos which follow Coghill's paradigm of continuously integrated behavior. In amniote embryos continuity is compatible with discontinuity, if one accepts the notion of incorporation of unspecific movements in newly established integrated patterns. And there is definitely continuity on the structural level, in neurogenesis, to which we shall turn next. The continuity–discontinuity problem is a complex one. Looked at close range, it dissolves itself into a number of specific problems. In this case, as in others, a dichotomous conceptualization, leading to an "either–or" position, does not do justice to the phenomena. In all such matters, the question is not "either–or" but rather: to what extent and in what respects does a specific sequence show one or the other alternative?

V. Incongruity of Neurogenesis and Development of Behavior

If complex integrated activities such as hatching behavior and many postnatal action systems make their appearance rather suddenly, *de novo*, in the sense indicated above, then one has to postulate that the neural machinery that makes such novelties in behavior possible must have been prepared in advance. The neural organization, including the circuitry that underlies all complex activities performed after birth, is created in a sequence of differentiation processess which are referred to as neurogenesis. This includes cellular and supercellular events and all aspects of fiber outgrowth and formation of central and peripheral nerve fiber patterns, and of course synaptogenesis. Whereas the early phases, such as proliferation, migration, axon and dendritic outgrowth, and some of the underlying mechanisms are fairly well known, we are much less well informed about the later phases, including the final fixation of synapses which is of crucial importance for function. Despite these gaps, the neuroembryologist is committed to the basic tenet of the continuity of all these process, in the sense that every step in progressive differentiation emerges from the preceding step, and that the increase in complexity is gradual and continuous.

There is nothing novel about all this. What has not been previously stated explicitly is the incongruity between neurogenesis and overt motility in amniote embryos. The unorganized movements in the chick embryo undergo little change, between 4 and 17 days, except that they become more frequent and more muscles get into the act. *In no way do they reflect the neurogenetic events which go on in the meantime, so to speak below the surface.* The gradual increase in the complexity of the circuitry finds no expression in the pattern of overt spontaneous motility; in fact, it is difficult to imagine how it could do so. Even if one admits that the resolution of our movement analysis is rather coarse, and that the structural and ultrastructural analysis is deplorably incomplete, I think that the lack of clear relationship between Type I motility and the observed or inferred neurogenetic processes is undeniable. Hence, the notion that neurogenesis fully "explains" or "determines" embryonic behavior development is not valid as a generalization. It does not apply to the chick and the rat embryos, though it may be valid for the salamander. In other words, even the most detailed knowledge of neural organization, including all significant synapses, in chick or rat embryos at a given stage would permit no prediction of the actual (Type I) movements performed at that stage. Nor would a progression in synaptogenesis from one stage to the other be reflected in the details of motility. However, in salamander embryos, the correspondence of progression in neurogenesis (especially synaptogenesis) and behavior is very close, indeed. It is one of Coghill's major achievements to have documented this in detail. But even in this form, there is a modicum of indeterminacy. For instance, in the "early flexure" stage, the head can move either to the left or to the right, though there is a high probability that it will move away from a unilateral stimulus. All one can say is that *the state of differentiation of the nervous system at a given stage delimits the range of behavioral potentialities.*

VI. Relation of Neurogenesis and Bioelectrical Phenomena

The electrophysiological events which are closely correlated with overt motility (Provine, this volume) cannot be expected to give a more precise reflection of ongoing neurogenesis than motility itself. The polyneuronal bursts, though they do have their own structure and though they undergo temporal changes (see Provine, this volume), appear as monotonous throughout the major embryonic period as the uncoordinated, jerky movements which they generate. While the pathways for integrated activity are being

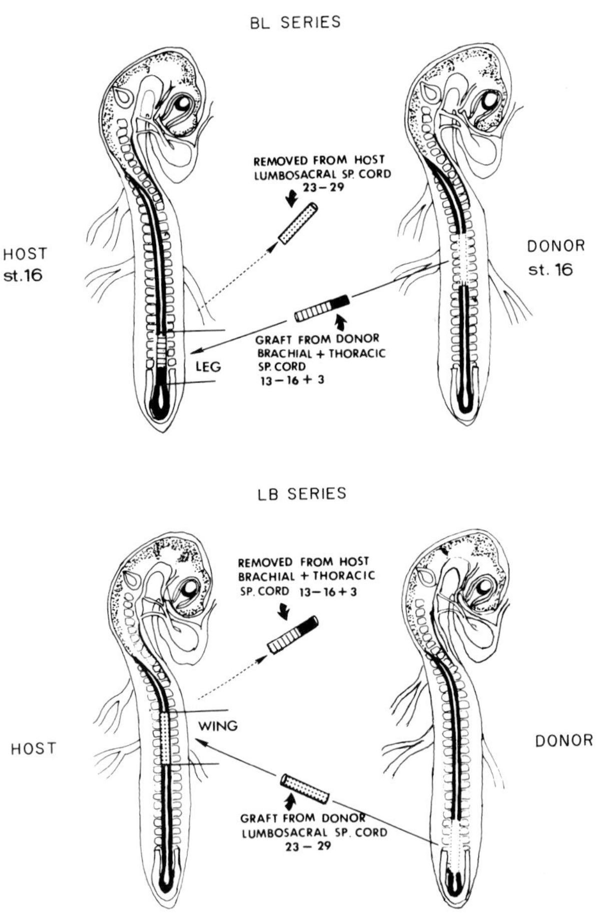

FIG. 1. Transplantation of brachial segments to lumbosacral level (Bl series) and of lumbosacral segments to brachial level (Lb series) in $2\frac{1}{2}$-day chick embryos. From Narayanan & Hamburger (1971).

created, very little, if any, of this structural refinement is translated into the ongoing bioelectrical activity. On the contrary, the double-electrode findings of Provine seem to substantiate my earlier notion that the discharges spread indiscriminately through the ventral cord, ignoring, as it were, any specific pathways. They sweep through the system as if it were an unorganized network.

The incongruity of burst activity and neurogenesis thus leads to an impasse which has to be resolved in some way. It is conceivable that at our present level of analysis of electrical burst activity a more subtle, more highly organized pattern of discharges along specific channels is not revealed. The following experiment performed by Dr. Narayanan suggests that this is, indeed, the case, and that while Type I motility goes on, preferential pathways are actually being used for impulse transmission. In 48-hour chick embryos, the brachial spinal cord segments were transplanted in the place of the lumbosacral segments of another embryo, and vice versa; thus, one set (Bl) had two brachial, and the other set (Lb) had two lumbosacral cords. No differences between normal and experimental embryos were found with respect to total activity and periodicity. However, when another parameter was recorded, namely the number of movements performed per time unit, it was found that in both experiments wings and legs moved more frequently together, in concert, than in normal embryos. This holds for the entire observation period from 9 to 17 days (Narayanan & Hamburger, 1971).

The tight linkage of wing and leg movements when they are both innervated by homonymous spinal cord segments presupposes a corresponding coupling of the underlying bioelectrical discharges: two homonymous nerve centers are more frequently excited near-simultaneously than are the heteronymous centers in the normal embryo. One can explain this in different ways: either particular longitudinal fiber tracts have preferential chemo-affinities for brachial, or lumbosacral neuron sets, respectively, and connect with them preferentially, or bioelectrical messages are directed specifically to brachial, or to lumbosacral, levels, and decoded simultaneously by both homonymous segments in the experimental embryo. No matter which explanation one accepts, the results show that in this case the discharge pattern has special features; our assumption that *all* discharges sweep indiscriminately through the ventral cord has to be qualified.

It should be mentioned that coupling of wing and leg movements is even tighter after hatching. In hatched chicks of the Bl series, the four extremities almost always moved together. A similar observation was made by Straznicky (1963), who explained the phenomenon much as we did. Incidentally, the legs in our Bl series never displayed alternating stepping movements; they always performed simultaneous ab- and adduction, as in wing flapping. This is additional evidence that region-specific differences are built into the spinal cord at very early stages, and that the circuitry for wing- or leg-specific coordination cannot be modulated by the appendages.

Returning to our main theme, if one accepts the notion that not all discharges in the spinal cord spread indiscriminately, and that impulse transmission along discrete pathways can also occur, then the *distinction between coordinated and uncoordinated motility becomes less sharp*. One would no longer consider an occasional coordinated movement such as wing flapping or alternating movements of legs, when it is interspersed with uncoordinated movements, as an exception; one would rather expect this to happen once in a while. The conceptual dichotomy of coordinated versus uncoordinated movements which I have stressed in the context of embryonic motility is perhaps too rigid. Once more, dichotomous thinking should give way to a more flexible scheme. This will become more obvious when we turn to reflexogenesis.

VII. Responses to Stimulation (Concurrence of Generalized and Local Responses)

In our search for clues to discharge patterns in the embryonic spinal cord, we now turn to *evoked responses*. Postnatal reflexes such as withdrawal, stretch, or sucking are performed in a stereotyped way, and it is assumed that specific, discrete pathways are involved, even if not all details of the reflex circuits have been identified. The precursors of reflexes should provide evidence for the existence of discrete pathways in the embryo.

A great deal of information on evoked responses is available in the literature. Most experiments were conducted during the 30's and 40's, using principally mammalian fetuses (review in Carmichael, 1970). The data were interpreted from different viewpoints, as, for instance, in support or refutation of Coghill's or Windle's theories, or in terms of fetal antecedents of postnatal reflexes, or with respect to the sequential order in which different modalities make their appearance (Gottlieb, 1971b). I shall consider only cutaneous tactile stimulation in birds and mammals, and I shall limit myself to a few selected data that are pertinent to the issues under discussion.

The earliest responses in birds and mammals are elicited by perioral stimulation, but in some forms (e.g., rat, cat, rhesus monkey) the palmar surface becomes sensitive at the same time. The responses consist first of a bending of the head, and thereafter expand from there to trunk, tail, and limbs; or of forearm flexion in response to palmar stimulation. In some instances, a close correlation between the early movements and the underlying structural differentiation has been established. This holds for the human (Humphrey, 1964), the monkey (Bodian, Melby, & Taylor, 1968), cat (Windle, 1934), and others. However, very soon this relationship is lost sight of in amniotes. This is due, in part, to the lack of information concerning details of structural maturation and synaptogenesis. However, the main reason is that the responses, rather than progressing toward clearly defined reflexes and coordinated patterns, become unstructured and generalized. All observers agree that soon after the initial period of local responses, stimulation, for instance of the perioral region, evokes what has been called a *generalized*, total pattern, or mass response, that is, the activation of several or all parts of the musculature. In other words, we are dealing with a *diffusion* or spreading or irradiation of the motor responses (not to be confused with the extension of reflexogenous areas which occurs at the same time). The generalized responses are the exclusive or predominant pattern for considerable periods of prenatal life. For instance, in the chick, the generalized response is the only motility type observed from the beginning of sensitivity at $7\frac{1}{2}$ days to 11 days (Hamburger & Narayanan, 1969). In the guinea pig with a gestation period of 68 days, the period of exclusive or predominant total activity extends from about 31 to 40–45 days (Carmichael, 1934), in the rat from 16 to 18 days (Angulo y Gonzalez, 1932; Narayanan *et al.*, 1971), and in the human fetus from about 8 to 13 weeks (Hooker, 1952, 1958; Humphrey, 1964, 1970). The initial stages of head bending and forearm flexion have been interpreted either as local responses or in terms of incipient generalized activity; this is a minor point. It should be understood that generalized activity does not always involve all parts of the body; the diffusion of the response may extend only to certain regions. For instance, stimulation of the palm or of other parts of the leg usually elicits only segmental responses, whereas stimulation of the trigeminal area more frequently results in mass action. The term "generalized" motility is therefore perhaps preferable to "total pattern."

The generalized response reminds one of the spontaneous Type I motility with which it shares the uncoordinated form of movements. In fact, an observer would find it difficult to distinguish between them. The diffusion of evoked responses in mammals suggests that we are dealing with a spread of bioelectrical activity through the spinal cord, similar to that established for the chick.

However, this is not the full story. It was found that in all forms, sooner or later more restricted, local responses occur side by side with generalized responses. There is a general trend in that local responses become more frequent as development progresses, and in late stages specific stereotyped reflexes can be obtained as the sole response. This is not the occasion to discuss in detail the origin and elaboration of reflexes. In the context of our discussion, special interest is focused on the intermediate stages in which *generalized and local responses can be elicited side by side.* We shall give a few examples. In the chick (Table I), all responses to beak stimulation are

TABLE I
RESPONSES TO TACTILE STIMULATION OF TRIGEMINAL AREAS IN THE CHICK (GENTLE STROKING)[a]

Age (days)	Age (hours)	Stimulated head areas	Responses
7	15	Beak	Weak total body movements
8	5–9	Beak	Total body movements
9	5–8	Beak	Same, including head, legs
10	5–9	All areas	Same, including head, wings, legs
11	5–12	Beak	Local head withdrawal only
		Other areas	Total body movements, including legs, wings
12	5–9	Beak	Local head withdrawal only
		Other areas	Total body movements, including wings, legs
13	5–8	All except posterior head	Local head withdrawal
		Posterior head	Same with occasional wing, leg movements
14	5–7	Same	Same as 13, 5–8
15	5–7	All areas	Local head withdrawal, leg kicking
16	5–6	All areas	Same
17	5–8	All areas	Same
18	5–8	All areas	Same, occasional rotatory head movements

[a]Hamburger & Narayanan (1969).

generalized up to 11–11½ days. At that stage, stimulation of the beak gives *local* head responses for the first time, whereas stimulation of posterior trigeminal head areas still gives the total response. At 15 days, one obtains consistently local head movement responses from all trigeminal areas, accompanied sometimes by kicking of the legs (Hamburger & Narayanan, 1969). In the guinea pig, the general trend is well illustrated by the stimulation of the lateral margin of the pinna. At 32 days, a generalized total pattern, including head, trunk, and extremities, is the only response. At 43 days, local contraction of the pinna is observed for the first time; it is accompanied by head, trunk, and limb movements. At 61 days, "a strictly localized twitch of the exact part of the pinna that had been touched [Carmichael, 1934, p. 438]" was recorded. In the human fetus, following perioral stimulation, the total pattern prevails up to 13 weeks. However, at 8½ weeks, the stimulation of the edge of the lower lip already evokes an active mouth opening, along with other body movements; likewise, from 10½ weeks on, stimulation of the upper eyelid elicits the contraction of the orbicularis oculi muscle, accompanied by mouth opening and movements of head and extremities (Hooker, 1958; Humphrey, 1970). *The local responses are so to speak embedded in the total pattern.* These few examples could easily be multiplied; they document the crucial point that generalized and localized responses occur side by side. The inference seems to be justified that exteroceptive sensory neurons which innervate the trigeminal and other cutaneous areas can discharge indiscriminately into neuron pools at different levels of the spinal cord, and at the same time along discrete pathways. The stimulation experiments thus lead to the same conclusion that was derived for spontaneous motility from the spinal cord transplantation experiment, namely, that *coordinated and uncoordinated behavioral activities are not mutually exclusive in embryos.* One can assume that while the reflex arcs are gradually refined by formation of the appropriate synaptic connections, they are preferentially used for channeling of impulses, though diffusion still continues up to late stages.

To summarize, as far as the structure ⟶ function relationship in evoked motility is concerned, in mammalian fetuses a close correlation has been established for the early stages of reflex responses (see p. 62). Again, at the end of the fetal period, the strictly localized responses which one obtains reflect the gradual perfection of the central reflex circuits. In the intermediate period, the picture is blurred by the phenomenon of diffusion of the motor responses, and no clear relationship can be established.

VIII. Inhibition

Generalized motility and the transition to coordinated activity cannot be properly understood without consideration of inhibition, inasmuch as all integrated activity depends on the subtle interplay of excitation and inhibition. Hence, the origin and elaboration of inhibition in the embryo is a matter of special interest. It is in the nature of inhibition that it cannot be demonstrated by observation of motility but only by experiment.

Data for the chick embryo are very limited. In our preparations of chronic gaps in the spinal cord, at the cervical or thoracic level, our observational methods have not given signs of inhibition up to 17 days; body regions caudal to the transection invariably show a quantitative reduction of motility (Hamburger, 1963; Hamburger, Balaban, Oppenheim, & Wenger, 1965). After 17 days, extirpation of the midbrain and of the otocysts raises the level of activity (Decker 1970; Decker & Hamburger, 1967). It is at this time, then, that the first inhibitory action of higher centers can be detected by these experiments. Yet, it is certain that inhibitory synapses become functional much earlier. A rise in activity was initiated by strychnine treatment at 14–15 days (R. Oppenheim and R. R. Provine, personal communications). Flat-vesicle synapses, which are considered as representing inhibitory synapses, were found in 11-day embryos (Foelix and Oppenheim, this volume). We assume that the inhibitory effects prior to day 17 are outweighed by excitation and therefore not detected by our methods. No further advance can be expected without application of electrophysiological and pharmacological methods.

Fortunately, we have much more precise information on the onset and development of inhibition in mammalian fetuses (review in Skoglund, 1969a, 1969b). We shall limit the discussion to the data on stretch reflex in sheep and guinea pig fetuses, though the extensive work on kittens is equally pertinent (see Purpura, 1971; Skoglund, 1969b). An essential feature of the Scandinavian studies is the use of electromyography.

TABLE II
STRETCH REFLEX IN THE FETUS

	Sheep (days)[a]	Guinea pig (days)[b]
Gestation period	150	65–68
First exteroceptive (plantar skin) response of leg	35–40	35
First sign of stretch reflex in gastrocnemius	60	45
Irradiation of excitation to antagonists	60–81	48–56
First observed antagonistic inhibition in leg	90 (no observations 82–89)	After 56

[a] Änggård *et al.* (1961).
[b] Bergström, Hellström, and Stenberg (1961), and Bergström *et al.* (1962).

In sheep, with a gestation period of 150 days, evoked responses to exteroceptive cutaneous stimulation can be obtained beginning at 35 days, whereas the proprioceptively mediated stretch reflex was first elicited at 60 days. This seems to be the stage at which maturation of muscle spindles occurs (see Table II). The most interesting finding, in our context, was that the excitatory response to stretch was not limited to the stretched muscle (in this case the gastrocnemius, an extensor), but it spread to an antagonist (the tibialis anterior, a flexor). This diffusion, or absence of antagonistic inhibition, goes on for an extended period, that is, up to 90 days. However, direct inhibition by another route, namely by stimulation of the ipsilateral leg skin, can be obtained earlier, i.e., from day 72 on (Änggård, Bergström, & Bernhard, 1961).

The corresponding data for the guinea pig are shown on Table II (Bergström, Hellström, & Stenberg, 1962). The gestation period, as mentioned, is 65–68 days. The first stretch reflex on a leg muscle was observed on day 45, and the diffusion of excitation to antagonists lasts to day 56. Plantar skin stimulation also results in a contraction of antagonistic muscles during the same period. After day 56, the plantar stimulation elicits only flexors (withdrawal), and the stretch only extensors (postural).

These experiments clarify several points: (*1*) Excitation precedes inhibition, at least in this particular system. (*2*) Lack of inhibitory mechanisms is responsible, at least in part, for the generalized diffusion, in evoked responses as well as in spontaneous motility. As stated by Änggård *et al.* (1961), "[the] pronounced lack of balance between the excitatory and inhibitory mechanisms . . . may explain the diffuse and widespread character of the reflex activity [p. 134]." (*3*) The transition from generalized to precisely localized response seems to be paralleled by the elaboration of the inhibitory mechanisms.

IX. Influence of Function on Structure

Gottlieb (1970 and Introduction, this volume) distinguishes between two different theoretical viewpoints. According to one view (predeterministic), the relationship is unidirectional (structure ⟶ function); according to the other (probabilistic), the relationship is bi-directional (structure ⇄ function), that is, there is a feedback of function on neurogenesis. Function is defined broadly, to include bioelectrical activity, movements, sensory input through stimulation, and other experiences.

In the previous sections, I have stressed the difficulties which one encounters when one tries to come to grips with the specifics of the unidirectional structure ⟶ function relations in amniote embryos, both with regard to spontaneous and to evoked motility.* With respect to the reverse relation, the scarcity of reliable data leaves most questions unresolved. Since Dr. Gottlieb has dealt with this matter in the Introduction, I shall limit myself to a few additional comments. For the purpose of analysis, the different aspects of function, that is, bioelectrical activity, motility, and sensory input, are considered separately.

A. Does Impulse Transmission Play a Role in Structural Differentiation?

The theories of neurobiotaxis (Ariëns Kappers, 1932) and of stimulogenous fibrillation (Bok, 1915) have postulated such influences. For instance, impulse transmission along particular fiber tracts was supposed to induce directional outgrowth of dendrites from adjacent neuroblasts toward the tract, and of axons in the opposite direction. These theories were not based on accurate observations nor on experimental evidence and are now discarded for good reasons (see Jacobson, 1970; Sperry, 1951). But this does not settle the problem. The very early start of motility in all vertebrate embryos is evidence that immature neuroblasts generate impulses, and

*It is clear from previous (and also following) sections that my viewpoint is not that of "predetermined epigenesis" as defined by Gottlieb elsewhere in this serial publication.

Provine (this volume) has recorded activity from the spinal cord of chick embryos as early as 4 days. Hence, there is ample opportunity for electrophysiological influences during a considerable part of neurogenesis. It is conceivable that electrical activity of a neuroblast or neuron plays a role in the regulation of its metabolism or, more specifically, in its differentiation processes; but the experimental design for tests *in vivo* will be difficult. Under the more favorable conditions of tissue culture, Crain, Bornstein, and Peterson (1968) have found that synapses can form in mammalian embryonic nerve tissue when electrical activity is blocked with Xylocaine (see Crain, Volume 2 of this serial publication). This observation does not support such a notion, but it is perhaps too early to generalize from this experiment.

B. *Does Motility as such Play a Role in Neurogenesis?*

A negative answer was obtained in a particular instance. In the frequently quoted experiments of Harrison (1904), Carmichael (1926), and Matthews and Detwiler (1926), salamander and frog embryos were kept in chloretone narcosis during the critical preswimming stages; they showed normal swimming after removal from the narcotic. The inference is that the sequence of synapse formations described by Coghill (1929), by which the swimming mechanism matures, can proceed normally in the absence of function. Similar experiments by Fromme (1941) in which he found impairment of swimming after removal from the narcotic are sometimes quoted as evidence to the contrary; however, this claim is not warranted. He found only that "the earliest appearance of swimming behavior does not compare favorably with that of the control group [p. 238]." He found normal swimming after full recovery and was probably dealing with a transient impairment. He used frog embryos, which have a much lower tolerance for anesthetics than salamander embryos. The fact that these time-honored but poorly controlled experiments have not been superseded to this day testifies to our ignorance in this matter. The performance of similar experiments on chick embryos is marred by troublesome side effects: paralysis by curare for 24–48 hours results in ankylosis and muscle atrophy (Drachman & Coulombre, 1962). Perhaps the deleterious effects can be avoided by applying intermittent treatment over longer periods which would reduce motility drastically, though not abolish it. There is another way of circumventing the difficulty. Since according to Provine (this volume), motility is closely correlated with burst activity, the monitoring of burst activity might give information on development of motility and, by inference, on synaptogenesis in embryos with ankylosis and myopathy.

But one should realize that even if such efforts were successful, any conclusions from this type of experiment concerning neurogenesis would be by inference only, and it seems altogether hopeless to expect precise information on function ⟶ structure relationships by this approach. It would be preferable to stay at one level of organization and ask either, whether experimental modification of bioelectrical activity at one stage influences such activity at a later stage, or whether manipulation of motility, including its suppression, during embryonic development affects later behavior. The assumption that this is the case is the basic tenet of Kuo's (1932) and Schneirla's (1965) theories, but the crucial experiments still have to be done.

We turn next to the question which, for a long time, has been in the center of theoretical considerations and controversy.

C. The Role of Sensory Input in Neurogenesis

Different aspects of this problem have been approached from different viewpoints and by a variety of experimental designs, such as: deafferentation in embryos, sensory deprivation or increase of stimulation postnatally, enrichment and impoverishment of the environment. I shall add only a few comments to those of Dr. Gottlieb in the Introduction.

The nervous system of birds and mammals is already at an advanced state of differentiation when the sensory-motor connections become functional. Our deafferentation experiments show that neurogenetic differentiation responsible for spontaneous (Type 1) motility in the chick embryo does not require sensory input (see Hamburger, 1968, 1970). This applies to the major part of the embryonic period up to 17 days. But, as was stated above, the question of whether or not subsequent pre- and postnatal coordinated action patterns and the concomitant neurogenetic events are dependent on sensory input at any stage remains unresolved.

As a broad generalization, one would assume that the central circuitry for those action patterns which are performed in a rather stereotyped way, such as locomotion, would be established according to an intrinsically programmed blueprint. On the other hand, those higher brain centers which subserve activities requiring a high degree of adjustment to environmental exigencies, such as the mammalian cortex as the seat of learning and memory, would retain a certain degree of plasticity up to postnatal stages.

The first point is illustrated by our spinal cord transplantations (p. 60). They show what other experiments had demonstrated before (e.g., Weiss, 1955), that the program for the circuitry underlying coordinated wing and leg movements, respectively, is built into the brachial and lumbosacral spinal cord regions at very early stages, and, furthermore, that the connectivities are not modified by sensory input from an atypical periphery. Concerning *plasticity*, two questions arise: Is it limited to functional specification of neurons or does it extend to the structural level? And, if the latter is the case, can sensory input control or modify structural differentiation, as for instance, synaptic linkage? Since higher centers in mammals do not complete their maturation until after birth, opportunities exist for a variety of sensory experiences to impinge on functional as well as structural specification of neurons and synapses. Of course, the major fiber tracts have long been established by then, and the fiber terminals have reached their assigned sites. For instance, the highly specific topographic projection of the retina fibers onto the optic tectum in chick embryos is completed long before visual perception begins (La Vail & Cowan, 1971). The mechanism by which this is accomplished is probably selective chemoaffinity between the nerve ending and the neuron with which it establishes contact (Sperry, 1963, 1965; see, however, Székely, Volume 2 of this serial publication). But this organizing mechanism may still permit a considerable degree of randomness on a smaller scale, such as modifications or shifts of synaptic links. It certainly does not exclude the possibility that in some systems the fine details of synaptic structure and linkage could be regulated by sensory experiences. In brain centers that retain plasticity after birth, connections of units may remain diffuse and uncommitted, and visual and other stimulation may play a determinative role in the specification of a synapse, or little-used synapses may disappear.

Experimental studies of the modifiability of neural structures by sensory stimulation are of rather recent date and not yet very numerous. A few examples may be cited. Several investigations have used visual deprivation in newborn mammals. The visual system has many advantages, but it is often difficult to decide whether one deals with regressive effects in an already differentiated structure or with the arrest of ongoing differentiation. Valverde (1967) has shown that in mice raised in darkness from birth, the number of dendritic spines on the apical dendrites of pyramidal cells in the visual cortex is markedly reduced, and it is very probable that the lack of sensory input causes the failure of normal spine formation. In a later investigation, Valverde (1971) has identified two populations of pyramidal cells: one in which dendritic spine growth is independent, and another one in which it is dependent on stimulation by light. Szentágothai and Hámori (1969) have described a special kind of dendritic spines in the lateral geniculate body of the dog which penetrate into deep invaginations of the surface of optic terminals, where they synapse. In the newborn dog these spines are not yet developed. If the eyes are sutured after birth, the spines fail to differentiate. They are completely absent at 2 months, when the adult pattern is established in normal puppies. In both instances, we are dealing with the arrest of a differentiation process, resulting from absence of sensory input. Finally, we present an example of enhancement of differentiation by stimulative experience. Rosenzweig and collaborators (see Rosenzweig, 1971) have shown that when young rats, after weaning, are exposed to an enriched or impoverished environment, respectively, a number of anatomical and biochemical parameters show significant changes. The investigations have now been extended to the synaptic level (Møllgaard, Diamond, Bennett, Rosenzweig, & Lindner, 1971). The axodendritic synapses in layer III of the occipital cortex (mostly visual) were used for quantitative comparison. In animals raised in the enriched environment, the mean length of synapses was 52% greater and the number of synapses per unit area of neuropil was 35% less than in animals raised in the impoverished environment. The total synaptic area was 40% greater in the former than in the latter. While the relation of these findings to behavior is not immediately obvious, it is of critical importance that the principle of regulation of at least some aspects of synaptogenesis by sensory input has been validated. The far-reaching consequences for behavioral performance in general, and learning and memory in particular, are obvious.

D. Performance Effect of Prenatal Sensory Experience on Postnatal Behavioral Performance

Finally, I shall comment briefly on this question, although this topic does not relate directly to the function ⟶ structure problem. Claims that such a role is of great significance have been made by Kuo (1967) and Schneirla (1965), but experimental tests have not been forthcoming, with the exception of the important experiments of Gottlieb (1971a) which will be reviewed briefly (see also Impekoven and Gold, this volume). The experiments are limited, so far, to acoustic stimulation of birds. As is known, birds begin to vocalize several days before hatching, and, as Gottlieb has shown, these late embryos are able not only to respond to acoustic stimulation, but to discriminate between the species-specific maternal call and other calls. The role of prenatal acoustic experience in postnatal responses was tested in deprivation experiments. Since acoustic isolation before hatching still leaves the embryo exposed to its own vocalization, it was necessary to devocalize the embryos. A simple but very effective procedure was devised which consists of coating the tympaniform membranes of the syrinx with nonflexible collodion, preventing the vibration of the membranes. The operation was done about 2 days before hatching, the embryos being partly pulled out of the shell. All experiments were done on mallard duck embryos, and their postnatal discriminative ability was tested using maternal mallard, duckling (sibling), maternal pintail, and maternal chicken calls. The individuals that had been reared prenatally in complete acoustic deprivation actually did show postnatally several deficiencies in their responses, as for instance a time lag in the development of the usual discriminatory abilities. The most significant deficiency is the inability of the experimental ducklings to discriminate between the mallard and the chicken maternal call, whereas they are capable of distinguishing between the duckling (sib) call and the pintail maternal call. Experiments are underway to identify the acoustic characteristic (rate, fundamental frequency, etc.) which makes the discrimination between the mallard and chicken maternal calls difficult. This is apparently the first instance in which experimental evidence has been provided for the theoretically important notion that stimulative events occurring normally before birth play a role in the perfection of species-specific perception after birth.

X. Concluding Remarks

The picture that I have presented of stucture-function relationships in amniote embryos will not satisfy a mind in search for clearly definable connections and broad generalizations. The only claim that my picture can make is that it is a fairly realistic portrait of the present situation. Part of its imperfection is due to our very limited knowledge in these matters. But there are other reasons. One is the fundamental difficulty inherent in every reductionist effort to "explain" phenomena at one level in terms of events occurring at lower levels. This becomes evident when one realizes that the different units with which one operates at different levels are incommensurate. On the behavioral level, a representative unit is the reflex; on the level of motility, the units are contractions of muscles and muscle groups; on the level of bioelectrical activity, we are dealing with single-unit and burst discharges; and on the structural level with differentiating neurons, neuron pools, connectivities, and synapses. What kind of relationship would one accept as an "explanation" of events at one level in terms of events at lower levels? In our search we have detected only two relationships that could make such a claim: that between bioelectrical bursts and activity phases in the chick embryo, and the relation between structural differentiation of reflex arcs and evoked responses, at the beginning of sensory competence and then again toward the end of fetal life, when local reflexes can be elicited. But why do we find incongruity of patterns in all other respects? I think this cannot be attributed to conceptual difficulties alone. A major difficulty is inherent in the phenomena themselves, or, to be more specific, in the prevalence of uncoordinated motility in amniote embryos and fetuses. The situation would look quite different if the embryos would make the gradual assemblage of embryonic motility units into integrated postnatal behavior patterns overtly manifest; or, in other words, if prenatal antecedents to postnatal action patterns were clearly recognizable. In this case, specific questions could be asked concerning relationships prevailing at specific stages of development. But the reality is different. We are, then, back at the question raised earlier: Why is there the prevalence of uncoordinated motility in amniotes? One could argue that autonomous discharge of electrical impulses is an elementary property of immature nerve tissue which generates automatically the observed motility type, as long as excitation prevails over inhibition, and in the absence of selection pressure against this form of overt motility, when the embryo and fetus is protected in the egg and uterus. Or, as I have indicated, the electrical discharges may play a facilitative or even indispensable role in the functional or structural maturation of neurons. The maintenance of articulations and musculature would then be a fringe benefit of the fetal exercise, as has been suggested by Eisenberg (1971). Be this as it may, the incongruities that I have spoken of are real.

Where do we go from here? This will depend largely on the preferences of the individual investigator. It seems to me that further studies of the forms and sequences of overt motility would not be particularly rewarding. At this moment, the best strategy would seem to be to confine oneself to one level of organization at a time, building on the beginnings that have been made, but with the problems posed by behavior development constantly on one's mind. On the behavioral level, the pioneer experiments of Gottlieb and others, which have been limited so far to the perinatal period, can be expanded to earlier stages, to different sensory modalities and different species. They can answer questions in which the developmental psychologist is particularly interested, without necessarily referring to lower levels of organization. The electrophysiological analysis is in its infancy; and the ultrastructural analysis, particularly of synaptogenesis, is likewise at its beginnings. Needless to say, in both areas answers to many questions will be available in the near future. If the focus of such investigations is on behaviorally critical stages, such as onset of sensory competence for different modalities, or initiation of specific integrated behavior patterns, then one can be confident that a more unified picture will emerge very soon.

References

Änggård, L., Bergström, R., & Bernhard, L. G. Analysis of prenatal spinal reflex activity in sheep. *Acta Physiologica Scandinavica*, 1961, **53**, 128–136.

Angulo y Gonzalez, A. W. The prenatal development of behavior in the albino rat. *Journal of Comparative Neurology*, 1932, **55**, 395–442.

Ariëns Kappers, C. U. Principles of development of the nervous system (Neurobiotaxis). In W. Penfield (Ed.), *Cytology and cellular pathology of the nervous system*. Vol. I. New York: Harper (Hoeber), 1932.

Balaban, M., & Hill, J. Perihatching behaviour patterns of chick embryos (*Gallus domesticus*). *Animal Behavior*, 1969, **17**, 430–439.

Bergström, R. M., Hellström, P. E., & Stenberg, D. Prenatal stretch reflex activity in the guinea pig. *Annales Chirurgiae et Gynaecologiae Fenniae*, 1961, **50**, 458–466.

Bergström, R. M., Hellström, P. E., & Stenberg, D. Studies in reflex irradiation in the foetal guinea-pig. *Annales Chirurgiae et Gynaecologiae Fenniae*, 1962, **51**, 171–178.

Bodian, D., Melby, E. C. & Taylor, N. Development of fine structure of spinal cord in monkey fetuses. II. Prereflex period to period of long intersegmental reflexes. *Journal of Comparative Neurology*, 1968, **133**, 113–166.

Bok, S. T. Die Entwicklung der Hirnnerven und ihrer zentralen Bahnen. Die stimulogene Fibrillation. *Folia Neurobiologica*, 1915, **9**, 475–565.

Carmichael, L. The development of behavior in vertebrates experimentally removed from the influence of external stimulation. *Psychological Review*, 1926, **33**, 51–58.

Carmichael, L. An experimental study in the prenatal guinea pig of the origin and development of reflexes and patterns of behavior in relation to the stimulation of specific receptor areas during the period of active fetal life. *Genetic Psychology Monographs*, 1934, **16**, 337–491.

Carmichael, L. The onset and early development of behavior. In A. Mussen (Ed.), *Carmichael's manual of child psychology*. Vol. I. New York: Wiley, 1970. Pp. 447–563.

Coghill, E. G. *Anatomy and the problem of behavior*. London & New York: Cambridge University Press, 1929.

Crain, S., Bornstein, M. B., & Peterson, E. R. Maturation of cultured embryonic CNS tissues during chronic exposure to agents which prevent bioelectric activity. *Brain Research*, 1968, **8**, 363–372.

Decker, J. D. The influence of early extirpation of the otocysts on development of behavior of the chick. *Journal of Experimental Zoology*, 1970, **174**, 349–364.

Decker, J. D., & Hamburger, V. The influence of different brain regions on periodic motility in the chick embryo. *Journal of Experimental Zoology*, 1967, **165**, 371–384.

Drachman, D. B., & Coulombre, A. J. Experimental club foot and arthrogryposis multiplex congenita. *Lancet*, 1962, **ii**, 523–526.

Eisenberg, L. Persistent problems in the study of the biopsychology of development. In E. Tobach, L. R. Aronson, & E. Shaw (Eds.), *The biopsychology of development*. New York: Academic Press, 1971. Pp. 515–529.

Fromme, A. An experimental study of the factors of maturation and practice in the behavioral development of the embryo of the frog, *Rana pipiens*. *Genetic Psychology Monographs*, 1941, **24**, 219–256.

Gottlieb, G. Conceptions of prenatal behavior. In L. R. Aronson, E. Tobach, D. S. Lehrman, & J. S. Rosenblatt (Eds.), *Development and evolution of behavior*. San Francisco: Friedman, 1970. Pp. 111–137.

Gottlieb, G. *Development of species identification in birds: An inquiry into the prenatal determinants of perception*. Chicago: University of Chicago Press, 1971. (a)

Gottlieb, G. Ontogenesis of sensory function in birds and mammals. In E. Tobach, L. R. Aronson, & E. Shaw (Eds.), *The biopsychology of development*. New York: Academic Press, 1971. Pp. 67–128. (b)

Gottlieb, G., & Kuo, Z.-Y. Development of behavior in the duck embryo. *Journal of Comparative and Physiological Psychology*, 1965, **59**, 183–188.

Hamburger, V. Some aspects of the embryology of behavior. *Quarterly Review of Biology*, 1963, **38**, 342–365.

Hamburger, V. Ontogeny of behaviour and its structural basis. In H. Waelsch (Ed.), *Comparative neurochemistry*. Oxford: Pergamon, 1964. Pp. 21–34.

Hamburger, V. Emergence of nervous coordination. Origins of integrated behavior. *Developmental Biology, Supplement*, 1968, **2**, 251–271.

Hamburger, V. Embryonic motility in vertebrates. In F. O. Schmitt (Ed.), *The neurosciences: Second study program*. New York: Rockefeller University Press, 1970. Pp. 141–151.

Hamburger, V. Development of embryonic motility. In E. Tobach, L. R. Aronson, & E. Shaw (Eds.), *The biopsychology of development*. New York: Academic Press, 1971. Pp. 45–66.

Hamburger, V., & Balaban, M. Observations and experiments on spontaneous rhythmical behavior in the chick embryo. *Developmental Biology*, 1963, **7**, 533–545.

Hamburger, V., & Narayanan, C. H. Effects of the deafferentation of the trigeminal area on the motility of the chick embryo. *Journal of Experimental Zoology*, 1969, **170**, 411–426.

Hamburger, V., & Oppenheim, R. Prehatching motility and hatching behavior in the chick. *Journal of Experimental Zoology*, 1967, **166**, 171–204.

Hamburger, V., Balaban, M., Oppenheim, R., & Wenger, E. Periodic motility of normal and spinal chick embryos between 8 and 17 days of incubation. *Journal of Experimental Zoology*, 1965, **159**, 1–14.

Hamburger, V., Wenger, E., & Oppenheim, R. Motility in the chick embryo in the absence of sensory input. *Journal of Experimental Zoology*, 1966, **162**, 133–160.

Harrison, R. G. An experimental study of the relation of the nervous system of the developing musculature in the embryo of the frog. *American Journal of Anatomy*, 1904, **3**, 197–220.

Hooker, D. *The prenatal origin of behavior*. Lawrence, Kans. University of Kansas Press, 1952.

Hooker, D. *Evidence of prenatal function of the central nervous system in man*. New York: American Museum of Natural History, 1958.

Humphrey, T. Some correlations between the appearance of human fetal reflexes and the development of the nervous system. *Progress in Brain Research*, 1964, **4**, 93–135.

Humphrey, T. Reflex activity in the oral and facial area of the human fetus. In J. F. Bosma (Ed.), *Second symposium on oral sensation and perception*. Springfield, Ill.: Thomas, 1970. Pp. 195–233.

Jacobson, M. *Developmental neurobiology*. New York: Holt, 1970.

Kovach, J. Development and mechanisms of behavior in the chick embryo during the last five days of incubation. *Journal of Comparative and Physiological Psychology*, 1970, **73**, 392–406.

Kuo, Z.-Y. Ontogeny of embryonic behavior in aves: V. The reflex concept in the light of embryonic behavior in birds. *Psychological Review*, 1932, **39**, 499–515.

Kuo, Z.-Y. *The dynamics of behavior development*. New York: Random House, 1967.

La Vail, J. F., & Cowan, W. M. The development of the chick optic tectum. I. Normal morphology and cytoarchitectonic development. *Brain Research*, 1971, **28**, 391–419.

Matthews, S. A., & Detwiler, S. R. The reactions of Amblystoma embryos following prolonged treatment with Chloretone. *Journal of Experimental Zoology*, 1926, **45**, 279–292.

Møllgaard, K., Diamond, M. C., Bennett, E. L., Rosenzweig, M. R., & Lindner, B. Quantitative synaptic changes with differential experience in rat brain. *International Journal of Neuroscience*, 1971, **2**, 113–128.

Narayanan, C. H., Fox, M. W., & Hamburger, V. Prenatal development of spontaneous and evoked activity in the rat (*Rattus norwegicus albinus*). *Behaviour*, 1971, **40**, 100–134.

Narayanan, C. H., & Hamburger, V. Motility in chick embryos with substitution of lumbosacral by brachial and brachial by lumbosacral spinal cord segments. *Journal of Experimental Zoology*, 1971, **178**, 415–432.

Oppenheim, R. Some aspects of embryonic behaviour in the duck (Anas platyrhynchos). *Animal Behaviour*, 1970, **18**, 335–352.

Purpura, D. Synaptogenesis in mammalian cortex: problems and perspectives. In M. B. Sterman, D. J. McGinty, & A. M. Adinolfi (Eds.), *Brain development and behavior*. New York: Academic Press, 1971. Pp. 23–41.

Rosenzweig, M. R. Effects of environment on development of brain and of behaviour. In E. Tobach, L. R. Aronson, & E. Shaw (Eds.), *The biopsychology of development*. Academic Press, 1971. Pp. 303–342.

Schneirla, T. C. Aspects of stimulation and organization in approach-withdrawal processes underlying vertebrate behavioral development. In D. S. Lehrman, R. A. Hinde, & E. Shaw (Eds.), *Advances in the study of behavior*. Vol. 1. New York: Academic Press, 1965. Pp. 1–74.

Skoglund, S. Growth and differentiation, with special emphasis on the central nervous system. *Annual Review of Physiology*, 1969, **31**, 19–42. (a)

Skoglund, S. Reflex maturation. In M. A. B. Brazier (Ed.), *The interneuron*. Berkeley: University of California Press, 1969. (b)

Sperry, R. Mechanisms of neural maturation. In R. Stevens (Ed.), *Handbook of experimental psychology*. New York: Wiley, 1951. Pp. 236–280.

Sperry, R. Chemoaffinity in the orderly growth of nerve fiber patterns and connections. *Proceedings of the National Academy of Sciences, U.S.*, 1963, **50**, 703–710.

Sperry, R. Embryogenesis of behavioral nerve nets. In R. L. DeHaan & R. Ursprung (Eds.), *Organogenesis*. New York: Holt, 1965. Pp. 161–186.

Straznicky, K. Function of heterotopic spinal cord segments investigated in the chick. *Acta Biologica (Budapest)*, 1963, **14**, 145–155.

Szentágothai, J., & Hámori, J. Growth and differentiation of synaptic structures under circumstances of deprivation of function and of distant connections. In S. H. Barondes (Ed.), *Symposia of the international society for cell biology*, 1969, Vol. 8. *Cellular dynamics of the neuron*. New York and London: Academic Press.

Tracy, H. C. The development of motility and behavior reactions in the toadfish (*Opsanus tau*). *Journal of Comparative Neurology*, 1926, **40**, 253–269.

Valverde, F. Apical dendrite spines of the visual cortex and light deprivation in the mouse. *Experimental Brain Research*, 1967, **3**, 337–352.

Valverde, F. Rate and extent of recovery from dark rearing in the visual cortex of the mouse. *Brain Research*, 1971, **33**, 1–11.

Visintini, F., & Levi-Montalcini, R. Relazione tra differenziazione strutturale e funzionale dei centri e delle vie nervose nell'embrione di pollo. *Schweizer Archiv für Neurologie und Psychiatrie*, 1939, **43**, 1–45.

Weiss, P. Nervous system (neurogenesis). In B. H. Willier, P. A. Weiss, & V. Hamburger (Eds.), *Analysis of development*. Philadelphia: Saunders, 1955.

Windle, W. F. Correlation between the development of local reflexes and reflex arcs in the spinal cord of cat embryos. *Journal of Comparative Neurology*, 1934, **59**, 487–505.

NEUROPHYSIOLOGICAL ASPECTS OF BEHAVIOR DEVELOPMENT IN THE CHICK EMBRYO

ROBERT R. PROVINE

Department of Psychology
Washington University
Saint Louis, Missouri

I. Introduction

One of the basic problems of behavior development is the analysis of the underlying neural mechanisms. Until recently, most attacks on this problem were purely behavioral in nature. Inferences concerning central nervous mechanisms were made on the basis of behavioral description. In some cases, the effects of various experimental manipulations such as deafferentation by extirpation or stimulation on behavior were examined. A basic, but not explicitly stated, tenet of many of these studies is that behavioral events should mirror the central nervous system (CNS) events giving rise to them. However, the assumption of a neurobehavioral parallelism is unwarranted, particularly in an immature system. A change in any one of the numerous neural, humoral, or muscular variables which intervene between CNS events and muscular contractions could radically modify the output of the developing motor system. Therefore, studies are needed which investigate simultaneously events within the CNS and the output of the CNS, behavior. This is one of the objectives of the work described here. Fortunately for the behavioral investigator, motility emerges as a convenient and rather accurate index of certain aspects of CNS motor development. However, we shall find that the CNS does not relinquish all of its secrets to the student of behavior.

The CNS structure receiving attention in this article is the embryonic spinal cord. The importance of this structure to the understanding of embryonic behavior was pointed out by Hamburger and his associates. In addition to giving rise to the "final common pathway," the spinal cord is a probable locus of origin of the autonomous discharges responsible for spontaneous embryonic motility in the chick. This is suggested by the finding that spontaneous motility continues in the absence of brain (Hamburger, Balaban, Oppenheim, & Wenger, 1965) and sensory input to the cord (Hamburger, Wenger, & Oppenheim, 1966). The electrophysiological studies to be presented here are a complement and extension of this behavioral work.

Before discussing the investigations of cord bioelectric activity, I would like to briefly review some of the main characteristics of embryonic motility in the chick. These data should help establish a perspective of the behavior in question.

II. Behavior of the Chick Embryo

A. Development of Motility

There is a generally good agreement among observers about the development of patterns of motility in the chick embryo (Corner & Bot, 1967; Hamburger, 1963; Hamburger *et al.*, 1965; Hamburger & Oppenheim, 1967; Preyer, 1885; Visintini & Levi-Montalcini, 1939). The following description relies primarily upon the reviews of Hamburger (1963, 1968) and Hamburger and Oppenheim (1967). Motility starts at $3\frac{1}{2}$–4 days of incubation with a slight flexion of the head. Movements of other body areas later become involved in a cephalocaudal sequence. At $4\frac{1}{2}$ days, S-flexures involving the somatic musculature start in the neck region and extend to the leg level, later involving the tail. Later, the waves increase in complexity; they may start in the trunk and spread to the tail or in both directions. At about $6\frac{1}{2}$ days, the wings and legs become motile. These limbs seem to move independently of each other, with no apparent correlation existing between left and right limb partners. The general pattern of cephalocaudal waves thus gives way to a generalized motility, with no apparent correlation existing between the movements of body parts. These movements eventually involve all body parts capable of moving at a given stage, including head, trunk, limbs, eyelids, tongue, and beak. These behavioral observations give us some important clues concerning underlying mechanisms, a topic which is pursued in later sections of this review.

Motility from its onset at $3\frac{1}{2}$–4 days up to about 13 days is intermittent, with activity phases alternating with inactivity phases. These activity phases occur at regular periods up to day 13. On the thirteenth day, the periodicity of motility is lost, and the embryo achieves its maximum amount of motility. After 13 days, the amount of motility declines. Up until late stages of incubation, most behavior consists of the jerky, apparently uncoordinated Type I motility, which has just been described (Hamburger & Oppenheim, 1967). Another similar but less frequent type of motility is present during development. It consists of whole body twitches or "startles." These movements are referred to as Type II. At approximately 17 days, prehatching and hatching movements appear which are different from the earlier Type I and II movements. These movements (Type III) are smooth and appear to involve the coordinated activity of body parts. The proportion of these Type III movements to Types I and II motility increases until hatching occurs on the twenty-first day.

B. *Spontaneous Nature of Embryonic Motility*

The chick embryo is motile under what appear to be stable environmental conditions. This suggests that motility may arise from endogenous physiological processes within the embryo. This possibility is particularly appealing for embryonic stages before 13 days when behavior is highly periodic (Hamburger, 1963). However, the assumption that patterned motor output must be structured by patterned sensory input is not well founded. For example, investigations of the insect nervous system have shown us that a random pattern of sensory input can yield a highly patterned motor output (Wilson & Wyman, 1965; Wyman, 1965). Preyer (1885) was the first to notice that motility appears in development before the time (day 7) when exteroceptive stimuli first become effective in evoking a behavioral response. This suggests the involvement of endogenous physiological processes up to 7 days. These and other lines of evidence supporting the concept of spontaneous embryonic motility are reviewed elsewhere (Hamburger, 1963, 1968, 1970; Hamburger *et al.*, 1966).

The most convincing evidence of the spontaneous (nonreflexogenic) nature of embryonic motility is provided by the previously mentioned study of Hamburger *et al.* (1966). These investigators removed several segments of the thoracic spinal cord of 2-day embryos. This isolated the lumbosacral cord from input from rostral cord regions and the brain. Simultaneously, a second operation was performed to remove the dorsal half of the cord caudal to the spinal gap. This second operation isolates the residual portion of the lumbosacral cord from brain and sensory input. Therefore, any movements of the legs of the operated embryos must arise within the totally isolated cord segment. The legs were highly motile up to 15–17 days, indicating that spontaneous embryonic motility may result from endogenous discharges within the embryonic spinal cord. Our discussion will now shift to the investigation of bioelectric phenomena within the embryonic spinal cord.

III. Ontogeny of Bioelectric Activity in the Embryonic Spinal Cord

The foremost problems encountered in studying a new system lie in determining what variables to investigate and at what level to investigate them. At first we must examine a system in a very general way and listen very carefully to what it has to say about its principles of operation. The variables studied and the methods used to study them must be appropriate to the system. With these considerations in mind, the first broad descriptive investigations of the embryonic cord were begun.

A. Exploratory Study of the Embryonic Spinal Cord

The first stage of our investigation was a general survey of electrical activity within the embryonic spinal cord (Provine, Sharma, Sandel, & Hamburger, 1970). Results from this study were used to direct subsequent more specific experiments. The study involved the dorsal-to-ventral probing of several areas of the lumbosacral spinal cord with glass pipette electrodes, filled with 3 *M* KCl solution in agar, with tip diameters of 3–5 μ. The lumbosacral region was examined because it forms the plexus innervating the legs. In order to facilitate the recording of unit activity, the embryos were immobilized with curare (2.5 μg/gm embryo wet weight). As will be discussed later, this drug had no noticeable effect on the neural activity in question.

Relatively mature 15-, 17-, and 19-day embryos were chosen for study because they were hardy and had large enough spinal cords to allow the depth of electrode penetration to be read accurately from a calibrated microdrive. In order to systematize the exploration, the mediolateral extent of the lumbosacral cord was divided into three regions: lateral (A), intermediate (B), and median (C). In each of these regions, the electrode was lowered in fixed increments of 68 μ. At each increment of electrode advance, general notes were taken concerning the local unit activity. In addition, ratings were given to certain aspects of cord activity to facilitate the quantification of data. The amount of unit activity (unit firing density) at a particular increment of electrode advance was assigned a 0–4 rating. Also, the presence or absence of burst activity was noted. Examples of cord activity equivalent to these ratings are given in Fig. 1. These data, when compiled, give a quantitative estimate of the amount of unit firing (mean unit firing density) and the percent of electrode runs showing burst activity at different cord depths and regions.

The results of this descriptive survey are presented in the activity profiles of Fig. 2. A high level of spontaneous unit activity was identified at most depths and regions of the lumbosacral cords of all embryos. The region corresponding to the dorsal columns and the ventral cord showed considerable activity, the ventral cord being the most active. In many embryos,

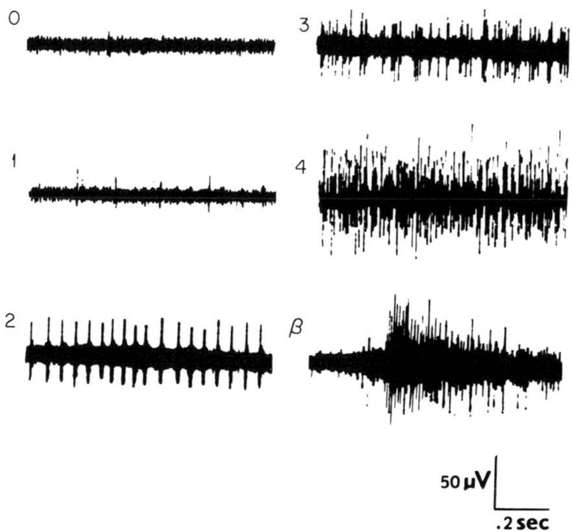

FIG. 1. Samples of unit activity representing different ratings of unit density: 0, no evidence of unit activity; 1, minimal unit activity with no sustained firing; 2, well-defined, sustained single-unit activity; 3, two or more well-defined single units with sustained firing; 4, dense multiunit activity; β, polyneuronal burst activity. From Provine *et al.* (1970).

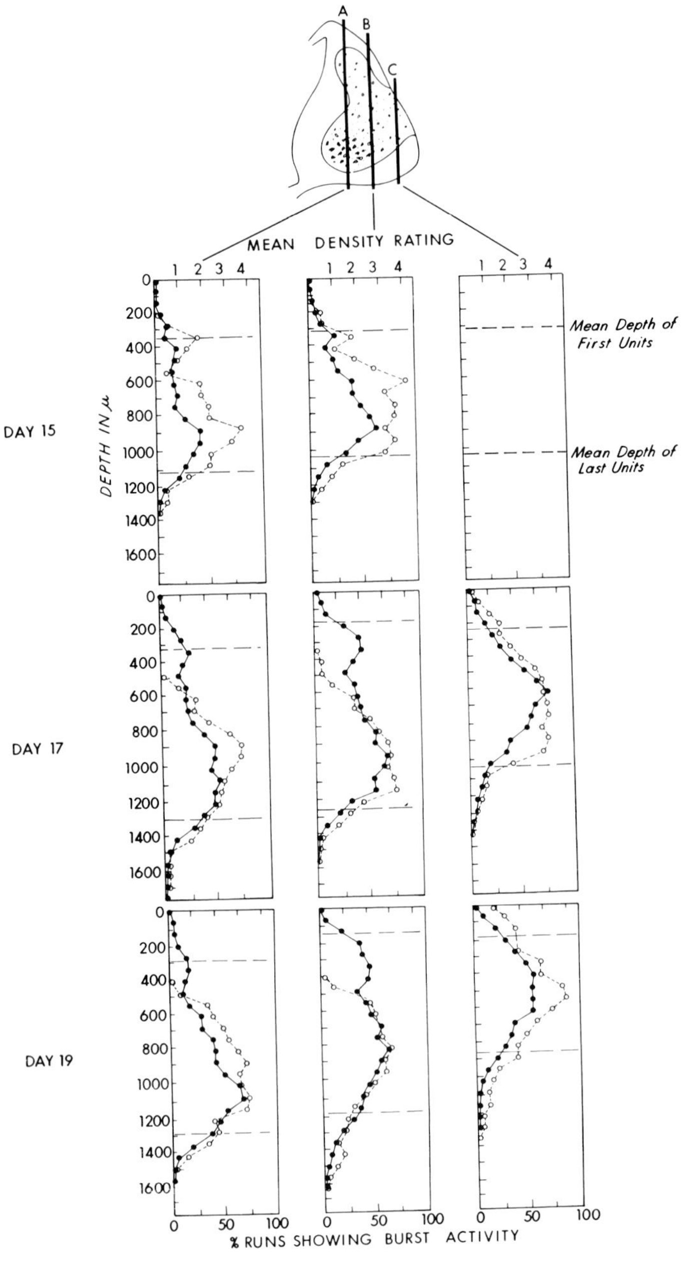
A
B
C
MEAN DENSITY RATING
1 2 3 4
Mean Depth of First Units
Mean Depth of Last Units
DAY 15
DAY 17
DAY 19
DEPTH IN μ
0
200
400
600
800
1000
1200
1400
1600
0 50 100
% RUNS SHOWING BURST ACTIVITY

an area of low activity was found at about one-third total cord depth. This region appears as a constriction in the activity profiles of regions A and B in Fig. 2. It is not clear whether this "quiet" region is due to low activity levels or simply consists of neurons which are difficult to monitor.

Prominent burst discharges were found in the ventral two-thirds of the cord. (Certain much shorter and less prominent discharges localized in the dorsal portion of the cord are mentioned in a later section.) The percentage of electrode passes showing burst activity at each increment of cord depth is shown by the broken line of Fig. 2. It is of interest that the bursts are restricted to the ventral cord region. This indicates that the dorsal and ventral portions of the embryonic cord possess region specific activity at least as early as day 15. This contrasts with the finding, to be discussed later, of considerable coupling of discharges in the ventral cord along the rostrocaudal axis.

The most significant result of this initial survey was the emergence of the burst discharge as an event of considerable bioelectric as well as behavioral interest. Bursts are noteworthy for several reasons. (*1*) They are massive polyneuronal discharges which are present throughout a considerable portion of the cord's cross section. (*2*) Bursts may represent a characteristic pattern of discharge for embryonic nervous systems. Such discharges have also been observed in cultures of embryonic spinal cord (Crain, 1970). (*3*) Bursts are of behavioral interest because they are a prime candidate for a neural correlate of embryonic motility. For these reasons, the burst discharges receive special attention in the following, more comprehensive, developmental study.

Ontogenetic Study of the 5- to 20-Day Spinal Cord

Using preliminary survey results to sharpen our focus, an extensive developmental study of the embryonic cord was initiated. This study had three primary objectives: (*1*) to establish the time of appearance of the first

---

FIG. 2. Profiles of unit density and burst activity in the lumbosacral cord medial to dorsal root 27. The activity profiles represent the distribution of unit activity at given depths of cord areas A, B, and C. The solid circles represent the density of unit activity found at each cord depth based on the 0–4 scale given at the top of the profiles. The open circles represent the percentage of electrode runs showing polyneuronal burst activity at each cord depth. The percentage scale is given at the bottom of the profiles. Zero depth represents the point of electrode cord contact for all profiles. Values were not computed for area C at 15 days. Profiles show ventral cord to have the highest unit density level. Burst activity is restricted to the ventral two-thirds of the cord in areas A and B. Area C is not an exception to this rule. Due to peculiarities of development, area C is equivalent to the ventral region of more lateral lumbosacral regions. From Provine *et al.* (1970).

unit activity; (*2*) to describe the ontogeny of the envelope and temporal distribution of burst discharges; and (*3*) to correlate these findings with behavioral data in order to establish a neural correlate of embryonic motility. The last objective will be explored more fully in the next section.

The present study was performed on curarized embryos. The embryos were prepared in essentially the same manner as those described before. One methodological difference involved the use of unsharpened 25-μ diameter tungsten wires insulated up to 5–10 μ of their tips with glass, instead of the KCl-agar pipettes which were used in the first study. The larger tungsten electrodes were ideal for observing the polyneuronal burst activity yet were usually small enough to allow observation of unit activity.

Electrical activity was present in the lumbosacral spinal cord by 5 days. (Activity was identified in the brachial region of one 4-day embryo.) Continuously firing and burst firing units have been identified. However, this distinction is somewhat blurred because many continuously firing units would accelerate or burst in synchrony with the polyneuronal burst discharges. These findings indicate that both simple spike activity and complex burst discharges are present very early in the developmental history of the spinal cord. At these stages, a high level of mitotic activity is present in the dorsal cord. This activity peaked several days earlier in the ventral cord (Corliss & Robertson, 1963; Hamburger, 1948; Langman & Haden, 1970; Levi-Montalcini, 1950). The existence of the bursts at early stages is particularly interesting because this activity suggests the presence of multisynaptic processes. In this regard it is noteworthy that Oppenheim and Foelix (1972 and this volume) have found ultrastructural evidence for synapses in the ventral cord of embryos younger than 4 and 5 days, the earliest observed in the present study.

Now that the presence of unit and multiunit (polyneuronal) burst activity has been established in young embryos, the discussion will shift exclusively to polyneuronal burst discharges, with emphasis being placed upon burst envelopes, burst periodicity, and the amount of burst activity present at different developmental stages. Integration is used to illustrate burst envelopes. By integration, I mean that a kind of moving average is being computed from the data in such a way that events of the recent past are weighted more heavily than events of the distant past. The amplitude of the integrated trace is a function of the unit firing density. This technique is a particularly useful means of rendering the low amplitude activity of young embryos. Burst discharges are compared with their integrated counterparts in Fig. 3. Examples of integrated burst discharges observed at different stages of development are shown in Fig. 4. These examples should be viewed as general steps in a developmental sequence and not as stage specific stereotypes because considerable variability in the form of burst envelopes was

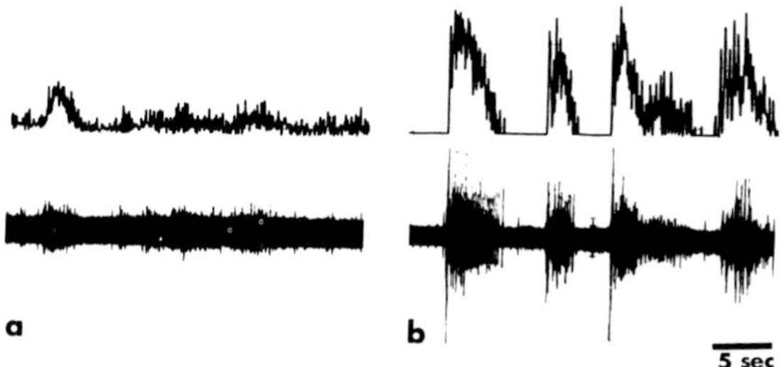

FIG. 3. Comparisons of raw (below) and integrated (above) records of identical burst discharges. (a) Five-day burst activity and (b) 17-day burst activity. From Provine (1972).

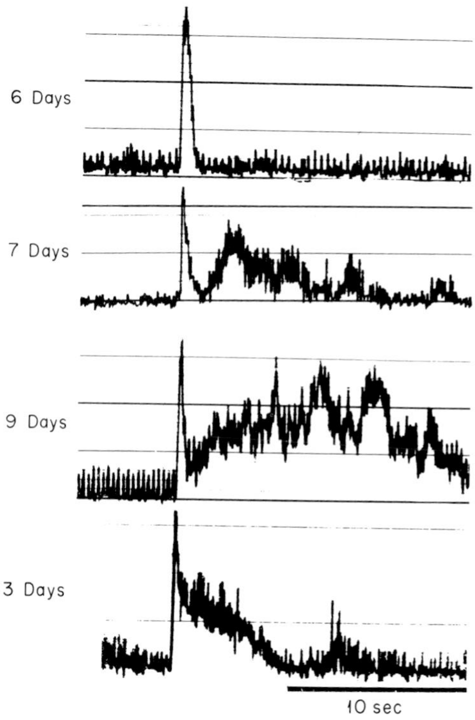

FIG. 4. Ontogeny of polyneuronal burst discharges. 6 days, simple discharge; 7 days, appearance of initiating and afterdischarge burst configuration and increase in burst duration; 9 days, decrease in latency of afterdischarge and further increase in burst duration; 13 days, essentially "mature" burst configuration with shortened afterdischarge. A gradual increase in the speed of burst onset is observed with development. The duration of the initiating discharge is shortened during the 6–13 day period. All records are integrated.

observed at all stages. The bursts observed at 6 days and before were simple accelerations in unit firing density. During the 7- to 8-day period, the duration and complexity of the burst envelopes increased due to the appearance of an *afterdischarge*. The afterdischarge is a trailing off of unit firing, usually several seconds in duration, which follows a shorter, more prominent, *initiating discharge*. However, not all short "initiating" type discharges are followed by afterdischarges. The duration of the afterdischarge increased up to 9–11 days. By 13 days, mature-appearing burst envelopes, which resembled those found at later stages, were observed (Fig. 4). These bursts have shorter afterdischarges and shorter lags between initiating and afterdischarges than were observed earlier.

The distribution of bursts in time varied as a function of development. Bursts tended to cluster together in groups of 2–4 at the 5- and 6-day stages. By 9 days, numerous bursts clustered together into regular periods of activity which recurred at regular intervals (Fig. 5). This regularity was lost when the maximum amount of burst activity was reached at 13 days (Fig. 5). These observations concerning burst periodicity have been substantiated in an extensive autocorrelational study of electrical activity in the embryonic spinal cord (J. S. Morrison & R. R. Provine, in preparation). In this latter study, the inability of autocorrelation procedures to detect regular periodic activity in the cord electrical data after 13 days indicates that regularity is lost and not simply masked by an overlay of irregular discharges. It should be emphasized that the periodicity in question is that of burst activity. At all stages, bursts were superimposed upon a relatively continuous level of unit activity.

The amount of burst activity observed showed reliable developmental trends. The quantity of burst activity present at a given stage is presented as percent burst activity. This is the percentage of total observation time occupied by burst discharges. Percent burst activity increased from 8% on day 5 up to a peak of 49% at 13 days. The amount of burst activity subsequently declined to only 14% on day 20, the day before hatching (Fig. 6).

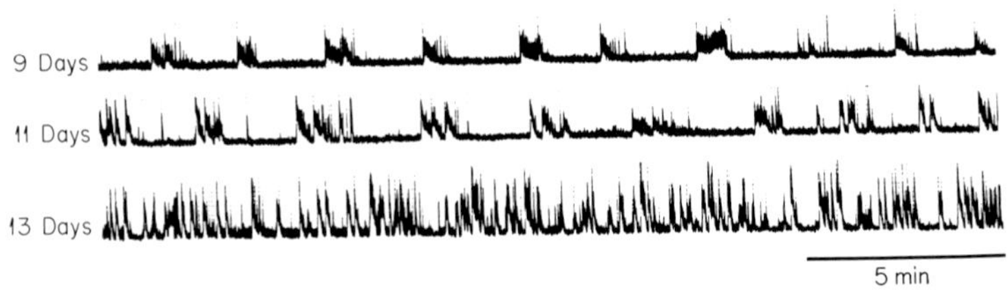

FIG. 5. Periodicity of burst discharges. Regular periods of activity and inactivity are present at 9 and 11 days. Regular periodicity is lost by 13 days. From Provine (1972).

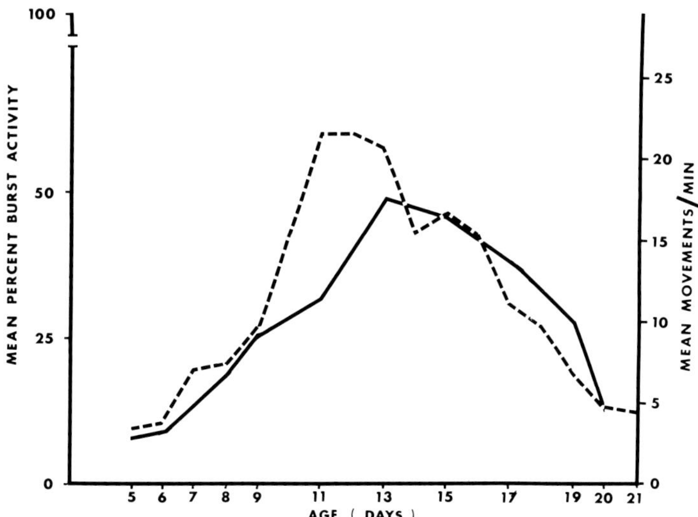

FIG. 6. Solid line represents the percentage of time during which polyneuronal burst discharges were present at different stages of development. Percent burst activity is compared with the mean number of body movements reported by Oppenheim (broken line) (1972). From Provine (1972).

These data concerning burst periodicity and percent burst activity, when considered together, suggest a mechanism which may underlie burst generation. Of particular interest are the observations that the maximum amount of burst activity is attained on day 13, the same day on which regular burst periodicity is lost. The coincidence of these events suggests that a threshold is being lowered upon a cyclic process. This could account for the increase in burst activity which occurs between 9 and 11 days while the period of burst discharges remains relatively stable. At 13 days, when burst periodicity is lost and the maximum amount of burst activity is observed, we may speculate that the threshold of our model system is sufficiently low to allow the spinal cord to be constantly above threshold. The system may be "freewheeling" at this time, no longer being constrained by the self-limiting periodic process which was present at earlier stages. The drop in percent burst activity which is observed after 13 days is probably not due to a raising of threshold to earlier levels but to another process, possibly a reduction in the capacity of the embryonic spinal cord for spontaneous discharge. This reduction may be due to active inhibitory processes. The presence of inhibition during this general period is indirectly indicated by an increase in the level of spontaneous unit activity in the ventral cord in response to strychnine, a drug thought to block postsynaptic inhibition (R. R. Provine, unpublished observations).

At this point, discussion will shift to the behavioral significance of the cord bioelectric activity. A fuller description and discussion of the ontogeny of bioelectrical activity is given elsewhere (Provine, 1972).

IV. Neural Correlates of Embryonic Motility

A remarkable parallelism is found between the polyneuronal burst discharges and embryonic motility. (*1*) The bursts are located in the ventral cord region which can sustain embryonic motility in absence of sensory, dorsal cord, and brain input (Hamburger *et al.*, 1966; Provine *et al.*, 1970). (*2*) The amounts of burst activity (percent burst activity) and motility are similar throughout development. Both increase up to a peak at about 13 days and then decrease until hatching at 21 days (Hamburger *et al.*, 1965; Provine, 1972). The close relationship between the number of embryonic movements per minute recorded by Oppenheim (1972) and percent burst activity is shown in Fig. 6. (*3*) The periodicity of burst activity and embryonic motility are similar. Both show regular periods of activity and inactivity before 13 days. This regular periodicity is lost at 13 days (Hamburger *et al.*, 1965; Provine, 1972). On the basis of these and other considerations (Provine, 1972), it seems probable that the polyneuronal burst discharge is a neural correlate of motility.

A. Simultaneous Observation of Cord Electrical Activity and Embryonic Motility

In order to establish directly the relationship between burst discharges and behavior, spontaneous motility and burst activity were simultaneously recorded from freely moving embryos. This project was undertaken with the collaboration of Mr. K. L. Ripley. Motility was monitored visually by an observer who activated an event marker in response to all discrete movements of the embryo. Cord electrical activity was simultaneously recorded with 25-μ glass-insulated tungsten "floating" electrodes which moved with the embryos. The concurrently recorded cord and motility data were taped for later analysis. At no time during the experiment did the observer of embryonic motility receive feedback concerning cord burst activity.

The results unequivocally establish the cord burst discharge as a neural correlate of embryonic motility. Burst discharges were correlated with movements in embryos ranging in age from 4 days until 21 days, the day of hatching (Fig. 7). The relationship between burst and movements was very high but not perfect. The relationship was particularly impressive in records in which the low number of movements and bursts simplify comparisons. Although we have not performed a correlational analysis upon our data, we

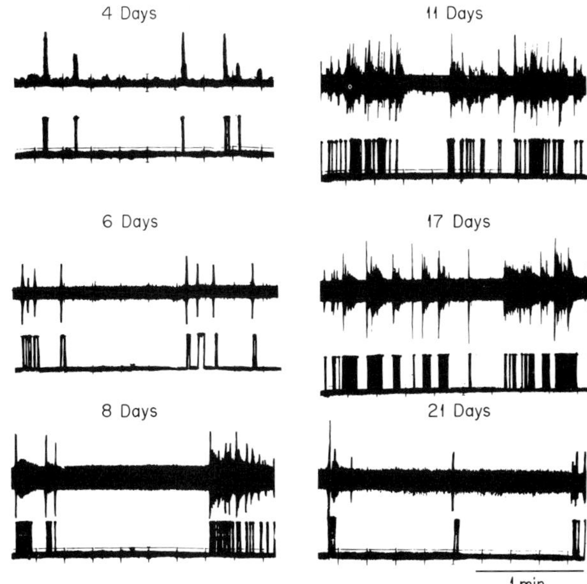

FIG. 7. Comparison of cord burst discharges (upper trace) with visually observed body movements (lower trace). Four-day cord activity was integrated to emphasize the low amplitude activity. Cord discharges were made from lumbosacral region, except at 4 days when brachial cord was monitored. From Ripley & Provine (1972).

can state with confidence that major burst discharges are always accompanied by the movement of some body part, and that only rarely do movements occur which are not in close temporal proximity to a burst.

B. Simultaneous Observation of Cord and Sciatic Nerve

A second series of experiments was performed with Mr. Ripley in order to show that burst discharges in curarized preparations result in motor nerve discharges. These experiments also control for the possibility that movements of the embryo might produce electrode movements which would evoke bursts. Such movement-produced artifacts could be responsible for the high movement-burst correlations. However, the previous observations of burst activity in curarized preparations make this possibility seem unlikely. In this experiment, the embryos were immobilized with curare and sciatic nerve discharges were used instead of motility as an index of cord motor output. For convenience of presentation, the concurrently recorded cord and sciatic nerve records were integrated and the sciatic nerve trace was inverted and placed base-to-base with the cord trace, thus forming a "cross-envelope." The symmetry of the resulting composite trace is an index of the similarity of envelopes and time relations of discharges from the two recording sites. This means of data presentation was suggested by Mr. J. S. Morrison and is described in Provine (1971) and Morrison and Provine (in preparation). A burst discharge is compared with its integrated counterpart in Fig. 8. The integrators used in the present study are different from the one used in the previous section.

The sciatic nerve discharges were correlated with cord burst discharges in the 15-, 17-, and 21-day embryos which were examined (Fig. 9). These results indicate that electrode movements do not evoke burst discharges in the floating electrode preparations. Most important, the present study provides further evidence that cord burst discharges are a reliable measure of embryonic "behavior" in the curarized preparation; the burst discharges must involve motoneurons which respond with volley discharges directed to the muscles through the motor nerves.

The present series of experiments establishes the spinal cord burst discharge as a neural correlate of embryonic motility. In doing so, it provides the first direct evidence of the *neurogenic* (in contrast to myogenic) basis of behavior in the chick embryo. The evidence of earlier investigators was indirect, being based upon the pharmacological action of curare and upon electrical stimulation of parts of the nervous system (see Hamburger, 1963, for a review). Investigators focusing on behavior should be encouraged by the finding that movements are a reasonable index of CNS bursts. Likewise, investigators with an electrophysiological orientation are fortunate to have a neural correlate which reliably reflects embryonic behavioral events.

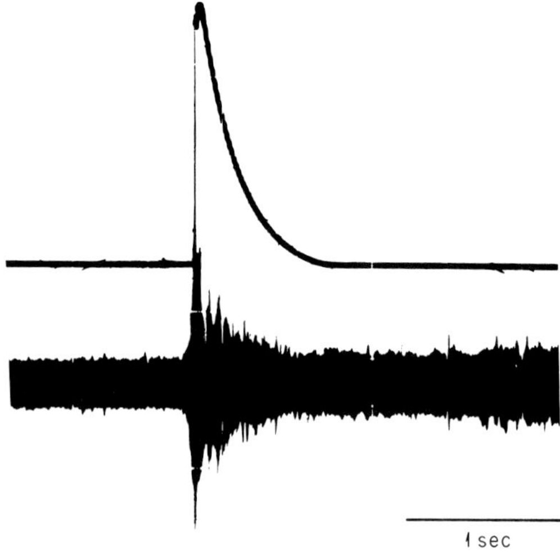

FIG. 8. Comparison of polyneuronal burst discharge (lower) with its integrated counterpart (upper). From Provine (1971).

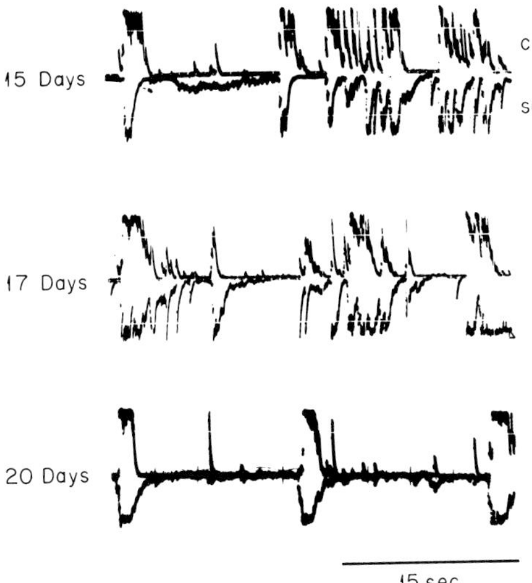

FIG. 9. Cross-envelopes comparing integrated burst activity simultaneously recorded from lumbosacral cord (c) and sciatic nerve (sn) of 15-, 17-, and 20-day embryos. From Ripley & Provine (1972).

The finding of a close correlation between bursts and behavior was not predictable, particularly at early stages. For instance, it would not have been surprising to find that the neuromuscular junction of young embryos was acting as a constraint on the ability of the CNS to express itself in terms of overt behavior, because only provisional neuromuscular contacts are present in the vicinity of muscles at the time of the first movements at $3\frac{1}{2}$–4 days (Hamburger, 1968; Mumenthaler & Engel, 1961; Visintini & Levi-Montalcini, 1939). It is not until 10–14 days, a week later, that the motor end plate approaches structural maturity (Mumenthaler & Engel, 1961). It is therefore interesting that the muscles appear to respond to the commands of the CNS from near the inception of motility. However, the apparent fidelity of response of the embryonic muscle to neural input may break down at a finer level of analysis; the present study only related gross cord discharges with gross movements of body parts.

A complete account of the material presented in this section is reported in Ripley and Provine (1972).

V. The Effect of Movement-Produced Stimulation on Embryonic Motility: An Electrophysiological Approach

One of the most interesting and controversial issues in behavior development concerns the relative importance of sensory stimulation and endogenous factors in the maintenance and shaping of embryonic motility. This matter is extensively discussed in the article by Viktor Hamburger, this volume, and in the Introduction (Section I) by Gilbert Gottlieb, this volume.

On the basis of descriptive evidence, several investigators contend that self-stimulation and other experiential factors play a significant role in evoking embryonic behavior (Gottlieb & Kuo, 1965; Kuo, 1932a, 1932b, 1932c, 1932d, 1967). For example, Kuo felt that the brushing of the legs against the beak and head constituted a major source of sensory stimulation. However, other investigators have drawn different conclusions from their descriptive evidence (Hamburger, 1963; Preyer, 1885; Visintini & Levi-Montalcini, 1939).More recent experimental studies which increased (Oppenheim, 1972) or decreased sensory input to the embryo (Hamburger & Narayanan, 1969; Helfenstein & Narayanan, 1969; Narayanan & Oppenheim, 1968; Oppenheim, 1966) or removed sources of sensory input altogether by radical surgical procedures (Hamburger *et al.*, 1966) give the impression that sensory input plays a minor, if any, role in embryonic motility, at least up to the last few days before hatching.

The present study tests the role of self-stimulation by comparing the amount of spinal cord burst activity (the motility correlate) observed before and after the immobilization of an embryo with curare. Curarization effectively breaks the feedback loop which may exist between motor outflow from the cord to the musculature and the possible sensory consequences of this activity. The curarization technique offers a methodological advance over previous studies and avoids the potential pitfalls of behavioral observation. Some of the limitations of visual studies of behavior were discussed by Gottlieb in the Introduction to Section I (this volume). This study was carried out with the assistance of Mr. K. L. Ripley.

Twenty 15-day embryos were used in this study. The 15-day stage was chosen because the embryos are large and hardy, being less likely to show adverse effects as a result of the experimental procedures than are the more delicate earlier stages. This is a critical consideration in an experiment such as the present one which has, by necessity, a fixed-order experimental design. Fifteen-day embryos were also selected because their behavior is typical of the jerky motility shown by younger embryos. Smooth, coordinated prehatching and hatching movements do not appear until day 17 (Hamburger & Oppenheim, 1967). The 15-day embryo was selected also because the exteroceptive (Gottlieb, 1968) and proprioceptive sensory systems were more likely to be mature and to influence behavior than at earlier stages.

Recording techniques and procedures were similar to those described for the previous study. Burst activity was recorded with "floating" glass-insulated 25-μ tungsten electrodes which were inserted into the ventral portion of the lumbosacral cord. Thirty minutes were allowed for the embryo to stabilize after preparation before an experiment was begun. Records of cord electrical activity observed during an experiment were stored on magnetic tape for later quantification. The amount of burst activity present in the control and experimental conditions were expressed in terms of percent burst activity as in previous experiments (p. 86). This is the percentage of time in an experimental period during which burst discharges were present. Double-blind procedures were used in all quantitative steps of the study.

The experimental design is illustrated in Fig. 10. The experiment was started by recording 15 minutes of baseline electrical activity from the spinal cords of untreated motile embryos. D-Tubocurarine (.03 mg) was then injected into the wing muscles, and 15 minutes were allowed for the curare to take effect. After this period, all embryos were completely immobile. The decline in the number of movements in a single embryo after the injection of curare is illustrated by the open circles in Fig. 10. Fifteen minutes of burst activity were then recorded from the immobile curarized embryos. In Fig. 10, the mean number of embryonic movements and the amount of cord burst activity (percent burst activity) are given for a single embryo at 2-minute intervals instead of the 15-minute intervals used in the study. This enables

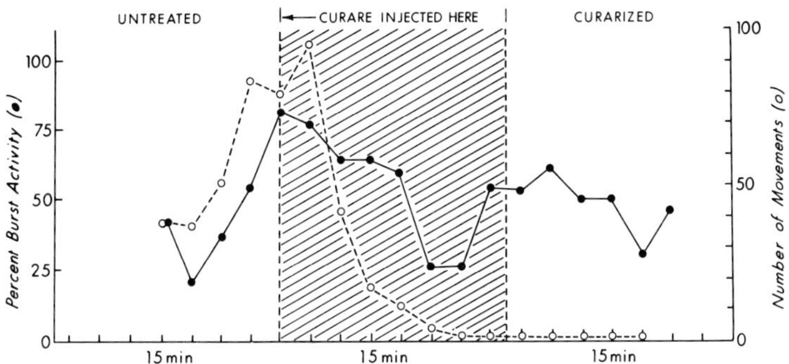

FIG. 10. The experimental design of the movement-produced stimulation study. The mean percentage of time that bursts were present in the spinal cords of 20 motile embryos during the 15-minute untreated condition was compared with the amount present in the same embryos during the 15-minute curarized (immobile) condition. The figure shows the amount of cord burst activity (closed circles) and simultaneously observed movements (open circles) which were observed in a *single* embryo at 2-minute intervals before and after the injection of curare. Cord burst activity is shown to continue at a high level after all movements have ceased. The presentation of the data for the embryo reported above differs from that reported in the actual study in that 10-minute instead of 15-minute periods of activity were recorded for the untreated and curarized conditions, and means are reported for 2-minute intervals instead of the entire 15-minute untreated and curarized conditions.

us to show short-term fluctuations in activity which emphasize the different effects curare has upon spinal cord electrical activity and embryonic motility. Motility was not a variable in the present experiments. Ten-minute periods of data for the untreated and curarized conditions are shown in Fig. 10 instead of the 15-minute periods used in the main body of the study. This was done only in the case shown in Fig. 10 in order not to exceed the storage capacity of our tape recorder.

The mean and standard deviation of the amount of burst activity recorded from 20 embryos during the 15-minute untreated control period before the injection of curare was 39.65 ± 15.63% *versus* 39.35 ± 16.69% for the 15-minute curarized condition. A two-tailed t test detected no significant differences between these means ($t = 0.08$, $df = 18$, $p > 00.05$). These electrophysiological results support the findings of the earlier behavioral experiments by Hamburger and associates which found that various sources of sensory input had little effect on motility (see above). However, the conclusion of the present study must be qualified because the dependent variable (percent burst activity) is only a rough index of the amount of motility and may not reflect subtle changes in the patterning of motor outflow to groups of muscles. The possibility also exists that the sensory events at one embryonic stage may influence behavior at another. However, until these later considerations receive experimental validation, we concur with Hamburger (this volume) that movement-produced exteroceptive and proprioceptive stimulation have little effect on ongoing embryonic behavior, at least up to late stages of incubation. It should also be pointed out that the present study deals with the short-term (less than 1 hour) effects of reduction of a potential class of sensory input, while the studies of Hamburger and associates have largely dealt with long-term or chronic effects of reduction or elimination of sensory input. This is an important distinction, because the long-term studies may allow the embryonic CNS to adjust to reduced levels of sensory input. The present short-term study would minimize such adjustments. For this reason, findings from short- and long-term studies should be considered complementary but not equivalent.

The present results are also important from a methodological point of view. They convincingly indicate that curare has no significant effect on the amount of spinal cord burst activity of 15-day embryos. That curare had no effect at other stages is suggested by the similarity of the amount of patterning of burst discharges and behavior at a given stage of development (Provine, 1972), and the quantitative and qualitative similarities of burst discharges observed before and after the injection of curare in a given embryo. The convenient technique of using immobile, curarized embryos for investigations of the development of spinal cord bioelectric activity thus seems justified.

VI. Synchrony and Spatial Distribution of Discharges: Communication within the Embryonic Spinal Cord

The initial experiment in this series (Provine *et al.*, 1970) demonstrated that prominent burst discharges are distributed throughout the ventral portion of a cross section of the lumbosacral spinal cord. These discharges were later shown to be correlated with embryonic movements (Provine, 1972; Ripley and Provine, 1972). The present study involves the description of the spatial distribution and synchrony of burst discharges occurring at two different spinal cord loci. This was done by concurrently recording activity with two electrodes. The results indicate whether the neural activity associated with embryonic motility involves the discharge of isolated islands of neurons or whether large expanses of the spinal cord discharge synchronously. Furthermore, by observing the spatial distribution of burst discharges at several stages of development, we may be able to determine if neurogenesis involves a progressive segregation or integration of segmental activity. As it turns out, neither process is an adequate description of the observed phenomena (Provine, 1971).

The method for preparing and recording from embryos is reported elsewhere (Provine, 1971). The only difference between this and previous recording procedures involved the making of two spinal incisions instead of one in some cases and the utilization of two, rather than one, 25-μ tungsten recording electrodes. Curare was used in all except two 6-day and two 17-day control embryos. Curare had no effect on the described phenomena. The correspondence between discharges occurring at different pairs of spinal loci was observed. A first experiment examined the coincidence of activity at various dorsal and ventral cord sites of four 17-day embryos. In a second experiment, the correspondence between discharges occurring at different points in the ventral portion of the cord along its rostrocaudal axis was observed in 34 embryos between 6 and 20 days of incubation. The following pairs of sites were examined: (*1*) ipsilateral: lumbosacral/lumbosacral, .5 mm electrode separation; (*2*) ipsilateral: lumbosacral/lumbosacral, 1.0 mm electrode separation; (*3*) bilateral: lumbosacral/lumbosacral; (*4*) ipsilateral; thoracic/lumbosacral; and (*5*) ipsilateral: brachial/lumbosacral. Integration was utilized to portray the acceleration in unit firing density which characterizes burst discharges. The input-output characteristics of this integrator were shown previously in Fig. 8. As before, the activity in two channels is presented with the cross-envelope technique in which one of the two simultaneously recorded integrated records is inverted and placed base-to-base with the activity in the other channel. The greater the symmetry of the cross envelope, the greater the similarity of the two signals.

A coincidence of activity was found between two ipsilateral dorsal column sites (260-μ and 265-μ electrode depth) which were separated by .5 mm along the cord's rostrocaudal axis (Fig. 11a). Note the effect of lowering one of these electrodes to a ventral column region (1400 μ) (Fig. 11b). The asymmetrical cross-envelope of Fig. 11b indicates that little, if any, correspondence was observed between spontaneous activity which was simultaneously observed in the dorsal and ventral columns of the cord. The short and rather infrequent discharges observed in the dorsal cord seem to be independent of the more frequent and lengthy discharges of the ventral region. Compare the asymmetrical cross-envelope for dorsal and ventral sites (Fig. 11b) with the highly symmetrical one representing the activity of two ventral sites (Fig. 11c). The dorsal electrode used to record the activity of Fig. 11b was lowered to a depth of 1700 μ to obtain the second channel of ventral cord activity shown in Fig. 11c. These findings indicate that a weak, if any, coupling exists between the spontaneous discharges observed in the dorsal and ventral spinal cord regions of 17-day embryos. In contrast, there seems to be highly coupled activity in the ventral cord. This coupling exists between sites along the rostrocaudal as well as mediolateral directions. Figure 11d shows the coupling between ventromedial and ventrolateral cord sites. The development and spatial extent of the coupling in the ventral cord will now be further explored at a wide range of developmental stages.

A remarkable correspondence was found between burst discharges occurring throughout the rostrocaudal axis of the ventral portion of spinal cords of embryos between 6 and 20 days of age. The similarities of time of onset and envelopes of bursts simultaneously recorded from different pairs of spinal cord loci are shown by the highly symmetrical cross-envelopes of Fig. 12. Almost identical discharges were observed at different loci within

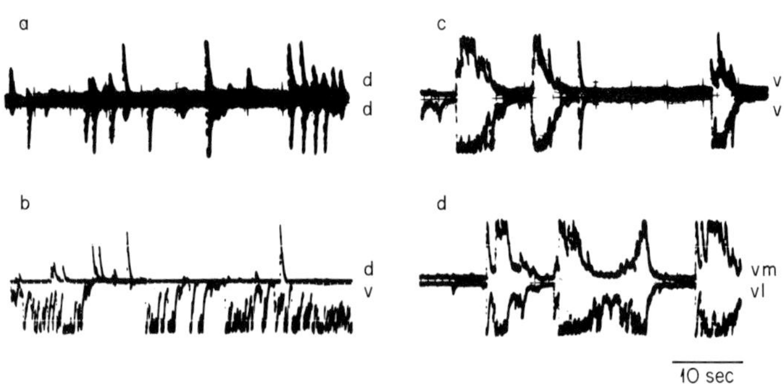

FIG. 11. Cross-envelopes comparing discharges simultaneously recorded from various ipsilateral pairs of regions of the lumbosacral spinal cord of a 17-day embryo. (a) dorsal/dorsal, d/d, (b) dorsal/ventral, d/v, (c) ventral/ventral, v/v, (d) ventrolateral/ventromedial, vl/vm.

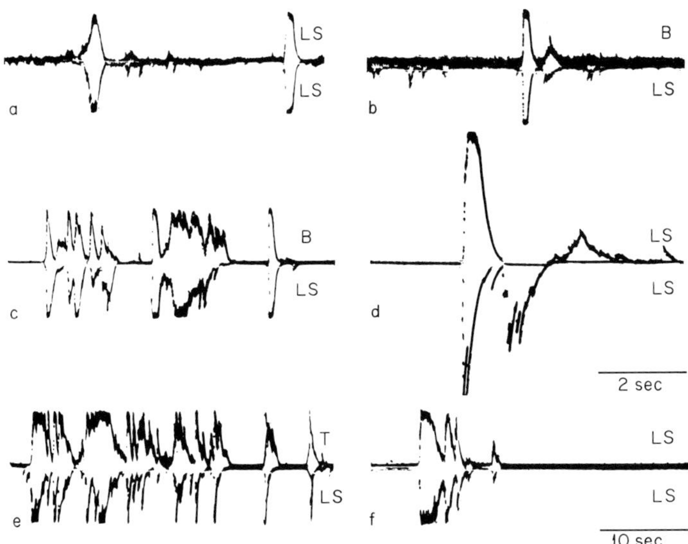

FIG. 12. Records of integrated polyneuronal burst discharges simultaneously recorded from pairs of spinal cord sites. Activity from one region is inverted and placed base-to-base with activity from the other region, so that symmetry of the resulting composite trace indicates the similarity of the activity from the two sites. Correspondence (symmetry) is shown between burst discharges appearing in various pairs of spinal cord loci recorded from embryos of the following ages: (a) 6-day, bilateral: lumbosacral/lumbosacral (LS/LS); (b) 6-day, ipsilateral: brachial/lumbosacral (B/LS); (c) 9-day, ipsilateral: brachial/lumbosacral (B/LS); (d) 9-day, bilateral: lumbosacral/lumbosacral (LS/LS); (e) 17-day, ipsilateral: thoracic/lumbosacral (T/LS); (f) 20-day, bilateral: lumbosacral/lumbosacral (LS/LS). The 9-day bilateral: LS/LS case (d) shows alternating region specific activity in the two cord halves after a common initial discharge. Records are from Provine (1971) except (d), which appears in J. S. Morrison and R. R. Provine (in preparation).

the spinal cords of 6- to 20-day embryos (Fig. 12). Note the impressive correspondence between discharges recorded in 6-day (Fig. 12a, b), 9-day (Fig. 12c), 17-day (Fig. 12e), and 20-day embryos (Fig. 12f). It is of particular interest that synchronous bursts were found in such functionally and anatomically dissimilar regions as the brachial and lumbosacral (Fig. 12b, c) and the thoracic and lumbosacral spinal cord (Fig. 12e). The similarity between discharges recorded from these different regions could be as great as those recorded within the same region which were separated by only .5 mm. Bilateral sites also showed a close correspondence (Fig. 12a, f). Although transregional coupling was the prevalent theme of these results, synchrony of discharges and similarity of burst envelopes were not always observed. In fact, most burst onsets were not precisely simultaneous. Lags varied from 0 to 100 msec or more. These lags do not show up in the slow sweep speed photos of Fig. 12. Also, region-specific activity was occasionally observed. This usually involved the afterdischarge. For example, after synchronous initiatory discharges, an afterdischarge may be present at only one site. In other cases, afterdischarges recorded at two sites would vary in duration. An interesting example of region-specific activity was detected in one 9-day record by Mr. J. S. Morrison (Morrison & Provine, in preparation). Note the alternation of burst activity at bilateral cord sites (Fig. 12d). This record suggests that waves of excitation are being passed back and forth across the cord's central commissure. Alternating leg movements may accompany this activity. In all embryos, the abrupt, high amplitude burst initiations were spatially coherent. Region-specific differences were usually confined to the longer and more complex afterdischarges which appeared after 6 days. Another less frequently observed form of regional activity had slow graded onsets in contrast to the explosive burst initiations typical of most spinal cord discharges.

Several factors indicate that the synchrony of the observed discharges is due to the propagation of activity through the cord and not to the recording of a single source through a volume conductor. (*1*) There was no decrement in burst amplitude as a function of the distance between recording electrodes. (*2*) Bursts occurring at different loci were rarely exactly synchronous. (*3*) Limited regional activity occurred. (*4*) Coupling was found along the rostrocaudal, but not the dorsoventral axis. (*5*) Results are generally compatible with behavioral evidence.

The results seem to explain the waves of contraction which pass through the bodies of young 4- to 6-day embryos. These S-movements are correlated with polyneuronal bursts which are probably propagated through the spinal cord. After the first few days of motility, considerable behavioral differentiation takes place. The limbs and various body parts begin to move independently of the somatic musculature. However, these later developments are not inconsistent with the finding of transregionally coherent burst activity. While body parts seldom twitch in synchrony, all body parts capable of moving seem to do so during an activity period. In general, the present findings are consistent with Hamburger's (1968) prediction that the neural discharge responsible for embryonic motility must "sweep through the whole system [p. 258]." Unfortunately, we are not yet in a position to decide whether the cord burst discharges are involved in the indiscriminate activation of all neuromuscular pathways, as was also suggested by Hamburger (1968, p. 258). To make a decision on this matter, we would require extensive data on the simultaneous activity of many specific muscles or motor nerves. The synchrony of cord burst discharges suggests only that large expanses of spinal cord tissue are being activated almost simultaneously. We are not yet in a position to carefully define what the product of this activity is in terms of the patterning of cord motor outflow. For example, discharges occurring at different cord loci may have very different behavioral consequences. Hamburger (this volume, p. 61) explores the latter point in his discussion of the work of Narayanan (Narayanan & Hamburger, 1971) and Straznicky (1963).

In concluding this section, it should be emphasized that the picture of spinal cord electrical activity which has been presented was painted in very broad brush strokes. Without doubt, information was lost when complex bioelectric phenomena were treated simply as burst "events." However, multielectrode procedures similar to those discussed in this section offer considerable potential in analyzing information flow in the embryonic CNS. For example, by analyzing sequences of discharges occurring at different points in an array of at least three electrodes placed throughout the embryonic nervous system, the origin and direction of propagation of discharges could be determined. Furthermore, shifts in the patterning of these sequences of events in a standard electrode array observed at different developmental stages may reveal the relative contributions of various cord and brain regions to cord burst activity and motility during ontogenesis. For example, a greater proportion of descending discharges originating in the brain as compared to ascending discharges arising within the cord may be observed in progressively older embryos. The development of communication between various brain regions may also be investigated using similar multielectrode techniques. Behavioral studies of embryos which have had parts of their CNS removed offer only an indirect and much less powerful approach to related problems (Decker & Hamburger, 1967; Hamburger, this volume; Hamburger *et al.*, 1965, 1966). The multielectrode technique may also be used to study the spread of excitation in response to sensory stimulation. Ideally, several cord output channels (muscles or motor nerves) could be monitored simultaneously with cord activity in order to give an accurate picture of input-output relationship in the developing nervous system. Our present rudimentary investigations suggest that such studies would be a promising next step in an electrophysiological investigation of the embryonic CNS.

VII. Summary

The present electrophysiological studies are a first step in the elucidation of the spinal mechanisms underlying the development of behavior in the chick embryo. Spinal cord bioelectric activity is traced from the fourth day of incubation until hatching on the twenty-first day. Of particular interest were polyneuronal burst discharges which are located in the ventral portion of the spinal cord. These discharges were synchronous with visually observed movements of the embryo from the inception of motility on day 4 until hatching on day 21. This finding establishes the polyneuronal burst discharge as a motility correlate and provides the first uniquivocal evidence that embryonic motility in the chick is *neurogenic*. The existence of a high correlation between CNS discharges and behavior indicates that behavioral studies provide a reliable means of studying certain aspects of neurogenesis in the chick embryo throughout development.

The special properties of the electrophysiological preparation were utilized to test the significance of movement-produced feedback to embryonic motility. This was done by comparing the amount of burst activity, the neural correlate of motility, before and after the immobilization of embryos with curare. No significant differences were found, indicating that movement-produced feedback has little, if any, effect on the amount of spontaneous embryonic motility at the 15-day stage tested.

A final study involved the examination of burst discharges occurring at two different spinal cord loci. This constituted a study of communication within the embryonic cord. The rather surprising result was that nearly synchronous burst discharges were observed along the rostrocaudal axis of the ventral cord from at least the sixth to the nineteenth day. This indicates that communication, which may be multisynaptic in nature, is present throughout most of embryogenesis.

Acknowledgments

I appreciate the comments and encouragement of Professor Viktor Hamburger. His contributions to the fields of neuroembryology and behavior development inspired the present research. I would also like to acknowledge the masterful engineering skills of Mr. Robert Loeffel and helpful discussions with Prof. T. Sandel and Mr. K. L. Ripley and Mr. J. S. Morrison. Much of the work described here was begun as a part of a doctoral dissertation which was submitted to the Department of Psychology, Washington University. The research was conducted in the Psychobiology Laboratory of Prof. T. Sandel and was supported by USPHS Grant GM 1900 and a traineeship under USPHS Grant P10 ES 00139 through the Center for the Biology of Natural Systems, Washington University.

References

Corliss, C. E., & Robertson, G. G. The pattern of mitotic density in the early chick neural epithelium. *Journal of Experimental Zoology*, 1963, **153**, 125–140.

Corner, M. A., & Bot, A. P. C. Developmental patterns in the central nervous system of birds. III. Somatic motility in the embryo and its relation to behavior after hatching. *Progress in Brain Research*, 1967, **26**, 214–236.

Crain, S. M. Bioelectric interactions between cultured fetal rodent spinal cord and skeletal muscle after innervation *in vitro*. *Journal of Experimental Zoology*, 1970, **173**, 353–370.

Decker, J. D., & Hamburger, V. The influence of different brain regions on periodic motility of the chick embryo. *Journal of Experimental Zoology*, 1967, **165**, 371–383.

Gottlieb, G. Prenatal behavior of birds. *Quarterly Review of Biology*, 1968, **43**, 148–174.

Gottlieb, G., & Kuo, Z.-Y. Development of behavior in the duck embryo. *Journal of Comparative and Physiological Psychology*, 1965, **59**, 183–188.

Hamburger, V. The mitotic patterns in the spinal cord of the chick embryos and their relation to histogenetic processes. *Journal of Comparative Neurology*, 1948, **88**, 221–284.

Hamburger, V. Some aspects of the embryology of behavior. *Quarterly Review of Biology*, 1963, **38**, 342–365.

Hamburger, V. Emergence of nervous coordination. *Developmental Biology, Supplement*, 1968, **2**, 258.

Hamburger, V. Embryonic motility in vertebrates. In F. O. Schmitt (Ed.), *The neurosciences: Second study program*. New York: Rockefeller University Press, 1970.

Hamburger, V., Balaban, M., Oppenheim, R., & Wenger, E. Periodic motility of normal and spinal chick embryos between 8 and 17 days of incubation. *Journal of Experimental Zoology*, 1965, **159**, 1–14.

Hamburger, V., & Narayanan, C. H. Effects of the differentiation of the trigeminal area on the motility of the chick embryo. *Journal of Experimental Zoology*, 1969, **170**, 411–426.

Hamburger, V., & Oppenheim, R. Prehatching motility and hatching behavior in the chick. *Journal of Experimental Zoology*, 1967, **166**, 171–204.

Hamburger, V., Wenger, E., & Oppenheim, R. Motility in the chick embryo in the absence of sensory output. *Journal of Experimental Zoology*, 1966, **162**, 133–160.

Helfenstein, M., & Narayanan, C. H. Effects of bilateral limb-bud extirpation on motility and prehatching behavior in chicks. *Journal of Experimental Zoology*, 1969, **172**, 233–244.

Kuo, Z.-Y. Ontogeny of embryonic behavior in Aves. I. The chronology and general nature of the behavior of the chick embryo. *Journal of Experimental Zoology*, 1932, **61**, 395–430. (a)

Kuo, Z.-Y. Ontogeny of embryonic behavior in Aves. II. The mechanical factors in the various stages leading to hatching. *Journal of Experimental Zoology*, 1932, **62**, 453–487. (b)

Kuo, Z.-Y. Ontogeny of embryonic behavior in Aves. III. The structural and environmental factors in embryonic behavior. *Journal of Comparative Psychology*, 1932, **13**, 245–271. (c)

Kuo, Z.-Y. Ontogeny of embryonic behavior in Aves. IV. The influence of embryonic movements upon the behavior after hatching. *Journal of Comparative Psychology*, 1932, **14**, 109–122. (d)

Kuo, Z.-Y. *The dynamics of behavior development*. New York: Random House, 1967.

Langman, J., & Haden, C. C. Formation and migration of neuroblasts in the spinal cord of the chick embryo. *Journal of Comparative Neurology*, 1970, **138**, 419–432.

Levi-Montalcini, R. The origin and development of the visceral system in the spinal cord of the chick embryo. *Journal of Morphology*, 1950, **86**, 253–284.

Mumenthaler, M., & Engel, W. K. Cytological localization of cholinesterase in developing chick embryo muscle. *Acta Anatomica*, 1961, **47**, 274–299.

Narayanan, C. H., & Hamburger, V. Motility in chick embryos with substitution of lumbosacral by brachial and brachial by lumbosacral spinal cord segments. *Journal of Experimental Zoology*, 1971, **178**, 415–432.

Narayanan, C. H., & Oppenheim, R. Experimental studies on hatching behavior in the chick. II. Extirpation of the right wing. *Journal of Experimental Zoology*, 1968, **168**, 395–402.

Oppenheim, R. W. Amniotic contraction and embryonic motility in the chick embryo. *Science*, 1966, **152**, 528–529.

Oppenheim, R. W. The embryology of behavior in birds: A critical review of the role of sensory stimulation in embryonic movement. *Proceedings of the XVth international congress of ornithology*. Leiden: Brill, 1972.

Oppenheim, R. W., & Foelix, R. F. Synaptogenesis in the chick embryo spinal cord. *Nature (London), New Biology*, 1972, **235**, 126–128.

Preyer, W. *Specielle physiologie des embryo*. Leipzig: Grieben's Verlag, 1885.

Provine, R. R. Embryonic spinal cord: synchrony and spatial distribution of polyneuronal burst discharges. *Brain Research*, 1971, **29**, 155–158.

Provine, R. R. Ontogeny of bioelectric activity in the spinal cord of the chick embryo and its behavioral implications. *Brain Research*, 1972, **41**, 365–378.

Provine, R. R., Sharma, S. C., Sandel, T. T., & Hamburger, V. Electrical activity in the spinal cord of the chick embryo *in situ*. *Proceedings of the National Academy of Sciences, U.S.*, 1970, **65**, 508–515.

Ripley, K. L., & Provine, R. R. Neural correlates of embryonic motility in the chick. *Brain Research*, 1972, **45**, 127–134.

Straznicky, K. Function of heterotopic spinal cord segments investigated in the chick. *Acta Biologica (Budapest)*, 1963, **14**, 145–155.

Visintini, F., & Levi-Montalcini, R. Relazione tra differenziazione strutturale e funzionale dei centri e delle vie nervose nell'embrione di pollo. *Schweizer Archiv. Für Neurologie und Psychiatrie*, 1939, **43**, 1–45.

Wilson, D. M., & Wyman, R. J. Motor patterns during random and rhythmic stimulation of locust thoracic ganglia. *Biophysical Journal*, 1965, **5**, 121–143.

Wyman, R. J. Probabilistic characterization of simultaneous nerve impulse sequences controlling dipteran flight. *Biophysical Journal*, 1965, **5**, 447–471.

SYNAPTOGENESIS IN THE AVIAN EMBRYO: ULTRASTRUCTURE AND POSSIBLE BEHAVIORAL CORRELATES

RAINER F. FOELIX AND RONALD W. OPPENHEIM

Neuroembryology Laboratory
Division of Research
North Carolina Department of Mental Health
Raleigh, North Carolina

> . . . in unwilling recognition of the truth—against which all seekers sooner or later stumble—that our great creative Mother, while she amuses us with apparently working in the broadest sunshine, is yet severely careful to keep her own secrets, and, in spite of her pretended openness, shows us nothing but results.
>
> *Nathaniel Hawthorne (1804–1864)*

I. Introduction

The systematic attempt to study both the morphology and function of the nervous system of embryos had its beginning with the classical studies of Coghill (1929) on the salamander, *Amblystoma punctatum.* In a long and carefully executed series of studies, Coghill was able to demonstrate that with the appearance of each new behavioral response there occurred a preceding, or concomitant change in the neuroanatomy which could be related to the behavioral response in a coherent and meaningful way. These studies of Coghill have not been surpassed for either their detailed and comprehensive nature or for the impetus that they had in stimulating a multitude of similar studies on other vertebrate forms during the 1920's, 30's and 40's. Some notable examples of correlated behavioral and neuroanatomical studies of embryos and fetuses conducted during that period of time are the studies of Barcroft and Barron (1939) on the sheep; Hooker and Humphrey on human embryos and fetuses (see Hooker, 1952); Tracy (1926) on the toadfish; Visintini and Levi-Montalcini (1939) on the chick embryo; and Windle and his co-workers on chick, rat, cat, and human embryos (Windle, 1940).

In spite of the fact that the silver staining techniques for nerve tissue (first used so fruitfully and elegantly by S. Ramón y Cajal) were available and utilized to great advantage by the above workers, including Coghill, there were, and still are, numerous problems inherent in these techniques that make it difficult to precisely relate neuroanatomical events with changes in embryonic behavior. The first disadvantage, of course, is not inherent to the staining techniques themselves, but rather is due to the limitations of the light microscope. Even with maximum resolving power small diameter neuronal processes simply can not be seen. Secondly, the silver techniques (especially the Golgi technique) are well known to be rather capricious in that they only stain a small percentage of all neurons. Thirdly, certain of these techniques work poorly during early developmental stages with avian embryonic nerve tissue (Ramón y Cajal, 1929). In spite of these obvious drawbacks the silver staining methods (especially the Golgi technique) have been, and continue to be, used quite fruitfully for the study of developing neuronal morphology (Marin-Padilla, 1970a, 1970b; Morest, 1969; Purpura & Pappas, 1968; Scheibel & Scheibel, 1970; Stelzner, 1971a).

While many of the problems inherent to light microscopical techniques have been alleviated with the advent of the electron microscope, ultrastructural studies introduce other, but equally frustrating, drawbacks. The most serious of these being that problems requiring quantitative answers are extremely recalcitrant to the electron microscope, due to the immense labor involved in attaining adequate sampling (Pease, 1964). Still, the use of the electron microscope, coupled with light microscopy and with the application of behavioral and neurophysiological techniques, provides a powerful approach to the correlated study of the functional and anatomical development of the embryonic nervous system. Bodian (1968, 1970b) and his colleagues have recently demonstrated the feasibility of such studies with the monkey fetus by showing that specific changes in synapse development are related to particular events in the ontogeny of behavior. Bodian's work represents the first attempt to apply combined ultrastructural and behavioral techniques to the embryonic nervous system, *in vivo.* Recent investigations of the ultrastructural and electrophysiological development of fragments of embryonic neural tissue, *in vitro*, conducted by Crain and his colleagues, also represent an elegant and significant contribution to our understanding of the morphological basis of the ontogeny of neural function (see Crain; Peterson; Bornstein; 1968; see also Crain; Volume 2 of this serial publication).

While there is a small, but growing literature on the fine structure of the developing nervous system of different vertebrate embryos and young animals (For excellent reviews of the literatures, see esp. Bunge, Bunge, & Peterson, 1967; Mugnaini, 1971; Tennyson, 1970), most workers have not combined both functional and ultrastructural studies in their experimental paradigm. Thus the studies of Bodian (1968) and Crain and his colleagues (1968) are especially significant not only for their pioneer efforts in this regard, but more importantly for their careful attempts to elucidate the relationship between the ontogeny of function (behavior or electrophysiology) and synaptogenesis. Indeed, on the basis of their efforts, synaptic development would seem to be the most compelling candidate for study in any serious attempt to relate the fine structure of the nervous system to the ontogeny of function. This is not to say, of course, that given a sufficient amount of information concerning synaptogenesis one should always expect to find a clear correlation with behavior. There may well be additional limiting factors that preclude the overt behavioral performance other than the formation of synaptic contact between neurons (e.g. maturation of muscle, liberation of sufficient transmitter substances, etc.).

The chick embryo is a particularly favorable subject for studying the simultaneous development of behavior and neuroanatomy. Firstly, there is a great deal of behavioral information available for the entire embryonic development of the chick (Corner & Bot, 1967; Gottlieb, 1968; Hamburger, 1963; Kuo, 1932; Oppenheim, 1972a). This information deals with both spontaneous and reflexogenic motor behavior and with the development of various sensory modalities. Secondly, the techniques for exposing and maintaining the embryo for behavioral observations are well worked out and provide a preparation that is far superior to that attained in mammalian embryos, where normalcy of the embryo is always in doubt (Windle, 1940). Thirdly, analytical techniques such as embryonic microsurgery (Hamburger, Wenger, & Oppenheim, 1966; Oppenheim & Narayanan, 1968), electrophysiology (Provine, 1971, and this volume), behavioral manipulations (Oppenheim, 1966, 1972b), neuropharmacological experiments (Sparber & Shideman, 1968), and neurochemical measures (Burt & Narayanan, 1970) can be utilized to an extent often not possible with mammalian embryos. And finally, with the exception of one brief report on the spinal cord (Glees & Sheppard, 1964), and a more extensive study of the cerebellum (Mugnaini, 1969), there is almost no information available on synaptogenesis in the chick embryo, *in situ* (however, other aspects of neurogenesis in the intact chick embryo have been examined with the electron microscope, e.g., Bellairs, 1959; Lyser, 1964, 1968).

The major aim of the research program that we have undertaken (the present article being the first *detailed* report of these studies; but see Oppenheim & Foelix, 1972) is to gain a better understanding of the morphological events, especially synaptogenesis, underlying the development of behavior in the chick embryo. Specifically, we want to determine: (*a*) when specific types of synapses are first formed in various regions of the spinal cord, e.g. axodendritic, axosomatic, excitatory and inhibitory synapes, etc; (*b*) the cellular derivation of these various synapses; (*c*) how these synapses change during development; (*d*) when, and in what order, myelination occurs both in different levels of the spinal cord and within a given level (dorsal, ventral, etc.); and (*e*) if and how the overt behavior of the embryo can be related to the above morphological events of synaptogenesis and myelination. We are well aware of the pitfalls that must be avoided if one is to convincingly demonstrate this last point. For example, until one can selectively eliminate a specific morphological structure (e.g., a certain type of synapse) and show that a particular aspect of behavior is modified, the establishment of mere correlations between a given structure and function will always rest on a rather questionable foundation. Even with such precautions, however, one must still be conservative in making inferences of causation between such grossly different levels of organization as ultrastructure and behavior.

II. General Morphological Development of the Spinal Motor System as Seen in the Light Microscope

The elegant and painstakingly accurate descriptions of histogenesis in the spinal cord of the chick embryo by S. Ramón y Cajal (1929) have not been surpassed up to the present day. These descriptions by Ramón y Cajal, based on the examination of histological sections treated with various silver staining techniques and beautifully illustrated by drawings from the light microscope, have served as an important source of information for much of the descriptive evidence discussed below, aided to some extent by our own reduced silver preparations.

It has become axiomatic in avian neuroembryology that neuromuscular development, including myelination and function, proceeds in a cephalo-caudal direction. Thus, within the spinal cord the cervical region appears to lead other regions in its maturation. It also seems to be generally true that within a given region of the spinal cord ventral areas (basal plate or anterior horn) mature prior to dorsal ones (alar plate or posterior horn). For example, Hamburger (1948) found that cells in the basal plate have their highest rate of mitotic activity at day 3, and that cell division steadily decreases after this time, whereas cells in the alar plate exhibit high rates of prolifera-tion between 3–6 days. Langman and Haden (1970), using autoradiographic techniques, have generally confirmed these findings; they report a complete cessation of *neuroblast* formation after 8 days. Because mitotic activity is still present after 8 days, these later produced cells are most likely *glioblasts* (Hamburger, 1948).

It is likely that as more and more detailed information becomes available on both neurogenesis and function in the avian spinal cord, this general scheme of rostral-caudal and ventral-dorsal development will have to be modified to accommodate the regional heterochronic development of specific functional systems as proposed in the theory of *systemogenesis* (Anokhin, 1964). For example, in view of our observation that synapto-genesis begins slightly later in the lumbosacral spinal cord but then rapidly attains a level of development comparable to the cervical region (see below), the statement of Anokhin (1964) that the lumbosacral spinal cord matures *prior* to the cervical region in the chicken embryo may be correct.

By 5–6 days of incubation, cell division in the spinal motor neurons and internuncial neurons has practically ceased. It is interesting, however, that many of the cells in these regions have already begun synaptogenesis and are functioning 2–3 days prior to this time (see below; Hamburger, this volume, Provine, this volume).

Between days 2 and 3 of incubation, neuroblasts in the ventral portion of the spinal cord can be seen to spin out axonal processes that are already directed toward, or in many cases, have already exited in the ventral root (Fig. 1). Except in very rare instances, these cells have not been found to possess dendrites at this time. Both Ramón y Cajal (1929) and Barron (1946) have reported that dendrites are first seen to develop from motoneurons at the time that the axon reaches the periphery. Thus, at this stage most of the neuroblasts are pyriform in shape and only somewhat later do they assume a bipolar geometry (Fig. 1). It is also at about this same time that a greater number of motor cells become impregnated with silver indicating

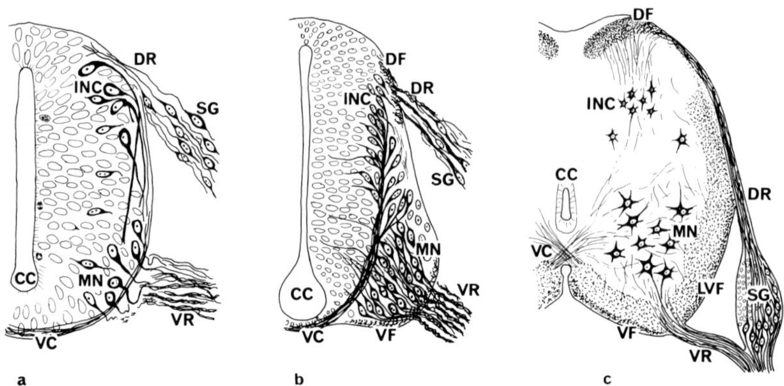

FIG. 1. Illustration of some of the developmental changes occurring in neuronal morphology in the spinal cord of the chick embryo. a, 3-day embryo, redrawn from Ramón y Cajal (1929). b, $4\frac{1}{2}$-day embryo, redrawn from Ramón y Cajal (1929). c, 12 to 13-day embryo, drawn from several consecutive sections of our own reduced silver material (Cajal-deCastro technique). Note the differentiation of neurons from the uni- and bipolar state in the early stages to multipolar neurons, as well as the increasing complexity in fiber growth with age. Abbreviations: CC, central canal; VC, ventral commissure; MN, motoneuron; VR, ventral root; SG, spinal ganglion; DR, dorsal root, INC, interneurons and commissural neurons; DF, dorsal funiculus; VF, ventral funiculus; LVF, lateral ventral funiculus. Drawing C is not to scale of A and B.

further differentiation. Although the first signs of *mature* myoneural junctions have not been found until about day 13 of incubation (Hirano, 1967), nerve fibers are already in close association with the cervical myotomes as early as 4 days (Visintini & Levi-Montalcini, 1939). The assumption that these early fibers have made functional contact seems well supported by curare experiments (Kuo, 1939; Visintini & Levi-Montalcini, 1939), and by the report that somites transplanted to the chorioallantoic membrane in the chick egg will not contract unless neural (spinal cord) tissue is included in the transplant (Alconero, 1965). Furthermore, Provine (this volume) has shown a clear correlation between central nervous system (CNS) bioelectric activity, peripheral nerve activity, and overt behavior from about day 4 of incubation. This evidence unequivocally supports the neurogenic (as opposed to myogenic) origin of the first embryonic movements. Thus, because overt contractions of somites in the cervical region can be observed already at $3\frac{1}{2}$ days in the chick embryo (Hamburger, 1963), it seems inexorably clear that motoneurons (and/or interneurons) are functioning long before they have completed differentiation. This fact raises the exceedingly interesting and theoretically important question of whether the early bioelectric activity plays any role in normal neurogenesis (Coghill, 1929; Gottlieb, 1971; Woodward, Hoffer, & Lapham, 1969).

By day 3 or 4 of incubation, the first dendrites can be seen arising from the soma opposite the exit of the axon, thus forming bipolar neurons. These dendrites are generally oriented obliquely in a dorsomedial direction. Between 5 and 6 days, the motoneurons in the ventral horn begin to take on a rather clear multipolar shape with dendrites arising from all sides of the soma (Fig. 1). Many of the dendrites from these early differentiated motor cells enter the ventrolateral funiculus.

Although internuncial (or association) neurons develop slightly later than motoneurons in the chick embryo, occasional axonal processes from these cells are already seen to impinge upon the motoneurons by 4 days. Since synapses, though rare, are found in the motor neuropil in the cervical region even earlier than 4 days (see below), it is possible that with the electron microscope small diameter internuncial axons will be found in this area before 4 days. According to Visintini and Levi-Montalcini (1939), these early association or interneuronal connections are primarily intersegmental, thus allowing for the integrated side-side *trunk and head* movements that begin to appear at about this time. On the other hand, some of these early synapses could arise from motoneuron collaterals or from fibers originating in the brain. Indeed, although the issue is not entirely settled, there is both anatomical and physiological evidence that fibers arising in the midbrain, and comprising the medial longitudinal fasciculus, have grown into the cervical region at the very inception of overt motor activity at 3–4 days, (Visintini & Levi-Montalcini, 1939). Because the ventral commissure system is present by about 4 days (Fig. 1), the anatomical basis for *bilateral* integration of the embryo's early movements also clearly exists.

According to both Windle and Orr (1934), and Visintini and Levi-Montalcini (1939), collaterals from sensory dorsal root fibers do not make contact with the internuncial neurons, thereby completing a multisynaptic cutaneous reflex arc, until 6–7 days of incubation. Although these data are based on silver techniques, and thus may have to be modified when ultrastructural evidence becomes available, the *functional* evidence strongly supports these reports; every investigator that has examined this question since the first report of Preyer (1885) agrees that artificially applied tactile stimulation does not evoke responses in the chick embryo until about day 7 of incubation.

After the first week of incubation, further neuroanatomical development in the spinal cord of the chick embryo as observed in the light microscope consists primarily of growth and further differentiation of features already present at that time (Fig. 1). Myelination in both the spinal cord and the peripheral nerves first begins at about 11–13 days of incubation and continues until after hatching (Bensted, Dobbing, Morgan, Reed, & Wright, 1957; El-Eishi, 1967; Mezei, Newburgh, & Hattori, 1971).

Although this brief summary of neuroanatomical events in the developing spinal cord of the chick embryo makes no claim for completeness (indeed many aspects of neurogenesis have been excluded altogether), it is hoped that it will at least serve the purpose of making the discussion below of synaptogenesis and behavior more understandable to the uninitiated.

III. General Discussion of Synapses

A. Historical

A synapse defines the contact region between two nerve cells, where nerve impulses are transmitted from one cell to the other. Synaptic structures were described in early light microscopical studies at the turn of the last century (Auerbach, 1898; Held, 1897; Ramón y Cajal, 1888), and it was Sherrington (Foster & Sherrington, 1897) who coined the term "synapse." At that time it was still quite controversial whether nerve cells were all interconnected by a protoplasmic continuity or whether each neuronal soma with its processes represented an individual and separate unit of the nervous system. While Ramón y Cajal (1888) and Harrison (1910) provided convincing anatomical evidence for the latter view, physiological support for the "neuron doctrine" (or for synaptic linkage) came from the observations of Wundt (1871) and later Jolly (1911) who demonstrated a minute, but distinct, delay (about 2 msec) in nerve impulse transmission within a simple reflex arc. The concept of a chemical transmission of nerve impulses at synaptic sites was already postulated by Elliott (1904) and Dale (1914), and later proven by the classical experiments of Loewi (1921) which showed the inhibitory effect of the *N. vagus* (acetylcholine) on heart rate.

Little was known of the detailed structure of synapses until the electron microscope became applicable to biological material in the early 1950's. De Robertis and Bennet (1954), and independently Palade and Palay (1954), found that nerve fibers contain accumulations of small vesicles in their presynaptic terminals. These electron-lucent, membrane bound vesicles of 300–600 Å diameter were termed "synaptic vesicles." It was soon proposed that they represent minute bags filled with neurotransmitter (e.g., acetylcholine), an idea which gained support from biochemical investigations of nerve cell fractions (De Robertis, Pellegrino de Iraldi, Rodriguez de Lores Arnaiz, & Salganicoff, 1962; Gray & Whitaker, 1960). It was thought that after arrival of a nerve impulse (action potential) at the axon terminal synaptic vesicles would liberate their content into the extracellular gap between the pre- and postsynaptic nerve fiber. The released neurotransmitter would then interact with the postsynaptic membrane and generate a new action potential. Although this concept is still basically accepted, certain modifications and restrictions of the details have been made (for extensive discussions of synaptic transmission, see De Robertis, 1964; Eccles, 1964; McLennan, 1970).

B. Structure of Mature Synapses

The major morphological criteria for a synapse (Plate I, Figs. 2, 3) are: (*1*) an aggregation of synaptic vesicles in the presynaptic terminal; (*2*) local, junctional densities on both pre- and postsynaptic membranes; (*3*) a distinct "synaptic cleft" (about 200 Å) between the parallel running membranes of the pre- and postsynaptic fiber, which usually contains electron-dense cleft material.

1. Synaptic Vesicles

The presynaptic terminal, which is usually an axon, is of variable shape (bouton/club/calyx) and is generally slightly expanded when compared to more proximal regions of the axon. It typically contains a few mitochondria and many synaptic vesicles which aggregate toward the thickened presynaptic cell membrane but is void of microtubules. The majority of synaptic vesicles are electron-lucent, spherical, and about 400 Å in diameter; they probably contain acetylcholine as a neurotransmitter (De Robertis, 1964). Electron-lucent, but flattened vesicles can also be found, especially after aldehyde fixation (Bodian, 1966, 1970; Dennison, 1971, Larramendi, Fickenscher, & Lemkey-Johnston, 1967; Uchizono, 1965; Valdivia, 1970; Walberg, 1965), and their presence has been related to inhibitory synapses. Other synaptic vesicles are electron-opaque (dense-cored) and may occur in two populations differing primarily in size (400–600 Å and 800–1200 Å diameter); these vesicles probably contain catecholamines, especially norepinephrine (De Robertis & Pellegrino de Iraldi, 1961).

Almost nothing is known about the *origin of synaptic vesicles.* It has been proposed that: (*1*) they are synthetized in the endoplasmic reticulum of the perikaryon (Palay, 1958; Stelzner, 1971b), and then transported by axoplasmic flow to the axon terminal; (*2*) that they are derivatives of mitochondria (Bloom, 1970), or microtubules (De Robertis, 1969), from within the axon; and (*3*) that they originate from pinocytotic vesicles which "bud off" the axonal membrane (Andres, 1964). It has been shown that the membrane of the synaptic vesicle is chemically different from the axonal membrane (Whittaker, 1965), and this fact would seem to argue against an origin from the axon membrane.

The commonly accepted concept of *transmitter release* in which the synaptic vesicle is thought to fuse with the presynaptic membrane and liberate

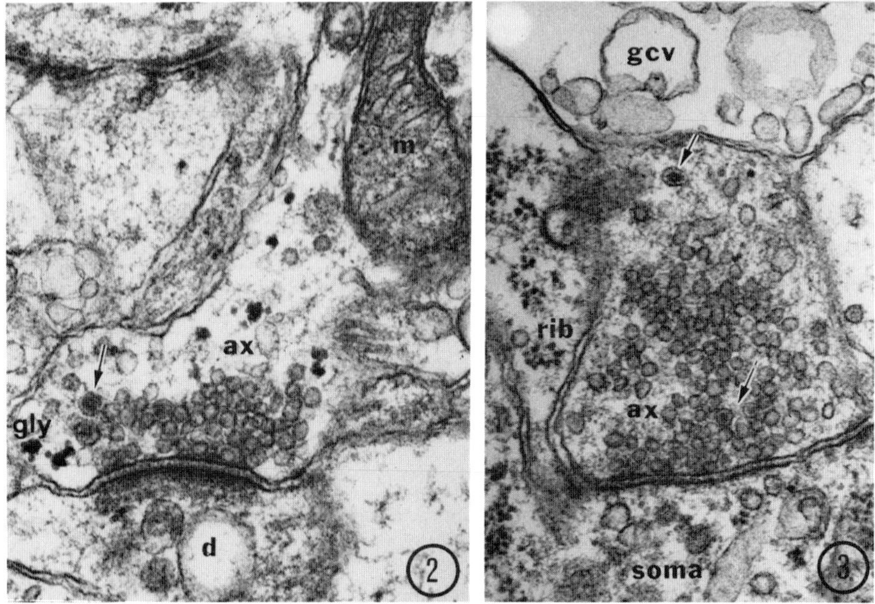

Plate I

FIG. 2. Mature axodendritic synapse from the cervical spinal cord of a 16-day pigeon embryo. The presynaptic nerve fiber (axon, ax) contains an aggregation of synaptic vesicles toward the cell membrane, some dense-core vesicles (arrow), glycogen granules (gly), and a mitochondrium (m). Note the filamentous material crossing the synaptic cleft and the distinct postsynaptic thickening of the dendrite (d) membrane. This micrograph pictures an asymmetric type synapse (Colonnier, 1968) or Type I, according to Gray (1959). × 46,750.

FIG. 3. Mature axosomatic synapse from the same spinal cord as in Fig. 2. Note the high density of synaptic vesicles in the axon terminal (ax) and a few dense-core vesicles (arrows). The membrane apposition between axon (ax) and soma is rather symmetrical in this case (Type 2, Gray, 1959). Part of a growth cone with growth cone vesicles (gcv) is seen in the upper part of the micrograph, rib, free ribosomes lying inside the perikaryon. × 42,500.

its content into the synaptic cleft has, in fact, never been proven. Although one can occasionally see micrographs clearly illustrating a vesicle "fusion" (Omega-shape, Plate VI, Fig. 13b), one can not decide whether this vesicle is moving "in" (pinocytosis) or "out" (exocytosis). Furthermore, such vesiculations are not restricted to the synaptic contact region but may be found anywhere on the furface of the axon terminal (Akert, Moor, Pfenninger, & Sandri, 1969). It is also not clear whether one synaptic vesicle releases all of its contents at once or in portions. A "quantal release" has been postulated by Fatt and Katz (1952) using physiological evidence and it was later suggested that the synaptic vesicles were the quantal unit (Del Castillo & Katz, 1956a, 1956b). Recently, Whittaker (1970) demonstrated two pools of acetylcholine in nerve terminals, one in the synaptic vesicles, and the other occurring free in the axoplasm. The released acetylcholine might well come from the free pool, but this is still an open question. The most recent hypothesis comes from Akert (1970, 1971) who postulates a temporary junction of synaptic vesicles at so-called "synaptopores" of 40–50 Å diameter in the presynaptic membrane. Such synaptopores have been discovered in Akert's laboratory by his co-worker Pfenninger.

Electron microscopical investigations have, with few exceptions (Hubbard & Kwanbunbumpen, 1968; Jones & Kwanbunbumpen, 1968, 1970a, 1970b), generally failed to demonstrate a decrease of synaptic vesicles after electrical stimulation. It has been proposed that the synaptic vesicles may simply represent a storage site that serves to replace any loss occurring in the free pool of neurotransmitter, with the latter being the actual neurotransmitter (Robertson, 1970).

2. Synaptic Densities

Most synapses exhibit asymmetric membrane thickenings, the postsynaptic membrane being wider and more conspicuous than the presynaptic. These thickenings ("synaptic material," see Bloom, 1970) are chemically different from the cell membrane as can be shown by selective staining techniques such as ethanolic phosphotungstic acid (E-PTA) (Bloom & Aghajanian, 1966; Gray, 1963), or Bismuth-Iodide (BI) (Akert & Pfenninger, 1969). The synaptic material on the presynaptic side appears as "dense projections" (Gray, 1963) which are arranged in a coherent, hexagonal network (Pfenninger, Sandri, Akert, & Eugster, 1969). This presynaptic grid is of a proteinaceous nature, rich in basic amino groups (Pfenninger, 1971a, 1971b), but its functional role is not yet understood. It should be mentioned that the presynaptic grid is usually not visible after standard preparation techniques for electron microscopy but only after E-PTA or BI application. It is also these presynaptic projections which differentiate a synapse from other junctional densities such as desmosomes (maculae adherentes).

The postsynaptic density is usually more pronounced than the presynaptic one. Short filamentous extensions of this density may project into the postsynaptic cytoplasm and have been termed "subsynaptic web" by De Robertis, Pellegrino de Iraldi, Rodriguez de Lores Arnaiz, and Salganicoff (1961). Occasionally several subjunctional dense bodies are found below the postsynaptic thickening (Taxi, 1961, 1965). It seems that the postsynaptic density is chemically different from the presynaptic dense projection, since it is clearly seen after OsO_4 fixation, in contrast to the dense projections.

3. Synaptic Cleft

A gap of about 200 Å appears between the pre- and postsynaptic fibers, while the proximity of nonsynaptic neuronal membrane appositions is usually around 100 Å. Some synaptic material occurs within the cleft, either in the form of crossing filaments or as a layer running parallel to pre- and postsynaptic membranes. It is thought that the cleft material has adhesive properties (Kelly, 1967), possibly due to mucopolysaccharides added to the proteinaceous cleft material (Akert, Gray, & Bloom, 1970). The adhesiveness of the cleft substance can be well demonstrated in "synaptosomes" (Whittaker, Michaelson, & Kirkland, 1964). Synaptosomes result from fractionating neuronal tissue and exhibit a presynaptic bag containing synaptic vesicles plus a firmly attached postsynaptic membrane.

C. Classification

Numerous attempts have been made to classify synapses both morphologically and physiologically. We want to present a brief list of several suggestions without trying to be complete or going into any details (Table I). For instance, beside the common axodendritic and axosomatic synapses, axoaxonic, dendrodendritic, somatosomatic and somatodendritic synapses (Peters, Palay, & de Webster, 1970) also exist, though much less frequently.

TABLE I
Classification of Mature Synapses

Axodendritic	*vs.*	Axosomatic	
Type I	*vs.*	Type II	Gray (1959)
Asymmetric (membrane thickening)	*vs.*	Symmetric	Colonnier (1968)
"S-type" (spherical synaptic vesicles)	*vs.*	"F-type" (flat synaptic vesicles)	Bodian (1966)
Excitatory	*vs.*	Inhibitory	Sherrington (1906)

Note: Often those characteristics in each of the left and right columns are all found in one synaptic type, but there are also numerous exceptions to this; for example, one may find axosomatic synapses with asymmetric membrane thickening and S-type vesicles.

D. Synaptogenesis

When nerve cells establish their first contacts during ontogeny, what are the morphological characteristics of these synapses? Are early synapses already equipped with clusters of synaptic vesicles and localized membrane thickenings, or do those characteristics appear sequentially? Many investigators of synaptogenesis hold the view that the first indication of a synapse are junctional thickenings while synaptic vesicles appear later (Glees & Sheppard, 1964, Hámorí & D'iachkova, 1964; Meller, 1964). Other authors (Bodian, 1968; Caley, 1971; Larramendi, 1969) report that membrane densities develop at about the same time, and only Ochi (1967) reported synaptic vesicles prior to membrane thickening.

One major problem in interpreting electron micrographs of developing nervous tissue is to assess whether a local membrane thickening is simply an adhesion site (desmosome) or possibly a precursor of a synapse.* Bearing this difficulty in mind, most authors are very cautious in their statements, but at least implicitly, the prevailing idea is still that local membrane densities are the first signs of synaptogenesis.

A wide variety of different animals and different regions of the nervous system have been studied with respect to developing synapses (see below). Many studies have used tissue culture techniques (Bunge *et al.*, 1967; James & Tresman, 1969; Meller, Breipohl, Wagner, & Knuth, 1969; Meller & Haupt, 1967) and apparently find that synaptogenesis proceeds as *in vivo*. In this connection it should be pointed out that the first appearance of synapses *in vitro* has been closely correlated with the onset of bioelectric activity (Crain & Peterson, 1967; Crain *et al.*, 1968), so it seems clear that nerve cells begin to interact electrically quite early in their development, whether *in vivo* (Provine, this volume), or *in vitro*.

IV. Observations of Synaptogenesis and the Ontogeny of Behavior in the Chicken and Pigeon Embryo

In our own investigations we have chosen to examine the spinal cord of the avian embryo as a kind of neuronal model in order to follow the events occurring in both synaptogenesis and behavior during the entire embryonic period. In this respect we are following the lead of Bodian (1970b) who is also studying the behavior and ultrastructure of the spinal cord of the monkey fetus. However, due to the advantages accruing to the avian embryo, as discussed in the Introduction to this article, we believe that an examination of these problems in the avian embryo may be more fruitful than similar studies with mammals.

*It is also possible that desmosome-like adhesion sites might initially serve the function of identifying and maintaining contact between neurons that are meant to be part of some later functional system, with synapse formation occurring at the same, or even another, site as the original adhesion; or in the case of cells not meant to be part of a functional system, the adhesion sites might be temporary. Altman (1971), for example, has suggested that coated vesicles in developing neurons may provide future synapses with either transient adhesion membranes or lasting dense membranes.

Since neurogenically mediated behavior, in the form of periodic muscular contractions, occurs first in the neck region of the avian embryo (Hamburger, 1963), we began our program of study by concentrating on the spinal cord in the cervical region. Furthermore, because it is now known that many aspects of the avian embryo's behavior are not initiated or controlled by sensory input (Hamburger *et al.*, 1966; Oppenheim, 1972a, 1972b), we have momentarily restricted our ultrastructural studies to the motor neuropil in the ventral or anterior portion of the spinal cord. In the future we plan to direct most of our attention to the lumbosacral spinal cord and to the development of hindlimb motor activity, and for this reason we include here preliminary observations on synaptogenesis in the lumbosacral spinal region of the chicken.

A. Material and Procedures

Each chicken embryo from 45 hours of incubation through 2, 3, 4½, 6, 7, 9, 11, 13, 16, and 19 days of age (hatching occurs on day 21) was observed (and often filmed) for 10–20 minutes to assess its behavioral repertoire, before the cervical or lumbosacral spinal cord was excised and processed for electron microscopy. After the embryo was staged according to the Hamburger-Hamilton normal stage series (Hamburger & Hamilton, 1951), the tissue was fixed in 5% phosphate-buffered glutaraldehyde, postfixed in 1% OsO_4, embedded in Epon, and the sections double-stained with uranyl acetate and lead citrate. Pigeon embryos were also observed and prepared for the electron microscope at 2, 3, 4, 7, 12, and 16 days of age (hatching occurs on day 16 or 17). Since few differences were noted between the pigeon and chicken, we will report primarily our findings on the chick spinal cord.

B. Results

1. Cervical Spinal Cord

a. First Phase. The youngest chick embryo (45 hours, stage 11) examined in our series did not show any distinct nerve fibers in the cervical spinal cord and, consequently, synapses were absent. Adhesion sites with symmetric membrane thickening do, however, occur between perikarya at this stage. Somatic motility was entirely absent.

At day 2 (60 hours, stage 16+), differentiated nerve fibers with microtubules, neurofilaments, mitochondria, vesicles, and occasional scattered ribosomes can be seen. Vesicles of about 400 Å are rarely encountered, and an aggregation of several vesicles toward the cell membrane is even less frequent. Among hundreds of inspected nerve fibers (axons, recognizable by their general lack of ribosomes), we found only two or three vesicle accumulations toward the axonal membrane (Plate II, Figs. 4a,b). Nonetheless, we consider these rare cases as possibly representing the first appearance of nascent synapses. The ultrastructural picture is similar in the pigeon, and neither species exhibits any motor behavior at this time.

At day 3 and $4\frac{1}{2}$, synaptic contacts are still rare. Only few synaptic vesicles aggregate toward the cell membrane, but now pre- and postsynaptic membranes often exhibit a slight thickening (Plate II, Fig. 5, Plate III, Figs. 6

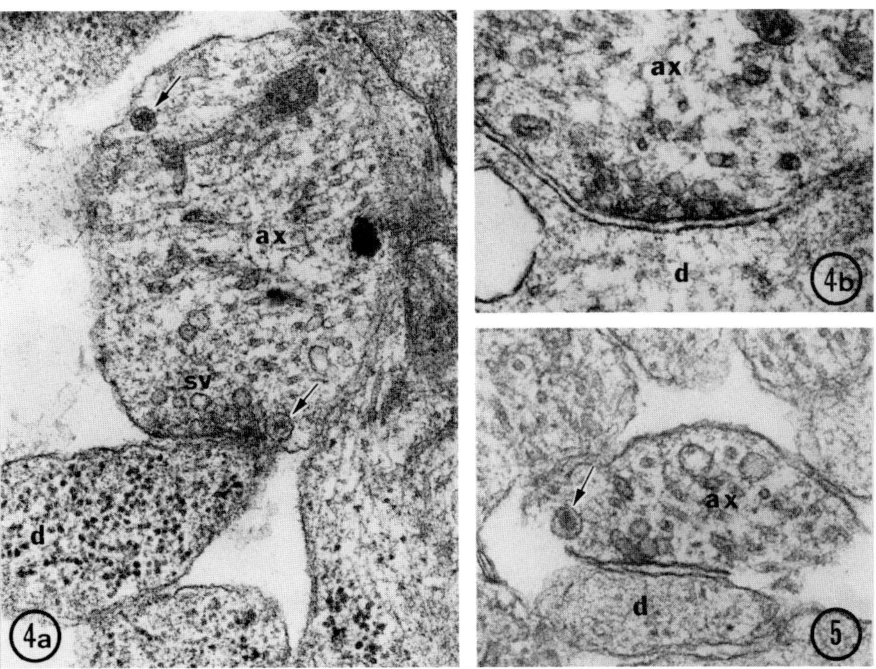

Plate II

FIG. 4a. The first indication of a synapse in the cervical spinal cord of a 2-day chicken embryo. The large axon (ax) contains several synaptic vesicles (sv) accumulated toward the membrane and two dense-core vesicles (arrows). The postsynaptic dendrite (d) exhibits numerous free ribosomes. × 41,650, by courtesy of *Nature*.

FIG. 4b. Corresponding to Fig. 4a, but in the 2-day pigeon embryo. Synaptic vesicles seem to be of variable size (as in Fig. 3), and some electron-dense substance is interspersed between them. × 50,150.

FIG. 5. Typical synapse of a 3-day chicken spinal cord. Only very few synaptic vesicles aggregate in the axon terminal (ax) and some larger dense-core vesicles (arrow) may also be present. × 50,150.

Note that all these early stages show virtually no membrane thickening in the synaptic contact region.

and 7). In a few cases an increased density was observed on the postsynaptic membrane (Plate III, Fig. 7).

All these early synapses are of the axodendritic type and exhibit only a short contact zone. Large dense-core vesicles (500–1000 Å in diameter) are quite common in the presynaptic terminal and occur more frequently than in later stages. Other cell organelles such as mitochondria and microtubules may also be present presynaptically.

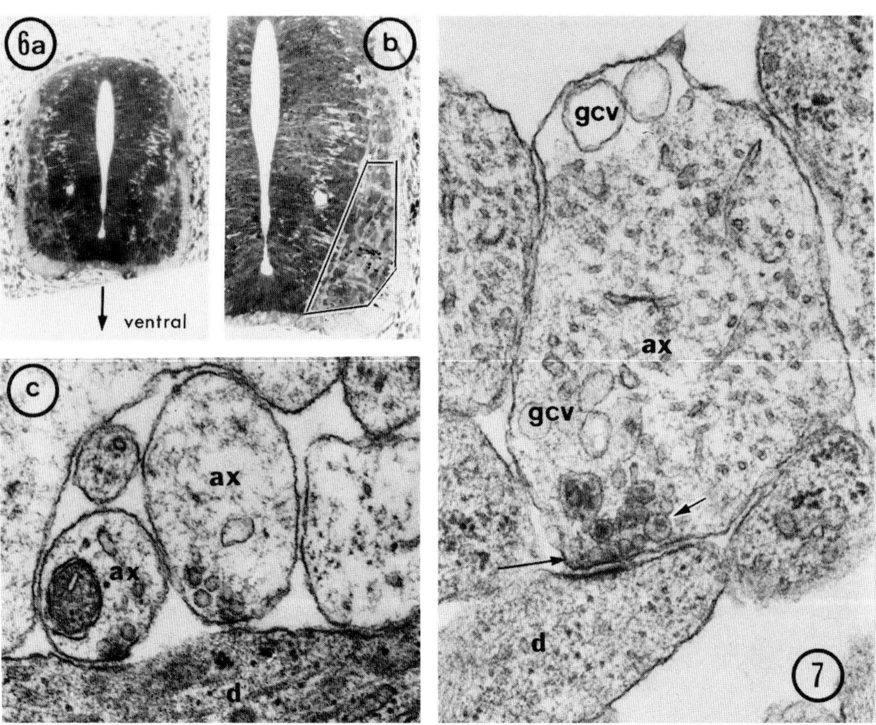

Plate III

FIG. 6a. Cross section of the midcervical spinal cord in a $4\frac{1}{2}$-day chick embryo. × 30. b. Higher magnification of the same section. The outlined area in the ventral horn region was examined with the electron microscope. × 150. c. Electron micrograph picturing two axodendritic synapses. Relatively few synaptic vesicles and hardly any membrane thickenings are seen. × 39,100.

FIG. 7. Exceptionally well-developed synapse from the spinal cord of a 3-day chick embryo. Immediately apposed to the presynaptic membrane lies a row of synaptic vesicles (long arrow), while several dense-core vesicles (short arrow) lie above. Only about one half of the synaptic contact zone shows distinct membrane thickening, which suggests that membrane densities appear later than vesicle aggregation. The axon (ax) is filled with numerous microtubules and large irregular growth cone vesicles (gcv) appear at two sites. × 40,800.

Motor behavior was first observed in the domestic chicken at 3 days, 12 hours and in the pigeon at 4 days, 12 hours (also see Tuge, 1937). In both cases the spontaneous movements consisted of three to five brief unilateral contractions of the neck musculature every 5 minutes. The remaining trunk region of the embryo only moved passively at this time. Although the synapses seen at these stages are still immature in appearance, the presence of overt movements suggests that they may be functional. It is also possible, however, that these early muscular contractions are produced by spontaneous firing of motoneurons and do not rely upon synaptic input from other neurons. By 4½ days when the cervical contractions become bilateral (side-side) and more integrated in appearance, it seems likely that functional synaptic input to motoneurons is occurring. Indeed, Provine (this volume) has found clear signs of complex bioelectric burst activity in the brachial spinal cord of the 4-day chicken embryo, which strongly argues for the synaptic interaction of neurons at this time.

b. Second Phase. Synapses at day 6 and 7 still possess relatively few synaptic vesicles, but pre- and postsynaptic membranes become denser and clearly differentiated (Plate IV, Figs. 8a,b). The postsynaptic membrane is usually more pronounced, and the synaptic contact zone is much longer than at day 3 or 4½; synaptic cleft material is clearly present. Axosomatic synapses were exceedingly rare, and therefore most synaptic contacts were axodendritic. While there are certainly more synapses by day 6 and 7 than during earlier stages, their number is still relatively low.

As in earlier stages, both the chicken and pigeon embryo perform regular and periodic side-side neck and trunk contractions but at a higher frequency (Fig. 9). Although limb movements first occur on day 6, they only accompany total body movements at this time; by days 7–8, one can detect the first independent limb movements. Tactile stimulation by stroking of the perioral region begins to evoke definite withdrawal responses by about day 7 in both chicken and pigeon embryos; this is somewhat later than Tuge's (1937) data indicate for the pigeon. Electrophysiological recordings from the ventral portion of the spinal cord at these ages indicates the appearance of so-called "after-discharges" (see Provine, this volume). These serve to lengthen the duration of individual electrical bursts. Furthermore, Provine (1971) finds that throughout much of the incubation period electrical burst discharges occur *in synchrony* in spatially separated regions of the cord (e.g., brachial and lumbosacral).

c. Third Phase. During the period between 9 and 13 days of incubation, the differentiation of synapses is characterized by: (1) an increase in the number of both actual synapses, and of synaptic vesicles (Plate V, Fig. 10; Plate VI, Fig. 12); (*2*) by the first appearance of flat synaptic vesicles (F-type,

Plate VI, Fig. 13b); and (*3*) by a noticable increase of axosomatic contacts (Plate V, Fig. 11). All these events begin to take place around day 9 and become more distinct by days 11–13. The majority of synaptic vesicles are spherical, some appear flattened (F-type) or pleiomorphic, and very few belong to the large dense core type, which are frequent in the early stages. Most synapses are axodendritic, and at day 13 several axons may be seen synapsing on one dendrite (Plate VII, Fig. 14). The synaptic contact zones are of variable length and there may even be a "regressive" tendency, as described by Larramendi (1969) for cerebellar synaptogenesis in the mouse. Some synapses in our 13-day material appear essentially mature as can be judged from the well differentiated pre- and postsynaptic membrane and the high number of synaptic vesicles (Plate VI, Fig. 12). The only difference to fully mature synapses as seen in the newly hatched chick may be a further increase of synaptic vesicles. The occurrence of other, less mature synapses suggests that synaptogenesis occurs in different cells at different times.

It should also be mentioned that at day 13 the first signs of myelination were observed in the ventral portion of the cervical cord, which agrees

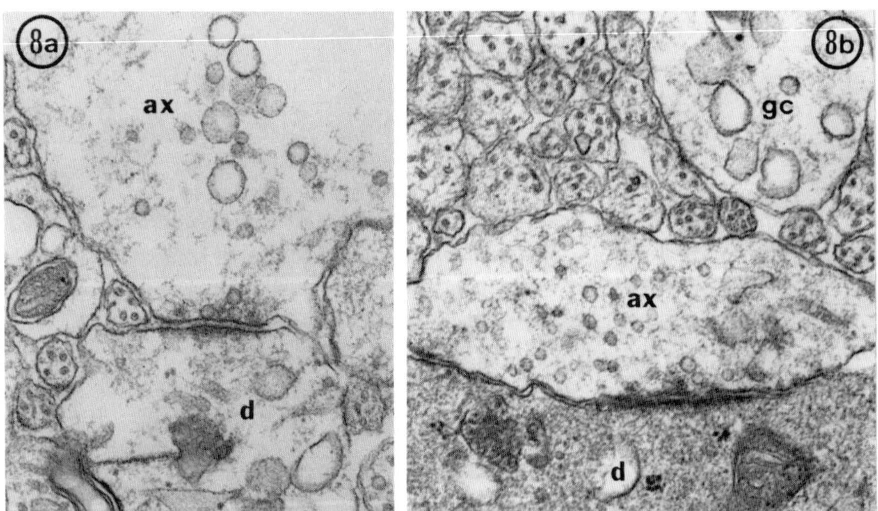

Plate IV

FIG. 8a. Axodendritic synapse in the spinal cord of a 6-day chick embryo still shows only a few synaptic vesicles but clearly differentiated pre- and postsynaptic membranes and synaptic cleft substance. Several large growth cone vesicles can be seen within the axon (ax) and additional synaptic vesicles are interspersed among them. × 36,550.

FIG. 8b. Another axodendritic synapse from the same 6-day embryo as in Fig. 8a. Several synaptic vesicles are distributed within the axoplasm, but only few aggregate toward the membrane. Note the relatively long synaptic contact zone. A growth cone (gc) of another axon is seen in the upper right corner. × 34,000.

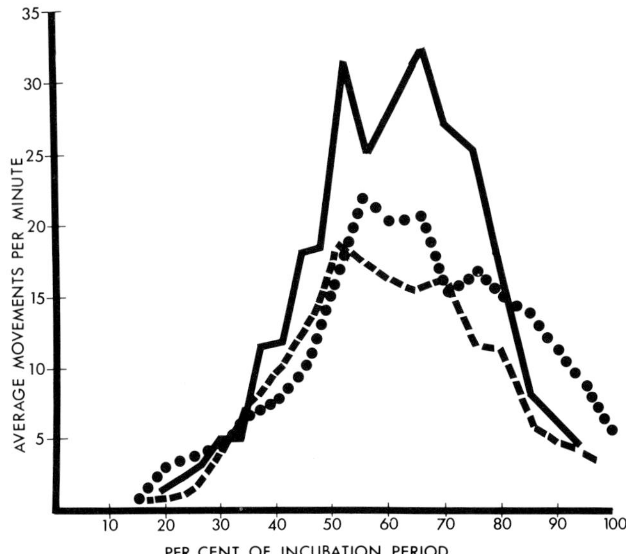

FIG. 9. Frequency of embryonic activity for chick, duck, and pigeon embryos at comparable stages of incubation. In all three species movements of any part of the embryo have been summed to provide a measure of *total activity*. Duck and chick data from Oppenheim (1970, 1972a); pigeon data from Dr. Marshall Harth (in preparation). Duck (—) n = 183; Chick (···) n = 1355; Pigeon (---) n = 243.

with histochemical evidence (see Section II). The fact that some synapses apparently conclude their morphological differentiation at the end of this phase is in good agreement with Bodian's (1970b) studies, where synaptogenesis terminates with the onset of glial ensheating (myelination).

Between 9 and 11 days of incubation the behavior of the chick embryo begins to change from the rather coordinated and stereotyped side-side contractions to more jerky, convulsivelike (and less coordinated) movements of all parts. Also by days 11–13 the frequency of movement reaches a peak (Fig. 9). However, up until about day 11 or 12 the temporal pattern of movement (cyclic motility) is still rather regular in its periodicity. It should be stressed at this point that the cyclic nature of motility in the avian embryo is one of the most characteristic features up to this time (Hamburger, 1963). Changes similar to those described above also occur in the pigeon embryo at these stages, although the peak of activity occurs slightly earlier (see Fig. 9).

Proprioceptive sensitivity in the chicken embryo has been reported to first appear on about day 10 or 11 (Visintini & Levi-Montalcini, 1939), and this coincides with both the differentiation of muscle spindles as seen in the

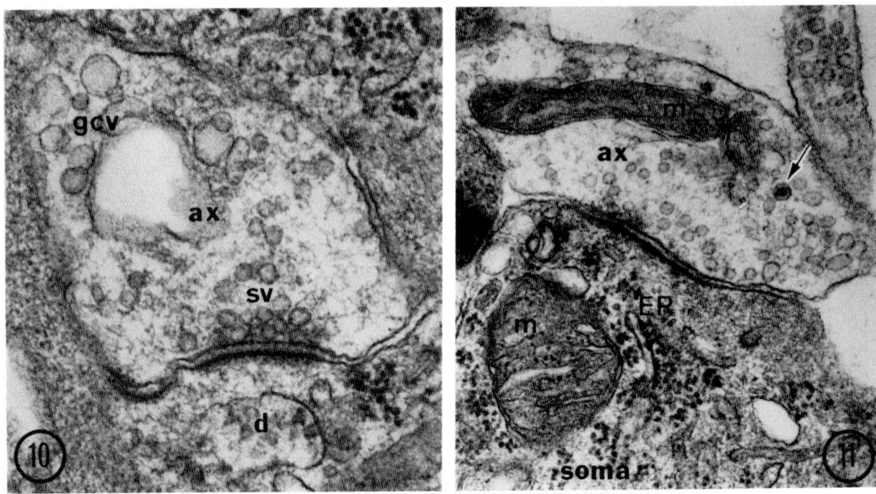

Plate V

FIG. 10. Axodendritic synapse of a 9-day chick embryo. Note the well-defined synaptic contact zone, clearly differentiated pre- and postsynaptic membranes, many synaptic vesicles (sv), and irregularly shaped growth cone vesicles (gcv). × 47,600.

FIG. 11. Axosomatic synapse in a 11-day chick embryo. The synaptic contact region is relatively long and slightly asymmetric. Note the high number of synaptic vesicles; a dense-core vesicle is indicated by an arrow. m, mitochondria; ER, rough endoplasmic reticulum. × 34,000.

light microscope (Tello, 1922) and with the closure of *monosynaptic* spinal reflex circuits (Visintini & Levi-Montalcini, 1939). The electrophysiological bursting which Provine (this volume) has recorded in the spinal cord of the chicken embryo during this period is exceedingly regular and in its temporal characteristics correlates quite well with the observed overt cyclic motility of the embryo.

d. Fourth Phase. After day 13 no dramatic changes occur with respect to synaptic fine structure. The number of synapses as well as the number of synaptic vesicles per terminal increases further. Many synapses appear mature (Plate VII, Fig. 15). F-type vesicles become more distinct and some synapses may contain exclusively flat vesicles. Compared to earlier stages, dense-core vesicles occur infrequently. The length of the synaptic contact zone seems to shorten further, especially in axodendritic synapses. Axosomatic contacts are considerably more numerous than in earlier stages and may be in close proximity to each other on one perikaryon (Plate VII, Fig. 16). Behaviorally the period between 13 and 19 days in the chicken embryo (and 14–17 days in the pigeon) is an exceedingly important one, as it is during this time that the embryo begins to prepare for hatching by assuming specific postures within the shell. The entire sequence of prehatching and hatching

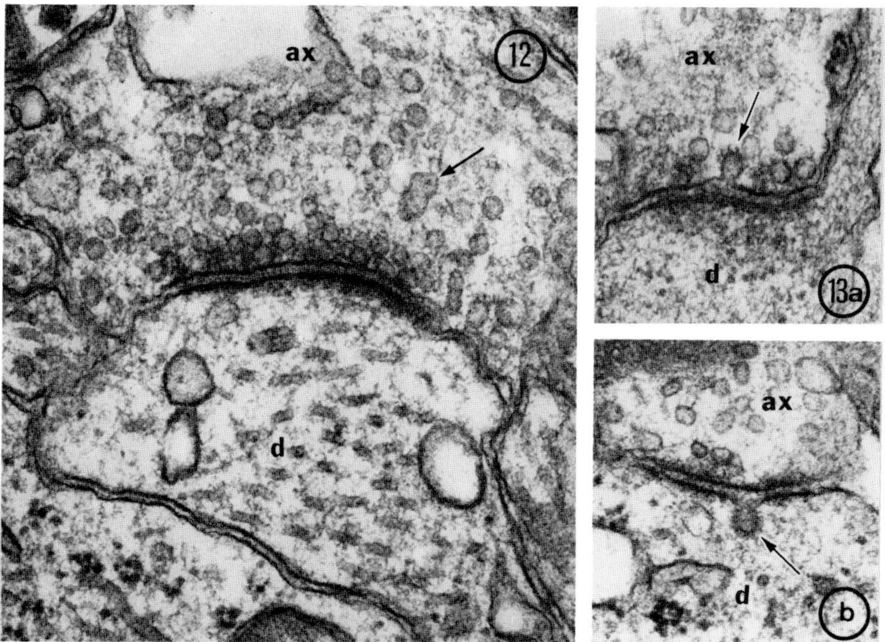

Plate VI

FIG. 12. Axodendritic synapse in a 13-day chick embryo. Most of the synaptic vesicles are clustered toward the presynaptic membrane. Synaptic cleft substance is seen as a dotted line between pre- and postsynaptic membranes. Arrow indicates a structure which may be in a fission process that will result in two synaptic vesicles. × 52,700.

FIG. 13a. Rarely an opening of a presynaptic vesicle into the synaptic cleft can be seen (arrow). It is, however, uncertain whether this is really a synaptic vesicle that would liberate a neurotransmitter. More likely we are dealing with an uptake of material (micropinocytosis) into the axon terminal (ax). Note the spiny extensions of the open vesicle on the axoplasmic side which are not present on the synaptic vesicles (13-day chick spinal cord). × 55,250.

FIG. 13b. A corresponding picture to Fig. 13a, but the vesiculation appears postsynaptically. The cytoplasmic surface of the vesicle is again coated and some material is presumably taken into the dendrite (d). Note the occurrence of flattened synaptic vesicles and symmetric membrane thickenings (13-day chick spinal cord). × 55,250.

behavior of both the chicken and pigeon embryo are reviewed elsewhere in this volume (see article by Oppenheim). However, two points should be mentioned here: (*a*) it is during this period that the movements of the embryo become less frequent, and more tonic in nature, and (*b*) in the chicken, at about 16–17 days of incubation, an entirely new behavior pattern appears that is more coordinated in its spatiotemporal pattern than the early embryonic motility; this behavior pattern, termed Type III activity, constitutes an important part of the prehatching and hatching activity of the chick.

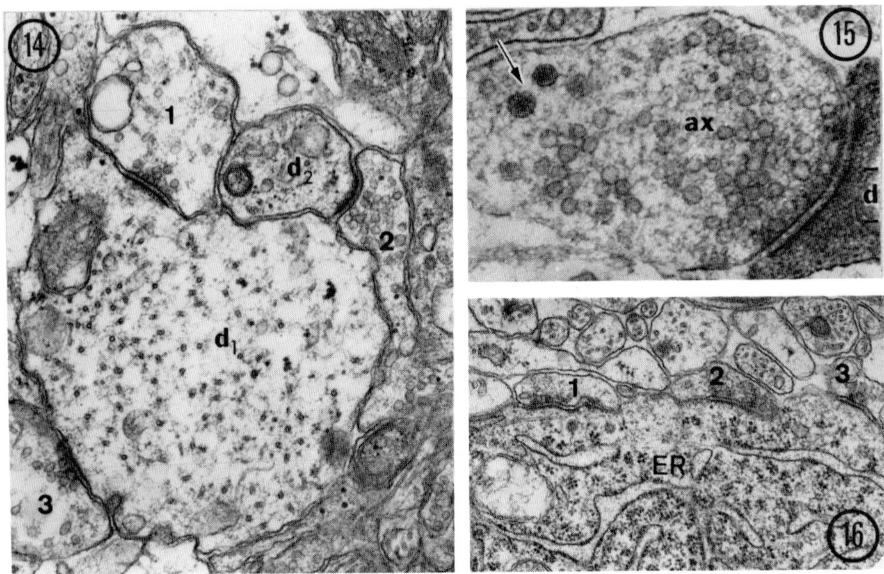

Plate VII

FIG. 14. The number of synapses is considerably increased in later embryonic stages. In this 13-day chick spinal cord three axons (1, 2, 3) synapse on two dendrites (d_1,d_2). Axons 1 and 3 synapse on the same dendrite (d_1). Synapse 3 has flattened synaptic vesicles, symmetrical membrane thickening, and also exhibits postsynaptic pinocytosis. × 25,500.

FIG. 15. Axodendritic synapse in a 16-day chick spinal cord. Note the high concentration of synaptic vesicles and two dense-core vesicles (arrow). × 50,150.

FIG. 16. Three axosomatic synapses (1, 2, 3) in a 19-day chick spinal cord. The contacts shown here are relatively small and contain only a few synaptic vesicles (compare Fig. 2) The ribosome-studded endoplasmic reticulum (ER) of the perikaryon is well developed. × 21,400.

The possibility that active inhibitory mechanisms may be functioning in the chick embryo has received some support from recent work in our laboratory in collaboration with Dr. Robert Provine. Sometime between day 7 and 9 of incubation, strychnine, an antagonist of the putative inhibitory transmitter glycine (Curtis, Duggan, & Johnston, 1969), when injected into the egg (2–5 μg), produces a brief but reliable increase in the frequency of ongoing embryonic movements. By 16–17 days of incubation this initial rise in the rate of movement is followed by the typical strychnine convulsions or seizures. These strychnine effects can also be seen in the electrophysiological recordings of polyneuronal bursts from the spinal cord of the embryo (see Provine, this volume); Picrotoxin, a substance thought to be a direct antagonist of the inhibitory neurotransmitter GABA (Engberg & Thaller, 1970; Galindo, 1969), has similar effects in the chick embryo, although the exact time of onset of this sensitivity has not yet been determined. These findings, while certainly consistent with the proposition that active inhibitory mechanisms are functional rather early during development in the chick embryo, are nevertheless only presumptive and will require final confirmation with intracellular techniques. In support of the thesis that active spinal inhibitory mechanisms are functioning by 9 days in the chick is the fact that monosynaptic reflexarcs can first be seen with reduced silver techniques by day 9, and from the observation that a local flexion reflex of the leg, presumably requiring synaptically mediated antagonistic inhibition, can first be elicited on day 9 (Visintini & Levi-Montalcini, 1939, and own personal observations). Furthermore, as we have mentioned earlier, both axodendritic and axosomatic synapses are present in the spinal cord of the chick embryo by 9 days. It is also important in this regard to note that in the rat spinal cord *in vitro* (Bunge *et al.*, 1967) predictable reactions to known pharmacological antagonists of inhibition (i.e. strychnine) roughly coincide with the appearance of axosomatic synapses. In our own material we have seen occasional, but rare, instances of axosomatic synapses as early as days 5–6, although no behavioral sensitivity to strychnine exists at this time. There does appear to be a correlation in our material between the appearance of synapses containing F-type vesicles on day 8–9 and the first signs of strychnine sensitivity. While these correlations may be no more than a temporal coincidence the possibility that, at least initially, postsynaptic spinal inhibition may be mediated at axosomatic junctions or by synapses containing F-type vesicles should be further examined.

Although previous behavioral studies have indicated that spinal inhibitory synapses arising from cells in the *brain* are not functional in the chick embryo until rather late stages of incubation (Hamburger, Balaban, Oppenheim, & Wenger, 1965), this possibility is currently being reexamined using more sensitive behavioral and electrophysiological techniques.

2. Lumbosacral Spinal Cord

In the Introduction to this article we pointed out that it is generally true that neuromuscular maturation proceeds in a rostrocaudal direction in the spinal cord of the chicken embryo. For example, the neural tube is formed 15–20 hours earlier in the cervical than in the lumbosacral region; overt muscular contractions first occur in the neck region and only later involve the trunk musculature; and the first responses to tactile stimulation also occur in the head-neck region, only later spreading to more caudal levels of the trunk. For the present discussion we are interested in determining (*a*) whether synaptogenesis is delayed in time in the lumbosacral cord, (*b*) if the sequence of synapse formation is similar to the cervical region, and (*c*) whether there is a close temporal relationship between the first appearance of synapses and the onset of overt motility, as was found in the cervical region.

Our youngest embryo in this series was 3 days of age (stage 20) and did not exhibit nerve fibers in the lumbosacral spinal cord, and hence synapses were absent. Yet, in this same embryo nerve fibers and some immature synapses were found in the cervical region. There also were *no* muscular contractions in *any* region of the embryo at this time. It is important to point out that in examining the lumbosacral cord in the electron microscope we have not limited ourselves to the prominent *lateral* motor region that primarily innervates the legs, but have also examined the more medial motor cells as well.

By about 5 days (stage 26) definite synapses (axodendritic only) were observed some of which were already relatively advanced in their differentiation (Plate VIII, Fig. 17). By this time synaptogenesis in the lumbosacral region appeared to have reached a stage comparable to that designated "phase 2" in the cervical region. However, we had the impression that the number of synapses was much lower in the lumbosacral region than in the cervical region of the same embryo. In any case, *after 5 days* we have so far been unable to detect any differences in the rate of maturation of synapses in the two regions. Although hindlimb movements do not occur on day 5, the *trunk* musculature in the lumbosacral region is already participating in the side-side contractions by this time.

On day 7 of incubation, synapse formation has advanced considerably over the 5-day stage, but in a manner similar to that already described for the cervical region, namely, an increased number of synapses, and a clearer differentiation of the pre- and postsynaptic membranes (Plate VIII, Figs. 18a,b,c). Synapses are almost entirely axodendritic. Behaviorally day 7 is marked by the first occurrence of active hindlimb movements. These almost invariably occur concomitant with trunk movements. All parts of the hindlimb are already capable of moving at this time.

By day 9 synaptogenesis in the lumbosacral region has reached the stage we term "*third phase*" in the cervical cord, and which is primarily characterized by the first appearance of axosomatic synapses (Plate VIII, Fig. 19). So far, however, we have not identified any F-type synapses in our 9 day lumbosacral material. By this time hindlimbs exhibit frequent *independent* movements primarily of a jerky nature. Although rare, short episodes of alternating flexions and extensions of the two legs are definitely seen by day 9, and both flexion and crossed extension reflexes can be elicited at this time (Oppenheim, 1972b).

As it is true of most other parts of the chick embryo, the frequency of leg movements is rather low at the time of their inception on day 6 or 7, and they then increase to a peak frequency on days 11–13. After this the movements gradually decline in frequency. Although it is tempting to relate this *rise* in activity to the increased number of axodendritic (excitatory?) synapses, and the later decline in movements with the increased number of axosomatic (inhibitory?) synapses observed by us during these stages, we are well aware of the present uncertainty concerning the morphological criteria for identifying excitatory and inhibitory synapses. For example, although the first synapses appearing in the cat neocortex and hippocampus are axodendritic they have also been clearly shown to be inhibitory in nature by electrophysiological techniques (Purpura, 1971).

As mentioned earlier, the above discussion is based upon rather preliminary observations of the lumbosacral region, and thus may have to be modified pending the outcome of more detailed morphological and behavioral observations which are presently being conducted. Yet, we have so far been impressed with the lack of any obvious differences in the temporal course of synaptogenesis between the cervical and lumbosacral spinal cord. It would be interesting to determine whether an altricial bird that does not walk until many days *after* hatching exhibits this same kind of temporal correspondence between the two spinal levels.

V. Summary and Conclusions

In summary (Fig. 20), we have found what we consider to be the first morphologically detectable synapses at 2–3 days of incubation in the chicken and pigeon cervical spinal cord, and on day 5 in the chicken lumbosacral region. These early synapses consist of a few synaptic vesicles, usually clustered toward the axonal cell membrane, and only rarely do they exhibit local membrane thickening. From our material, membrane thickening seems to be the second step in the differentiation of a synapse and occurs around day 6; the number of synaptic vesicles per terminal is only slightly increased at this time. The next phase of differentiation yields (*1*) a much higher number of synaptic vesicles, (*2*) two different types of vesicles (S- and F-type) and, (*3*) an increase in axosomatic contacts. Fully mature synapses appear first after day 13, coincidental with the onset of myelination. After day 13 more synapses attain the mature state, synaptic density (or number) increases, and the proportion of axosomatic to axodendritic synapses also increases, although the latter are always in the majority.

Two of our findings should be emphasized with respect to previous observations on synaptogenesis: (*1*) Synaptogenesis was associated with synaptic vesicles from the very beginning. The common concept of membrane densities appearing first apparently does not hold true in our material, al-

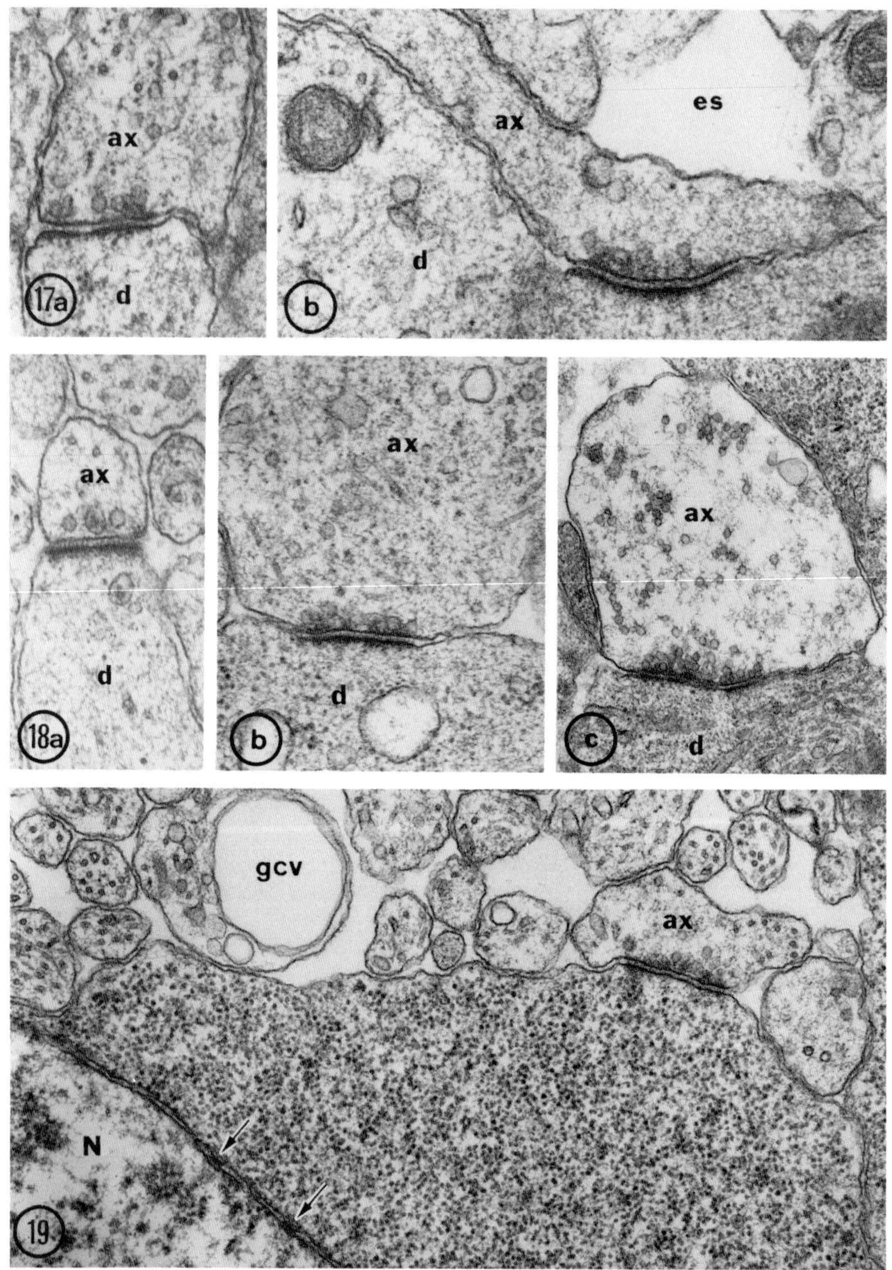
ax
d
17a
ax
es
d
b
ax
d
18a
ax
d
b
ax
d
c
gcv
ax
N
19

though to absolutely prove this it would be necessary to follow the same synapse from stage to stage, a procedure unlikely to be technically feasible for some time. It was pointed out before that membrane densities per se are a very questionable criterion for a synapse, since they could as well be pure adhesion sites (provided that they are symmetrical). Besides, there is no need to interpret local membrane densities as precursors of a synapse if membrane densities associated with synaptic vesicles can already be found very early during ontogeny (also see Caley, 1971). (*2*) Two other studies of synaptogenesis in the developing chick spinal cord (Glees & Sheppard, 1964; Wechsler, 1966) describe membrane thickenings as the first indication of a synapse (at day 5) and find the first definite synapse with synaptic vesicles around day 10. The reason why no synapses were found on earlier days by these investigators is most probably due to the fact that they are rare indeed, and easy to overlook. Another contributing factor may be the orientation problem in the electron microscope, especially when using conventional, fine-meshed specimen grids. If one does not search in the ventral horn area of the spinal cord, the chances are very slim of detecting any synapses at these early stages. Independent from our studies, Dr. D. Stelzner (personal communication) was also able to find synapses in the cervical spinal cord of the 3-day chicken embryo. There is, therefore, substantial evidence that

---

Plate VIII

Synaptogenesis in the lumbosacral spinal cord of the chicken embryo.

FIG. 17. 5-day embryo. a. Early axodendritic synapse with few synaptic vesicles and slight membrane thickening. × 44,200, b. Extremely well-developed axodendritic synapse with clearly differentiated pre- and postsynaptic membrane but relatively few synaptic vesicles. es, extracellular space. × 50,150.

FIG. 18. Variation of synaptic development in a 7-day embryo. a. Short axodendritic contact with very few synaptic vesicles but asymmetrically differentiated membrane thickenings. Note the dense cored vesicle in pre- and postsynaptic fiber. × 42,500, b. The typical 7-day synapse exhibits few synaptic vesicles but well-differentiated membrane thickenings. ×36,550. c. Exceptionally well-developed axodendritic synapse with numerous synaptic vesicles in the axon terminal; this increased number of synaptic vesicles usually only becomes apparent after day 9 of incubation. Note that membrane thickenings appear symmetrical, in contrast to a and b. × 24,650.

FIG. 19. Early axosomatic synapse in a 9-day embryo. A moderate number of synaptic vesicles aggregates toward the presynaptic membrane. Membrane thickening is slight and symmetrical. The soma is densely filled by ribosomes; part of the nucleus (N) is seen in the lower left corner, exhibiting two pores (arrows) in its nuclear envelope. gcv, Growth cone vesicles. × 28,900.

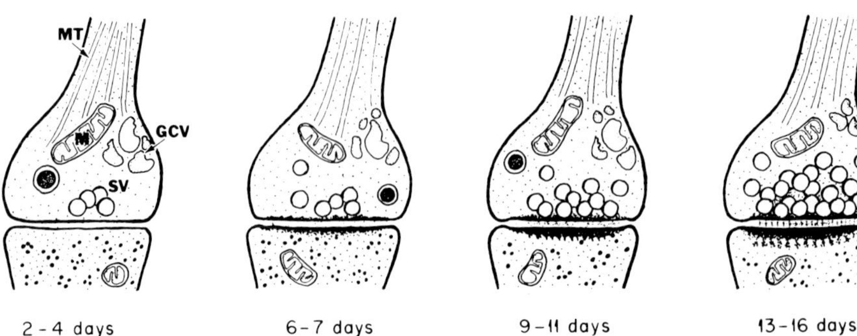

FIG. 20. Diagrammatic representation of synaptogenesis in the embryonic spinal cord of the chick. In the early stages (2–4 days of incubation) the presynaptic terminal (axon) contains only a few synaptic vesicles (SV) and several irregularly shaped growth cone vesicles (GCV). Other presynaptic cell organelles are dense-core vesicles, mitochondria (M), and microtubules (MT). The postsynaptic element (dendrite) is characterized by numerous free ribosomes (black dots) as well as mitochondria and other cell organelles. Usually no membrane thickening is observed in the contact region during the very early stages. The second phase of synaptogenesis (6–7 days) yields a definite membrane differentiation, often with a more pronounced postsynaptic membrane. In comparison to early stages, the number of synaptic vesicles increases only slightly. During the third phase of synaptogenesis (9–11 days) the number of synaptic vesicles rises distinctly, while further membrane thickening occurs. The synaptic cleft material, which could already be detected in previous stages, is now more regularly arranged. During the completion phase (13–16 days) the number of synaptic vesicles and the membrane thickenings increase further. Dense-core vesicles and growth cone vesicles diminish and the microtubules do not seem to penetrate the axon terminals as far as in previous stages. Many synapses after days 13–16 appear morphologically mature.

synaptogenesis in the avian embryo begins much earlier than previously assumed.

From extensive studies on cerebellar synaptogenesis in the mouse, Larramendi (1969) has reached the following conclusions: (*1*) Synaptogenesis begins after a certain *asynaptic period* (after the completion of cell migration); (*2*) the *synaptic period* continues for 1–2 weeks after its initiation and hence has a relatively short duration; and (*3*) synaptogenesis also shows regressive phenomena such as shortening of synaptic contacts ("synaptic adhesion waning").

From our observation, the *asynaptic period* of Larramendi must be extremely compressed in the avian embryo, since the first synapses are seen shortly after the first nerve fibers appear. (One could hardly speak of an *asynaptic period* during the first 2 days, when neuroblasts have not even developed any processes.) The duration of the *synaptic period* does seem to be quite similar in mouse cerebellum and chick spinal cord. According to our observations, synaptogenesis starts at day 2–3 of incubation, and the first fully mature synapses were seen by day 13. This means a minimal time of about 10 days is necessary for synaptic development. Of course, new synaptic contacts become established after day 3 and will consequently terminate their development later. In later stages (16 and 19 days) immature synapses in the spinal cord are rarely seen, indicating that most synapses have completed their differentiation, which in turn supports the idea of a limited *synaptic period* for specific neural regions. As a logical consequence of the *synaptic period* concept, Larramendi (1969) suggests that adult synapses may be remnants of early established synapses. The evidence for *synaptic adhesion waning* is less clear in our material and would require a quantitative evaluation of numerous micrographs. From our limited observations, however, we have the impression that many synaptic contacts do indeed shorten in the later stages. Larramendi (1969) suggests a postsynaptic pinocytosis as the mechanism being responsible for the shortening process. Our Fig. 13b may picture such an event.

The first synapses found by Bodian (1968, 1970b) in the spinal cord of the macaque embryo were all axodendritic, and the first observed synaptic vesicles were all spherical (S-type). This suggests an interesting functional correlation, namely that in many systems the first synapses appearing during development may be exclusively excitatory. Supporting this idea are the observations that the development of synapses possessing those morphological characteristics thought to be related to *excitation* also appear first during development in the mouse cerebellum (Larramendi, 1969), the cat spinal cord (Conradi & Skoglund, 1969), the rat cerebral cortex (Johnson & Armstrong-James, 1970), the rat spinal cord, *in vitro* (Bunge *et al.*, 1967), and our own studies on the avian spinal cord *in vivo* show the same tendency.

In this regard it is perhaps useful to once again point out that although we have attempted, in the present paper, to discuss the major functional changes that occur *pari passu* with synapse development in the chick spinal cord, we recognize that any suggested correlations between these two events are still highly tentative and must await analytic confirmation. For it is not only the fine structural changes of a developing synapse that should be related to behavior but, of course, also the number, type, and location (on a given cell) of functioning synapses, all of which lead to a much more complex system. For example, in agreement with all previous investigators of synaptogenesis, Aghajanian and Bloom (1967) have also noted a conspicuous increase of synapses during a certain synaptogenic period (postnatal day 14–26 for the rat cortex), and they report that this increase seems to correspond well with the onset of certain behavioral and EEG patterns.

Finally, for comparative reasons we find it interesting that our preliminary results concerning synaptogenesis and embryonic behavior in the pigeon (M. Harth, personal communication; Oppenheim, 1972c), an altricial species that hatches 3–4 days earlier than the chicken, is remarkably similar to the precocial chicken. However, such similarities may only apply to lower levels of the nervous system (e.g., spinal cord or medulla).

VI. Present and Future Goals

Although we began our studies of synaptogenesis in the chick embryo by studying the cervical spinal region (primarily because motility begins in the neck musculature thus maximizing the possibility of our finding synapses at the earliest stages), we are now directing all our attention to the lumbosacral spinal cord. Our primary reason for this is that by concentrating on this region we may be able to examine in detail the ontogeny and neuroanatomical substrate of locomotion which is an exceedingly important behavior pattern of many terrestrial vertebrates. A few of the problems we hope to pursue in the future are:

1. What is the cellular origin of the synapses appearing during normal ontogeny and what are the behavioral consequences of elimination of these sources? A related question is whether elimination of one source of synapses during ontogeny elicits a compensatory mechanism (e.g., collateral sprouting) in other potential source cells.
2. Is there any relationship between the appearance of a new type of synapse and the behavioral sensitivity of the embryo to pharmacological agents known to be involved in synaptic function?
3. Do synapses formed by cells destined to die during normal ontogeny differ in any way from other synapses? Or, can synapses even be made upon such cells?
4. Can synaptic contacts be made upon motoneurons, the axons of which have failed to reach their appropriate peripheral termination (e.g., after early limb bud removal)?
5. Do experimental modifications in either motor activity or sensory input play any role in synaptogenesis and behavior during the embryonic period?

Although we are under no illusions that the techniques and methods that we have outlined in the present article will permit a definitive answer to any of these questions, we are confident that such a multifaceted approach, if patiently and judiciously applied, will allow the embryo to reveal some of its well-kept secrets and thereby increase our understanding of both behavioral and neuroanatomical events occurring during this important period of development.

Acknowledgments

The research reported in this article has been supported in part by NSF grant GB 31874; by grant 16598-03 from the National Institute of Mental Health; by a grant from the United Health Services of North Carolina; and by the continuing generous support of the North Carolina Department of Mental Health. We thank Rubenia Daniels for competent clerical assistance.

References

Aghajanian, G. K., & Bloom, F. E. The formation of synaptic junctions in developing rat brain–a quantitative electron microscope study. *Brain Research*, 1967, **6**, 716–727.

Akert, K. Synaptopores—An alternative? *Neurosciences Research Program*, *Bulletin*, 1970, **8**, 416–420.

Akert, K. Struktur and Ultrastruktur von Nervenzellen und Synapsen. *Klinische Wochenschrift*, 1971, **49**, 509–519.

Akert, K., Gray, E. G., & Bloom, F. E. Structure of specialized junctions. *Neurosciences Research Program*, *Bulletin*, 1970, **8**, 336–360.

Akert, K., Moor, H., Pfenninger, K., & Sandri, C. Contributions of new impregnation methods and freeze-etchings to the problems of synaptic fine structure. *Progress in Brain Research*, 1969, **31**, 223–240.

Akert, K., & Pfenninger, K. Synaptic fine structure and neural dynamics. In S. H. Barondes (Ed.), *Symposia of the international society for cell biology*. Vol. 8. *Cellular dynamics of the neuron*. New York: Academic Press, 19.

Alconero, B. B. The nature of the earliest spontaneous activity of the chick embryo. *Journal of Embryology and Experimental Morphology*, 1965, **13**, 255–266.

Altman, J. Coated vesicles and synaptogenesis. A developmental study in the cerebellar cortex of the rat. *Brain Research*, 1971, **30**, 311–322.

Andres, K. H. Mikropinozytose in Zentralnervensystem. *Zeitschrift für Zellforschung und Mikorskopische Anatomie*, 1964, **64**, 63–73.

Anokhin, P. K. Systemogenesis as a general regulator of brain development. *Progress in Brain Research*, 1964, **9**, 54–86.

Auerbach, L. Nervenendigungen in den Centralorganen. *Neurologisches Zentralblatt*, 1898, **17**, 445–454

Barcroft, J., & Barron, D. H. The development of behavior in foetal sheep. *Journal of Comparative Neurology*, 1939, **70**, 477–502.

Barron, D. H. Observations on the early differentiation of the motor neuroblasts in the spinal cord of the chick. *Journal of Comparative Neurology*, 1946, **85**, 149–169.

Bellairs, R. The development of the nervous system in chick embryos studied by electron microscopy. *Journal of Embryology and Experimental Morphology*, 1959, **7**, 94–115.

Bensted, J. P. M., Dobbing, J., Morgan, R. S., Reed, R. T. W., & Wright, G. P. Neuroglial development and myelination in the spinal cord of the chick embryo. *Journal of Embryology and Experimental Morphology*, 1957, **5**, 428–437.

Bloom, F. E. Correlating structure and function of synaptic ultrastructure. In F. O. Schmitt (Ed.), *The neurosciences*: *Second study program*. New York: Rockefeller University Press, 1970.

Bloom, F. E., & Aghajanian, G. K. Cytochemistry of synapses: selective staining for electron microscopy. *Science*, 1966, **154**, 1575–1577.

Bodian, D. Electron microscopy: two major synaptic types on spinal motoneurons. *Science*, 1966, **151**, 1093–1094.

Bodian, D. Development of fine structure of spinal cord in monkey fetuses. II. Pre-reflex period to period of long intersegmental reflexes. *Journal of Comparative Neurology*, 1968, **133**, 113–166.

Bodian, D. An electron microscopic characterization of classes of synaptic vesicles by means of controlled aldehyde fixation. *Journal of Cell Biology*, 1970, **44**, 115–124. (a)

Bodian, D. A model of synaptic and behavioral ontogeny. In F. O. Schmitt (Ed.), *The neurosciences: Second study program*. New York: Rockefeller University Press, 1970. (b)

Bunge, M. B., Bunge, R. P., & Peterson, E. R. The onset of synapse formation in spinal cord cultures as studied by electron microscopy. *Brain Research*, 1967, **6**, 728–749.

Burt, A. M., & Narayanan, C. H. Effect of extrinsic neuronal connections on development of acetylcholinesterase and choline acetyltransferase activity in the ventral half of the chick spinal cord. *Experimental Neurology*, 1970, **29**, 201–210.

Caley, D. W. Differentiation of the neural elements of the cerebral cortex in the rat. In D. C. Pease (Ed.), *Cellular aspects of neural growth and differentiation*. Los Angeles: University of California Press, 1971.

Coghill, G. E. *Anatomy and the problem of behavior*. Cambridge, Eng.: Cambridge University Press, 1929.

Colonnier, M. Synaptic patterns on different cell types in the different laminae of the cat visual cortex: an electron microscope study. *Brain Research*, 1968, **9**, 268–287.

Conradi, S., & Skoglund, S. Motoneurons synaptology in kittens. An electron microscopic study of the structure and location of neuronal and glial elements on cat lumbo-sacral motoneurons in the normal state and after dorsal root section. *Acta Physiologica Scandinavica, Supplementum*, 1969, **333**.

Corner, M. A., & Bot, A. P. C. Developmental patterns in the central nervous system of birds. III. Somatic motility during the embryonic period and its relation to behavior after hatching. *Progress in Brain Research*, 1967, **26**, 214–236.

Crain, S. M., & Peterson, E. R. Onset and development of functional interneuronal connections in explants of rat spinal cord-ganglia during maturation in culture. *Brain Research*, 1967, **6**, 750–762.

Crain, S. M., Peterson, E. R., & Bornstein, M. B. Formation of functional interneuronal connexions between explants of various mammalian central nervous tissues during development *in vitro*. In G. E. W. Wolstenholme & M. O'Connor (Eds.), *Ciba foundation symposium: Growth of the nervous system*. Boston: Little, Brown, 1968.

Curtis, D. R., Duggan, A. W., & Johnston, G. A. R. Glycine, strychnine, picrotoxin and spinal inhibition. *Brain Research*, 1969, **14**, 759–762.

Dale, H. H. The action of certain esters and ethers of choline, and their relation to muscarine. *Journal of Pharmacology and Experimental Therapeutics*, 1914, **6**, 147–190.

Del Castillo, J., & Katz, B. Biophysical aspects of neuro-muscular transmission. *Progress in Biophysics, and Biophysical Chemistry*, 1956, **6**, 121–170. (a)

Del Castillo, J., & Katz, B. Localization of active spots within the neuromuscular junction of the frog. *Journal of Physiology (London)*, 1956, **132**, 630–649. (b)

Dennison, M. E. Electron stereoscopy as a means of classifying synaptic vesicles. *Journal of Cell Science*, 1971, **8**, 525–539.

De Robertis, E. *Histophysiology of synapses and neurosecretion*. Oxford: Pergamon, 1964.

De Robertis, E. Structural and chemical studies on storage and receptor sites for biogenic amines in the central nervous system. In S. H. Barondes (Ed.), *Symposia of the international society for cell biology*. Vol. 8. *Cellular dynamics of the neuron*. New York: Academic Press, 1969.

De Robertis, E., & Bennett, H. S. Submicroscopic vesicular component in the synapse. *Federation Proceedings, Federation of American Societies for Experimental Biology*, 1954, **13**, 35.

De Robertis, E., & Pellegrino de Iraldi, A. Plurivesicular nerve endings in the pineal gland of the rat. *Journal of Biophysical and Biochemical Cytology*, 1961, **1**, 47–58.

De Robertis, E., Pellegrino de Iraldi, A., Rodriguez de Lores Arnaiz, G., & Salganicoff, L. 1961. Electron microscope observations on nerve endings isolated from rat brain. *Anatomical Record*, 1961, **139**, 220–221.

De Robertis, E., Pellegrino de Iraldi, A., Rodriguez de Lores Arnaiz, G., & Salganicoff, L. Cholinergic and non-cholinergic nerve endings in rat brain. I. Isolation and subcellular distribution of acetylcholine and acetylcholinesterase. *Journal of Neurochemistry*, 1962, **9**, 23–35.

Eccles, J. C. *The physiology of synapses*. Berlin Springer-Verlag, 1964.

El-Eishi, H. I. Biochemical and histochemical studies on myelination in the chick embryo spinal cord. *Journal of Neurochemistry*, 1967, **14**, 405–412.

Elliott, T. R. On the action of adrenalin. *Journal of Physiology* (*London*), 1904, **31**, 20P.

Engberg, I., & Thaller, A. On the interaction of picrotoxin with GABA and glycine in the spinal cord. *Brain Research*, 1970, **19**, 151–154.

Fatt, P., & Katz, B. Spontaneous subthreshold activity at motor nerve endings. *Journal of Physiology* (*London*), 1952, **117**, 109–128.

Foster, M., & Sherrington, C. S. *A textbook of physiology*, Part III. *The central nervous system*. (7th ed.) London: Macmillan, 1897.

Galindo, A. GABA—picrotoxin interaction in the mammalian central nervous system. *Brain Research*, 1969, **14**, 763–767.

Glees, P., & Sheppard, B. L. Electron microscopical studies of the synapse in the developing chick spinal cord. *Zeitschrift für Zellforschung und Mikroskopische Anatomie*, 1964, **62**, 356–362.

Gottlieb, G. Prenatal behavior of birds. *Quarterly Review of Biology*, 1968, **43**, 148–174.

Gottlieb, G. Ontogenesis of sensory function in birds and mammals. In E. Tobach, L. R. Aronson, & E. Shaw (Eds.), *The biopsychology of development*. New York: Academic Press, 1971.

Gray, E. G. Axo-somatic and axo-dendritic synapses of the cerebral cortex: an electron microscope study. *Journal of Anatomy*, 1959, **93**, 420–433.

Gray, E. G. Electron microscopy of presynaptic organelles of the spinal cord. *Journal of Anatomy* 1963, **97**, 101–106.

Gray, E. G., & Whittaker, V. P. The isolation of synaptic vesicles from central nervous system. *Journal of Physiology* (*London*), 1960, **153**, 35P–37P.

Hamburger, V. The mitotic patterns in the spinal cord of the chick embryo and their relation to histogenetic processes. *Journal of Comparative Neurology*, 1948, **88**, 221–284.

Hamburger, V. Some aspects of the embryology of behavior. *Quarterly Review of Biology*, 1963, **38**, 342–365.

Hamburger, V., & Hamilton, H. L. A series of normal stages in the development of the chick embryo. *Journal of Morphology*, 1951, **88**, 49–92.

Hamburger, V., Balaban, M., Oppenheim, R., & Wenger, R. Periodic motility of normal and spinal chick embryos between 8 and 17 days of incubation. *Journal of Experimental Zoology*, 1965, **159**, 1–14.

Hamburger, V., Wenger, E., & Oppenheim, R. Motility in the chick embryo in the absence of sensory input. *Journal of Experimental Zoology*, 1966, **162**, 133–160.

Hámorí, J., & D'iachkova, K. N. Electron microscope studies on developmental differentiation of ciliary ganglion synapses in the chick. *Acta Biologica* (*Budapest*), 1964, **15**, 213–230.

Harrison, R. G. The outgrowth of the nerve fiber as a mode of protoplasmic movement. *Journal of Experimental Zoology*, 1910, **9**, 787–846.

Held, H. Beiträge zur Struktur der Nervenzellen und ihrer Fortsätze. *Archiv für Anatomie und Physiologie, Physiologische Abteilung*, 1897, **21**, 204–294.

Hirano, H. Ultrastructural study on the morphogenesis of the neuromuscular junction in the skeletal muscle of the chick. *Zeitschrift für Zellforschung und Mikroskopische Anatomie*, 1967, **79**, 198–208.

Hooker, D. *The prenatal origin of behavior*. Lawrence, Kans.: University of Kansas Press, 1952.

Hubbard, J. I., & Kwanbunbumpen, S. Evidence for the vesicle hypothesis. *Journal of Physiology (London)*, 1968, **194**, 407–420.

James, D. W., & Tresman, R. L. Synaptic profiles in the outgrowth from chick spinal cord *in vitro. Zeitschrift für Zellforschung und Mikroskopische Anatomie*, 1969, **101**, 598–606.

Johnson, R., & Armstrong-James, M. Morphology of superficial postnatal cerebral cortex with special reference to synapses. *Zeitschrift für Zellforschung und Mikroskopische Anatomie*, 1970, **110**, 540–558.

Jolly, W. A. The time relations of the knee-jerk and simple reflexes. *Quarterly Journal of Experimental Physiology and Cognate Medical Sciences*, 1911, **4**, 67–87.

Jones, S. F., & Kwanbunbumpen, S. On the role of synaptic vesicles in transmitter release. *Life Sciences*, 1968, **7**, 1251–1255.

Jones, S. F., & Kwanbunbumpen, S. The effects of nerve stimulation and hemicholinium on synaptic vesicles at the mammalian neuromuscular junction. *Journal of Physiology (London)*, 1970, **207**, 31–50.

Jones, S. F., & Kwanbunbumpen, S. Some effects of nerve stimulation and hemicholinium on quantal transmitter release at the mammalian neuromuscular junction. *Journal of Physiology (London)*, 1970, **207**, 51–61.

Kelly, D. E. Fine structure of cell contact and the synapse. *Anesthesiology*, 1967, **28**, 6–30.

Kuo, K.-Y. Ontogeny of embryonic behavior in aves. I. The chronology and general nature of the behavior of the chick embryo. *Journal of Experimental Zoology*, 1932, **61**, 395–430.

Kuo, K.-Y. Studies in the physiology of the embryonic nervous system. I. Effects of curarization on somatic movement. *Journal of Experimental Zoology*, 1939, **82**, 371–396.

Langman, J. & Haden, C. C. Formation and migration of neuroblasts in the spinal cord of the chick embryo. *Journal of Comparative Neurology*, 1970, **138**, 419–432.

Larramendi, L. M. H. Analysis of synaptogenesis in the cerebellum of the mouse. In R. Llinás (Ed.), *Neurobiology of cerebellar evolution and development*. Chicago: American Medical Association, 1969.

Larramendi, L. M. H., Fickenscher, L., & Lemkey-Johnston, N. Synaptic vesicles of inhibitory and excitatory terminals in the cerebellum. *Science*, 1967, **156**, 967–969.

Loewi, O. Über humorale Übertragbarkeit der Herznervenwirkung I. Mitteilung. *Pflügers Archiv der Gesamten Physiologie des Menschen und der Tiere*, 1921, **189**, 239–242.

Lyser, K. M. Early differentiation of motor neuroblasts in the chick embryo as studied by electron microscopy. I. General aspects. *Developmental Biology*, 1964, **10**, 433–466.

Lyser, K. M. Early differentiation of motor neuroblasts in the chick embryo as studied by electron microscopy. II. Microtubules and neurofilaments. *Developmental Biology*, 1968, **17**, 117–142.

McLennan, H. *Synaptic transmission.* Philadelphia Saunders, 1970.

Marin-Padilla, M. Prenatal and early postnatal ontogenesis of the human motor cortex: A Golgi study. I. The sequential development of the cortical layers. *Brain Research*, 1970, **23**, 167–183. (a)

Marin-Padilla, M. Prenatal and early postnatal ontogenesis of the human motor cortex: A Golgi study. II. The basket-pyramidal system. *Brain Research*, 1970, **23**, 185–191. (b)

Meller, K. Elektronenmikroskopische Befunde zur Differenzierung der Rezeptorzellen und

Bipolarzellen der Retina und ihrer synaptischen Verbindungen. *Zeitschrift für Zellforschung und Mikroskopische Anatomie*, 1964, **64**, 733–750.

Meller, K., Breipohl, W., Wagner, H. H., & Knuth, A. Die Differenzierung isolierter Nerven- und Gliazellen aus trypsinisiertem Rückenmark von Hühnerembryonen in Gewebekulturen. *Zeitschrift für Zellforschung und Mikroskopische Anatomie*, 1969, **101**, 135–151.

Meller, K., & Haupt, R. Die Feinstruktur der Neuro-, Glio- und Ependymoblasten von Hühnerembryonen in der Gewebekultur. *Zeitschrift für Zellforschung und Mikroskopische Anatomie*, 1967, **76**, 260–277.

Mezei, C., Newburgh, R. W., & Hattori, T. Cholesterol, cholesterol esters and cholesterol esterase in the sciatic nerve during development of the chick. *Journal of Neurochemistry*, 1971, **18**, 463–468.

Morest, D. K. The growth of dendrites in the mammalian brain. *Zeitschrift für Anatomie und Entwicklungsgeschichte*, 1969, **128**, 290–317.

Mugnaini, E. Ultrastructural studies on the cerebellar histogenesis. II. Maturation of nerve cell populations and establishment of synaptic connections in the cerebellar cortex of the chick. In R. Llinás (Ed.), *Neurobiology of cerebellar evolution and development*. Chicago: American Medical Association, 1969.

Mugnaini, E. Developmental aspects of synaptology with special emphasis upon the cerebellar cortex. In D. C. Pease (Ed.), *Cellular aspects of neural growth and differentiation* . Los Angeles: University of California Press, 1971.

Ochi, J. Elektronenmikroskopische Untersuchungen des Bulbus Olfactorius der Ratte während der Entwicklung. *Zeitschrift für Zellforschung und Mikroskopische Anatomie*, 1967, **76**, 339–348.

Oppenheim, R. W. Amniotic contraction and embryonic motility in the chick embryo. *Science*, 1966, **152**, 528–529.

Oppenheim, R. W. Some aspects of embryonic behavior in the duck (*Anas platyrhynchos*). *Animal Behavior*, 1970, **18**, 335–352.

Oppenheim, R. W. The embryology of behavior in birds: A critical review of the role of sensory stimulation in embryonic movement. In K. H. Voous (Ed.), *Proceedings of the XVth international ornithological congress*. Leiden: Brill, 1972. (a)

Oppenheim, R. W. An experimental investigation of the possible role of tactile and proprioceptive stimulation in certain aspects of embryonic behavior in the chick. *Developmental Psychobiology*, 1972, **5**, 71–91. (b)

Oppenheim, R. W. Prehatching and hatching behavior in birds: A comparative study of altricial and precocial species. *Animal Behavior*, 1972, **20**, 644–655. (c)

Oppenheim, R. W., & Foelix, R. F. Synaptogenesis in the chick embryo spinal cord. *Nature (London)*, 1972, **235**, 126–128.

Oppenheim, R. W., & Narayanan, C. H. Experimental studies on hatching behavior in the chick. I. Thoracic spinal gaps. *Journal of Experimental Zoology*, 1968, **168**, 387–394.

Palade, G. E., & Palay, S. L. Electron microscope observations of interneuronal and neuromuscular synapses. *Anatomical Record*, 1954, **118**, 335–336.

Palay, S. L. The morphology of synapses in the central nervous system. *Experimental Cell Research, Supplement*, 1958, **5**, 275–293.

Pease, D. C. *Histological techniques for electron microscopy*. (2nd ed.) New York: Academic Press, 1964.

Peters, A., Palay, S. L., & de Webster, H. F. *The fine structure of the nervous system. The cells and their processes*. New York: Harper, 1970.

Pfenninger, K. H. The cytochemistry of synaptic densities. I. An analysis of the Bismuth Iodide impregnation method. *Journal of Ultrastructure Research*, 1971, **34**, 103–122. (a)

Pfenninger, K. H. The cytochemistry of synaptic densities. II. Proteinaceous components and mechanism of synaptic connectivity. *Journal of Ultrastructure Research*, 1971, **35**, 451–475. (b)

Pfenninger, K., Sandri, C., Akert, K., & Eugster, C. H. Contribution on the problem of structural organization of the presynaptic area. *Brain Research*, 1969, **12**, 10–18.

Preyer, W. *Specielle Physiologie des Embryo*. Leipzig: Grieben, 1885.

Provine, R. R. Embryonic spinal cord: synchrony and spatial distribution of polyneuronal burst discharges. *Brain Research*, 1971, **29**, 155–158.

Purpura, D. P. Synaptogenesis in mammalian cortex: problems and perspectives. In M. B. Sterman, D. J. McGinty, & A. M. Adinolfi (Eds.), *Brain development and behavior*. New York: Academic Press, 1971.

Purpura, D. P., & Pappas, G. D. Structural characteristics of neurons in the feline hippocampus during postnatal ontogenesis. *Experimental Neurology*, 1968, **22**, 379–393.

Ramón y Cajal, S. Estructura de los centros nerviosos de las aves. Con dos laminas litograficas. *Revista Trimestral de Histologia Normal y Patologica*, 1888, May 1.

Ramón y Cajal, S. *Études sur la neurogénèse de quelques vértébrés*. Madrid, 1929. Translated by L. Guth, *Studies on vertebrate neurogenesis*. Springfield, Ill.: Thomas, 1960.

Robertson, J. D. The ultrastructure of synapses. In F. O. Schmitt (Ed.), *The neurosciences: Second study program*. New York: Rockefeller University Press, 1970.

Scheibel, M. E., & Scheibel, A. B. Developmental relationship between spinal motoneuron dendrite bundles and patterned activity in the hindlimbs. *Experimental Neurology*, 1970, **29**, 328–335.

Sherrington, C. S. *Intergrative action of the nervous system*. New Haven: Yale University Press, 1906.

Sparber, S. B., & Shideman, F. E. Prenatal administration of reserpine: effect upon hatching, behavior and brainstem catecholamines. *Developmental Psychobiology*, 1968, **1**, 236–244.

Stelzner, D. J. The normal postnatal development of synaptic endfeet in the lumbo-sacral spinal cord and of responses in the hind-limbs of the albino rat. *Experimental Neurology*, 1971, **31**, 337–357. (a)

Stelzner, D. J. The relationship between synaptic vesicles, golgi apparatus, and smooth endoplasmic reticulum: a developmental study using the zinc-iodide-osmium technique. *Zeitschrift für Zellforschung und Mikroskopische Anatomie*, 1971, **120**, 332–345. (b)

Taxi, J. Étude de l'ultrastructure des zones synaptiques dans les ganglions sympathiques de la grenouille. *Comptes Rendus Hebdomadaires des Séances de l'Academie des Sciences*, 1961, **252**, 174–176.

Taxi, J. Contribution a l'étude des connexions des neurones moteurs du système nerveux autonome. *Annales des Sciences Naturelles. Zoologie et Biologie Animale, Series 12*, 1965, **7**, 413–674.

Tello, J. F. Die Entstehung der motorischen und sensiblen Nervenendigungen. I. In dem lokomotorischen System der höheren Wirbeltiere, *Zeitschrift für Anatomie und Entwicklungsgeschichte*, 1922, **64**, 348–440.

Tennyson, V. M. The fine structure of the developing nervous system. In W. A. Himwich (Ed.), *Developmental neurobiology*. Springfield, Ill.: Thomas, 1970.

Tracy, H. C. The development of motility and behavior reactions in the toadfish, *Opsanus tau*. *Journal of Comparative Neurology*, 1926, **40**, 253–369.

Tuge, H. The development of behavior in avian embryos. *Journal of Comparative Neurology*, 1937, **66**, 157–179.

Uchizono, K. Characteristics of excitatory and inhibitory synapses in the central nervous system of the cat. *Nature (London)*, 1965, **207**, 642–643.

Valdivia, O. Methods of fixation and the morphology of synaptic vesicles. *Anatomical Record*, 1970, **166**, 392.

Visintini, F., & Levi-Montalcini, R. Relazione tra differenziazione strutturale e funzionale dei centri e delle vie nervose nell'embrione di pollo. *Schweizer Archiv für Neurologie und Psychiatrie*, 1939, **40**, 1–45.

Walberg, F. Elongated vesicles in terminal boutons of the central nervous system, a result of aldehyde fixation. *Acta Anatomica*, 1965, **65**, 224–235.

Wechsler, W. Elektronenmikroskopischer Beitrag zur Nervenzelldifferenzierung und Histogenese der grauen Substanz des Rückenmarkes von Hühnerembryonen. *Zeitschrift für Zellforschung und Mikroskopische Anatomie*, 1966, **74**, 401–422.

Whittaker, V. P. The application of subcellular fractionation techniques to the study of brain function. *Progress in Biophysics, and Molecular Biology*, 1965, **15**, 39–96.

Whittaker, V. P. The investigation of synaptic function by means of subcellular fractionation techniques. In F. O. Schmitt (Ed.), *The neurosciences: Second study program.* New York: Rockefeller University Press, 1970.

Whittaker, V. P., Michaelson, I. A., & Kirkland, R. J. A. The separation of synaptic vesicles from nerve-ending particles ("synaptosomes"). *Biochemical Journal*, 1964, **90**, 293–303.

Windle, W. F. *Physiology of the fetus: Origin and extent of function in prenatal life.* Philadelphia: Saunders, 1940.

Windle, W. F., & Orr, D. W. The development of behavior in chick embryos: spinal cord structure correlated with early somatic motility. *Journal of Comparative Neurology*, 1934, **60**, 287–308.

Woodward, D. J., Hoffer, B. J., & Lapham, L. W. Postnatal development of electrical and enzyme histochemical activity in Purkinje cells. *Experimental Neurology*, 1969, **23**, 120–239.

Wundt, W. M. *Untersuchungen zur Mechanik der Nerven und Nervencentren. Abt. 2. Über den Reflexvorgang und das Wesen der centralen Innervation.* Erlangen: Enke, 1871.

THE EMBRYONIC BEHAVIOR OF CERTAIN CRUSTACEANS

MICHAEL BERRILL

Biology Department
Trent University
Peterborough, Ontario, Canada

I. Introduction

The behavior of invertebrate embryos is unknown in any detail. Though Davis (1968), in his review of hatching mechanisms in aquatic invertebrates, included the brief observations that have been made on embryonic behavior, most accounts of crustacean embryonic development have been morphological descriptions. (See Oppenheim, this volume, for a summary of hatching behavior in invertebrates.) Accounts of the early embryology of crustacean nervous systems are old ones (reviewed by Korschelt, 1944) and appear unreliable. There is, however, current work on adult decapod transmitter substances (Orkand & Kravitz, 1971) and neuronal circuitry (Zucker,

Kennedy, & Selverston, 1971) which may provide some insight into the embryonic situation. The Class Crustacea should prove to be a profitable group to examine for correlation of embryonic behavior with nervous system development, as well as for evidence as to the nature of the control of embryonic movements. The initial step, though, must be a description of the variation in organization of embryonic movements among representative forms.

Most crustacean females carry their developing eggs in brood chambers or attach them to their abdominal appendages. The embryos may hatch as simple naupliar larvae (characterized by having only three pairs of appendages and a single median eye), or they may develop through one or more distinct but suppressed larval-like stages while still embryos, emerging at the end of their development looking like miniature replicas of their parents. For present purposes, embryonic development in crustaceans is defined as that period in which the developing individual remains passively attached to the maternal swimmerets or in the maternal brood chamber. The shedding of the membranes may or may not precede the termination of this period.

II. Methods

Six crustacean species of four orders were studied: the lobster *Homarus americanus* of the Decapoda; the skeleton shrimp *Caprella unica* of the Amphipoda; *Daphnia pulex* of the Cladocera; and *Mysis relicta*, *Neomysis americana*, and *Praunus flexuosus* of the opossum shrimp, the Mysidacea. Of these, *M. relicta* and *D. pulex* were collected from the Kawartha Lakes of Southern Ontario, *C. unica* from the east coast of Newfoundland, and *H. americanus*, *N. americana*, and *P. flexuosus* from the shores of Maine and New Brunswick.

Embryos of the six species were removed at various stages of development from their mothers, and were maintained and observed at a temperature range that depended on the species: *H. americanus* and *D. pulex* at 20–23° C; *P. flexuosus*, *N. americana*, and *C. unica* at 10–15° C; and *M. relicta* at 4–10° C. The fresh water species were kept in pond water, the salt water species in water from the area they were captured in, and apart from controlling the temperature and maintaining aeration, no other control of the medium was attempted.

Observations in the 4–15° C range were made in controlled-temperature rooms. All observations were made visually, not electronically, on individuals in 30-cc observation dishes under the dissecting microscope. Pertinent movements were recorded on a manually stimulated event recorder. All embryos, with the exception of *H. americanus*, were raised to the end of their embryonic development following observation or experimentation, and wherever applicable brood mates were used as controls for individuals subjected to experimentation. Fine glass needles were used to remove embryonic membranes. Attempts to elicit reflex response to tactile stimulation were made by lightly stroking the embryos at various places with the tip of a short adult human hair.

III. Results

A. Mysids (Opossum Shrimp)

1. Reproductive Biology

Of the three mysids, *M. relicta* lives on the bottom mud of temperate and northern lakes, though a certain per cent of a population may be planktonic, rising off the bottom at night (Beeton, 1960). Lasenby (1971) has compared the energetics of arctic and temperate populations. *P. flexuosus* lives near and among the rocky shore seaweed, usually *Fucus* and *Ascophyllum* sp., in protected marine bays. *N. americana*, both marine and estuarine, usually forages on sandy bottoms, though parts of its populations are also planktonic (Herman, 1963). The female of each species may carry up to 40 eggs in a midventral brood pouch (Fig. 1), and the young emerge from their brood pouches as juveniles, looking and behaving much like miniature adults.

On the coast of Maine, *P. flexuosus* and *N. americana* breed repeatedly throughout the summer, each brood taking approximately 3 weeks to develop at 12–15°C. Within several hours after the emergence of a brood, the female molts and lays a new set of eggs. *M. relicta*, on the other hand, breeds but once during the year in the Kawartha Lakes. The females extrude their eggs in December or January, and the young emerge in April or May, about the time the ice breaks up and the phytoplankton bloom begins. According to Nair (1939) in his work with *Mesopodopsis orientalis*, the female is fertilized within 10 minutes after molting, perhaps chemically attracting males to her, and this is probably true of most mysid species.

The embryology of *Mesopodopsis orientalis* (Nair, 1939) and *Hemimysis lamornae* (Manton, 1928) has been described in detail. After a mysid embryo sheds its two egg membranes, it develops through two suppressed larval-like stages, molting at the end of each stage. The second molt terminates embryonic development. Davis (1966) described the shedding of these four embryonic membranes by *Mysidium columbiae*, and his description appears to be representative of the order. The embryonic period can be divided conveniently into three major stages: the prenaupliar stage, which ends with the shedding of the second egg membrane; the naupliar stage, which ends with the first embryonic molt; and the postnaupliar stage, which ends with

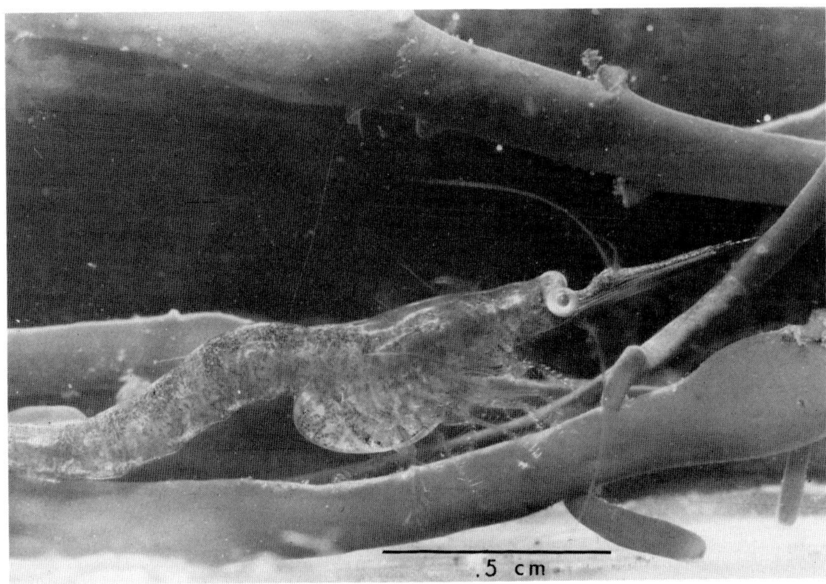

FIG. 1. *Praunus flexuosus*, a pregnant female clinging to seaweed. She has 25–30 young embryos in her mid-ventral brood pouch.

the second embryonic molt (Fig. 2). Table I summarizes the duration of these stages in *P. flexuosus*, *M. relicta*, and *N. americana*.

2. EMBRYONIC BEHAVIOR

The development of behavior of the embryos of *P. flexuosus*, *N. americana*, and *M. relicta* is similar in its general organization (Berrill, 1969, 1971). At

TABLE I
COMPARISON OF THE LENGTH OF THE NAUPLIAR AND POSTNAUPLIAR STAGES OF THE EMBRYOS OF THE THREE MYSID SPECIES

Species	Temperature (°C)	Age	Stage
M. relicta	6	1.5–2 months	Shed second egg membranes,
P. flexuosus	15	6–7 days	begin first larval-like stage
N. americana	15	5–6 days	(naupliar stage)
M. relicta	6	3–3.5 months	Shed naupliar membranes,
P. flexuosus	15	14–15 days	begin second larval-like stage
N. americana	15	13–14 days	(postnaupliar stage)
M. relicta	6	4.5–5 months	Shed postnaupliar membranes,
P. flexuosus	15	21–22 days	emerge as juvenile mysids
N. americana	15	20–21 days	

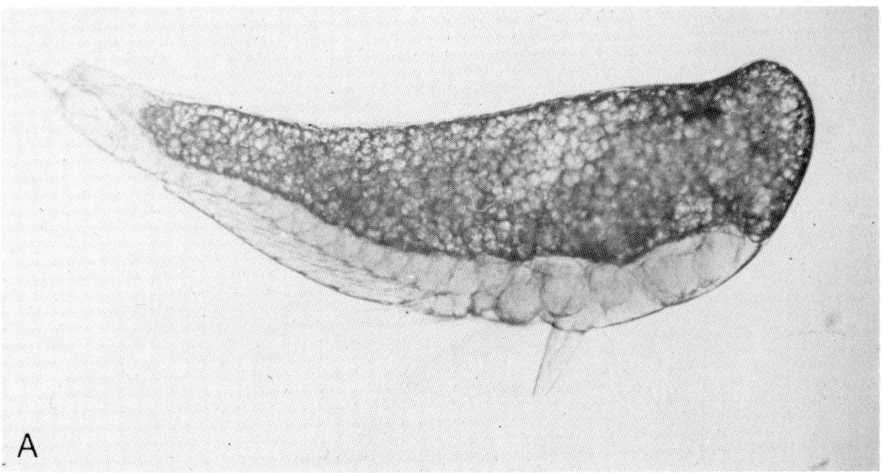

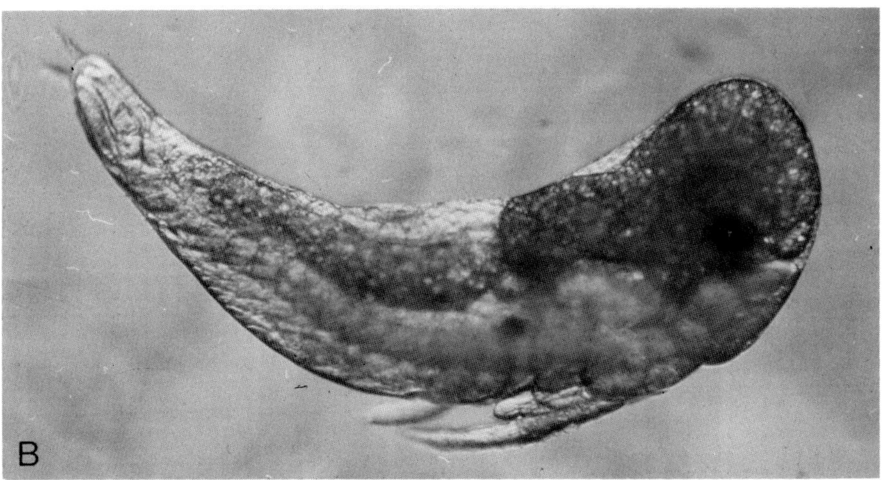

FIG. 2. *Mysis relicta* embryos. (A) Early naupliar stage, still immobile. (B) Mid-naupliar stage, time of the first heart beats and body flinches. (C) Postnaupliar stage, free of its naupliar membrane and near the end of its embryonic development (see p. 146).

approximately the middle of the naupliar stage, the heart begins to beat and the yolk begins to be moved by rhythmic contractions of the intestine. By the end of this stage, occasional contractions of the whole abdominal and thoracic region occur within the naupliar membrane that is about to be shed. No other movements are visible under dissecting microscope.

When the embryo sheds the naupliar membrane, the thoracic appendages and the abdomen become free to move. The appendages flutter weakly at first but with increasing strength as the postnaupliar stage progresses, and

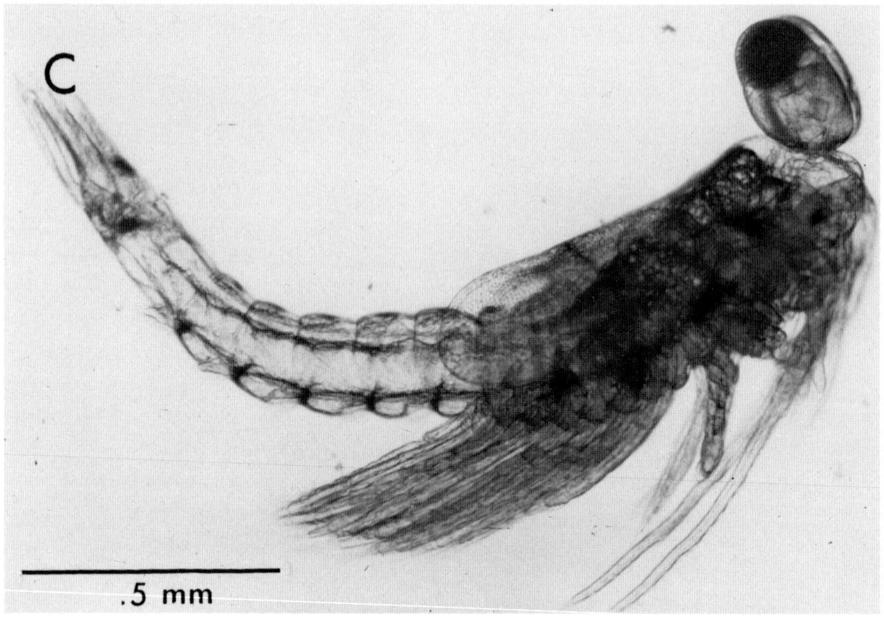

FIG. 2. (*Continued*)

the abdomen repeatedly bends gently dorsoventrally. The antennas also flutter for the first time, apparently independently of the thoracic appendages. Violent intermittent twitches of the abdomen occur throughout this stage, increasing dramatically in amplitude and gradually in frequency in an apparent absence of external stimulation. Such twitches can be observed to develop progressively into the abrupt avoidance movements that characterize postembryonic mysids, and they may represent an elaboration of the occasional general contractions of the naupliar embryos. Certain of the movements of the naupliar and postnaupliar stages occur with an organization that is rhythmic or nearly rhythmic, for instance, the early heartbeat and the apparently spontaneous violent twitches and gentle bending of the abdomen (Fig. 3A,D,E). Others appear to occur unpredictably, for instance the flinches of the naupliar embryo and the fluttering of the thoracic appendages of the postnaupliar embryo (Fig. 3C).

At the end of its postnaupliar stage of development, the embryo suddenly starts to twitch its abdomen more rapidly and with more strength than previously. The violence of this activity propels the embryo out of its brood chamber and away from its mother who is potentially cannibalistic on her young; at the same time it molts its postnaupliar membrane and becomes a

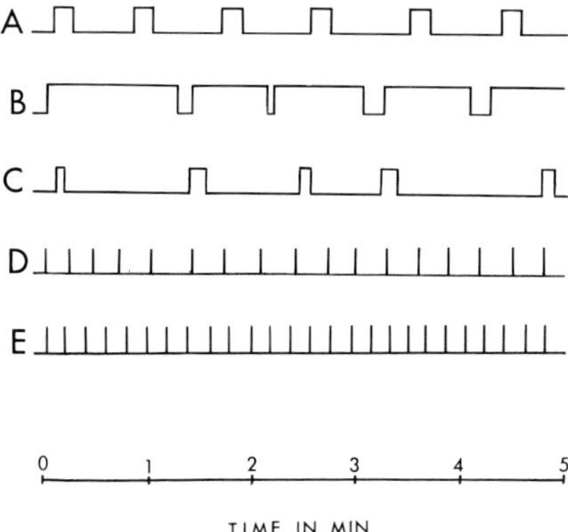

FIG. 3. Sample recordings of the embryonic movements of *Praunus flexuosus*. (A) Day 13 (naupliar stage), bouts of heartbeats. (B–E) Day 19 (postnaupliar stage). (B) Bouts of heart beats. (C) Bouts of fluttering of thoracic appendages. (D) Apparently spontaneous twitches of the thorax and abdomen. (E) Gentle dorsoventral bending of the abdomen.

juvenile mysid. For about 90 seconds thereafter, the new juvenile continues its violent abdominal twitches, until the swimming legs of the thorax hesitantly start their synchronous beating. Within another 30 seconds their beat becomes continuous, the abdominal twitches (or avoidance movements) cease, and the juvenile swims off (Fig. 4).

Lacking histological evidence of the timing of completion of neural reflex loops, it is difficult to distinguish myogenic (nonneural, muscular) response to tactile stimulation from neurogenic or neural response. However, an embryonic response which involves much of the thoracic and abdominal musculature and which appears to develop progressively into a reflex response of the postembryonic organism, is at least likely to be neurogenic (and therefore reflexogenic) as well. The first unquivocal response of this type to stimulation by hair tip in *N. americana* and *P. flexuosus* occurs during the first day of the postnaupliar stage. Even the response, as indicated by immediate twitching of the abdomen, is weak and infrequent, but it increases in strength and frequency throughout the stage. *P. flexuosus* responds with greater frequency earlier than *N. americana* (Fig. 5), suggesting perhaps that its neuromuscular development is more advanced. The lack of similar responsiveness to tactile stimulation by naupliar stage embryos must remain inconclusive evidence for the earlier development of spontaneous movements than of reflex movements, for the constricting naupliar membrane may prevent all but gross or internal movements of an otherwise responsive embryo. Experimental removal of this membrane, without harm to the embryo, has not yet been successful.

The female ventilates her brood throughout its development by rhythmic-

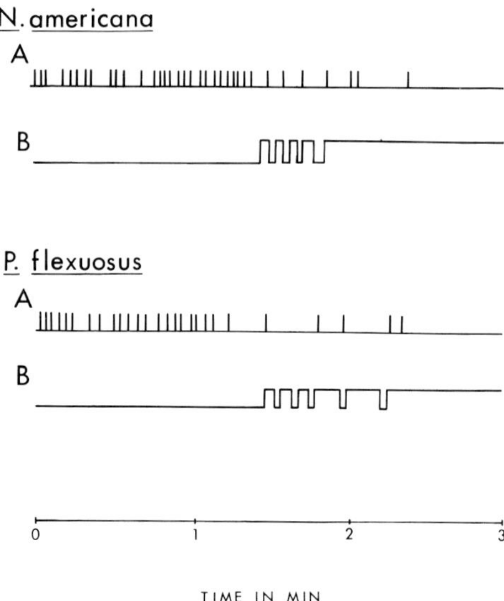

FIG. 4. Sample recordings of two new juveniles of *Praunus flexuosus* and *Neomysis americana* escaping from their brood chambers and mothers. Their repeated and violent avoidance reflexes cease as they start to swim. (A) Avoidance reflexes. (B) Onset of swimming.

ally moving the appendages (oostegites) which form the brood chamber, and such movements continually jostle the embryos. Moreover, once the embryos molt into their postnaupliar stage, they may possibly be stimulated by each other's movements. Though isolated embryos exhibit apparently spontaneous movements, their natural environment is potentially a highly stimulating one.

B. Caprellid Amphipod (Skeleton Shrimp)

Caprella unica is a commensal caprellid that scavenges for food among the spines of the starfish *Asterias vulgaris* and *A. forbesi*, and has been found from Newfoundland to Cape Hatteras (McCain, 1968). The embryology of caprellids seems to have been largely ignored, and the accounts of the development and hatching of gammarid amphipods (Davis, 1959; Le Roux, 1933) may be only generally applicable since gammarids are so unlike caprellids in their morphology and behavior. Both describe the gammarid embryo as actively rupturing its surrounding egg membranes at the end of its development, but other details of its embryonic behavior are not known, and the organization of the embryonic behavior of *C. unica* (Berrill, 1971) may or may not be typical of gammarids as well.

Each mature female *C. unica* breeds repeatedly during the summer months, each time laying a set of 10–15 eggs in a mid-ventral brood pouch where they are fertilized. Raised at 11°C, the young emerge as juveniles 16 days after fertilization to settle and scavenge on the surface of the host starfish, and the female then molts and lays a new set of eggs. Like the female mysids, the female *C. unica* ventilates her brood throughout its development, but there is no indication of potential cannibalism of emerging juveniles as there is in the mysids.

The *C. unica* embryo curls increasingly around itself as it develops encased in its two egg membranes, and it remains apparently motionless until day 11 of its development at 11°C (Fig. 6), when rhythmic contractions of the gut begin. On day 12, slight local twitches of the integument and appendage tips occur. On day 13 the first heartbeats are visible under dissecting microscope, and the posterior thorax bends repeatedly, seeming to curl the embryo even more tightly within its egg membranes. The outer egg membrane usually bursts on day 13. On day 14, the repeated bending of the posterior thorax increases in amplitude and appears to rupture the inner egg membrane and assist the juvenile in freeing itself of it. All of these movements appear to occur rhythmically or almost rhythmically (Fig. 7);

Within 2 hours after hatching, the antennae flutter independently of each other, and the pereiopods (hind legs) can weakly grasp fine algal filaments.

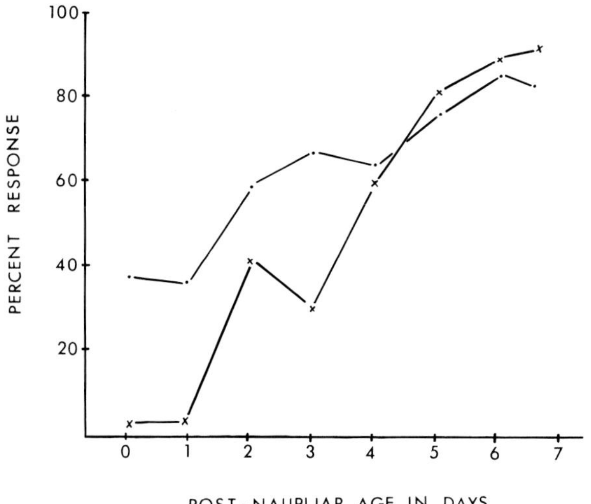

FIG. 5. Percent of postnaupliar embryos of *Praunus flexuosus* (●) and *Neomysis americana* (x) that twitched at least once in response to tactile stimulation. The embryos of both species spend 7 days in their postnaupliar phase of development. Five hundred embryos of each species were tested.

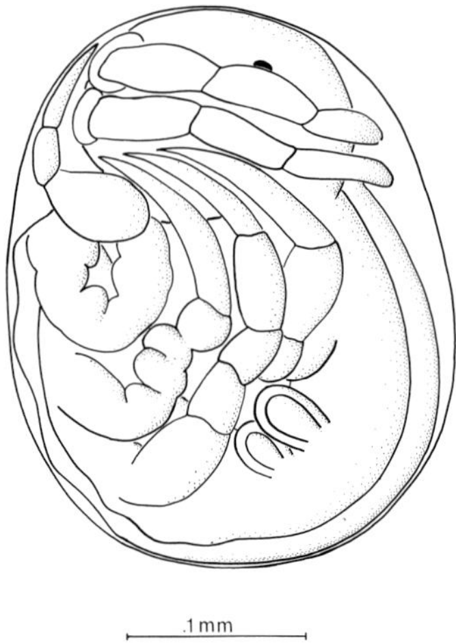

FIG. 6. *Caprella unica* embryo 11 days old and about to start its first slight movements. The outer egg membrane has been removed for a clearer view of the embryo.

FIG. 7. Sample recordings of the activities of individual embryos of *C. unica*. (A-E) Activities of the 13-day-old embryo. (A) Heartbeat. (B) Twitching of the dorsal thoracic integument. (C) Gut contractions. (D) Twitching of a gnathopod tip. (E) Curling of the posterior thorax. (F) Curling of the posterior thorax of the 14-day-old embryo 3 hours before shedding the second egg membrane. (G) Curling of the posterior thorax 1 hour after shedding the second egg membrane.

Within another 2 hours, the juvenile can grasp such filaments securely, clean itself with its gnathopods (large anterior pair of thoracic appendages), and avoid objects by ducking away from them with a motion that appears to be an elaboration of the curling and uncurling actions that began on day 13 (Fig. 8). Twelve hours after hatching, the juveniles appear as responsive and agile as their parents, yet they remain in their brood chambers for a further 25–40 hours. The time of their emergence is correlated with the disappearance of the last yolk in their intestines.

Though apparently rhythmic and spontaneous movements of the embryos of *C. unica* begin on day 11 and 12, no response to stimulation by hair tip is observed in embryos freed of their egg membranes until day 14, several hours before hatching. Though spontaneous movement appears to precede reflex response in this species, the amplitude of the first reflex and spontaneous movements may be too slight for certain recognition, and neural histological evidence again is necessary to help identify the relative times of onset of the two kinds of movements. The ventilation movements by the female are a source of continual stimulation to her embryos, should evidence of earlier reflex responsiveness be forthcoming.

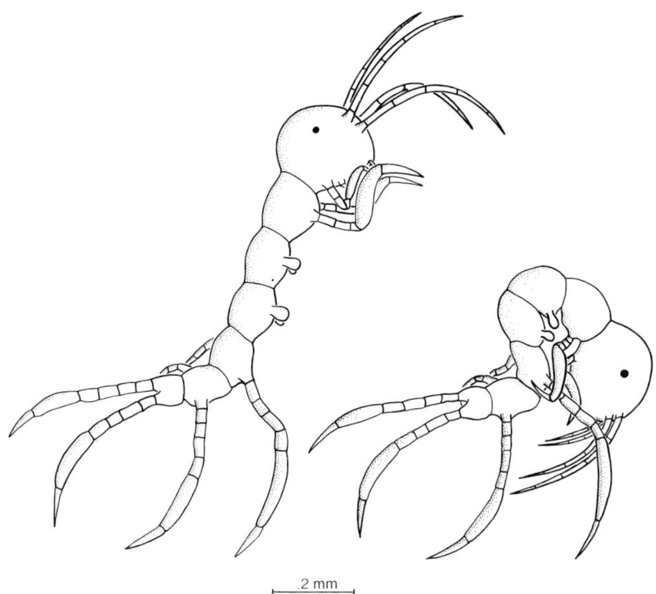

FIG. 8. *C. unica*, 4 hours after shedding its second egg membrane, able to avoid objects by ducking away from them, but not yet able to walk easily.

C. Cladoceran

The development of the embryos of *Daphnia pulex*, a water flea, has been described by Esslova (1959) and Wotzel (1937). The embryos develop in a dorsal brood chamber, 1–12 in each brood, and the parent female molts and lays a new set of eggs within 1–2 hours after the brood emerges. The embryo sheds its first egg membrane early in its development, sheds its second with about one-third of its development time left, and then finally molts to terminate embryonic development and become a juvenile daphnid. The embryos of other cladocerans appear to develop in much the same way (Obreshkove & Fraser, 1940). The embryonic behavior of *D. pulex* has been described by Berrill and Henderson (1972).

At 21°C, the *D. pulex* embryo takes 46–48 hours to develop. It remains inactive until 30 hours old, 2–3 hours before it sheds its second egg membrane. Then its heart and gut begin activity, and muscles at the base of the antennae (and elsewhere under the integument) start to twitch apparently rhythmically (Fig. 9). During the next 2–3 hours the antennal contractions increase in strength, giving the embryo the appearance of shrugging within its egg membrane. The membrane then ruptures and the embryo appears to wriggle free of it by means of its antennal movements.

For the next 10 hours the antennas flutter intermittently, with increasing strength, but they remain against the sides of the embryo when not fluttering. Then approximately 4 hours before the end of its development, the embryo raises its antennas permanently, and they become increasingly active until the embryo is in almost continuous motion (Fig. 10). If the embryo is isolated in a water column at this time, its antennal activity is enough to keep it off the bottom and swimming jerkily, though it lacks the avoidance flights typical of the fully developed organism, and its thoracic appendages have yet to be freed to begin their rhythmic beating. With the molt terminating its development, the embryo becomes a juvenile with the agility it needs to escape its brood chamber.

Reflex responses to tactile stimulation are first visible in the 30-hour-old embryo at the time the first apparently rhythmic and spontaneous movements also occur. Maternal ventilation and the movements of crowded embryos again provide a potential source of continual stimulation for the embryos.

D. Decapod (Lobster)

The reproductive biology and embryology of the American lobster *Homarus americanus* was reviewed in detail by Herrick (1911). The sexually mature female exudes up to 20,000 eggs after she has mated, fertilizing them as she glues them on to her abdominal swimmerets. The following mid-

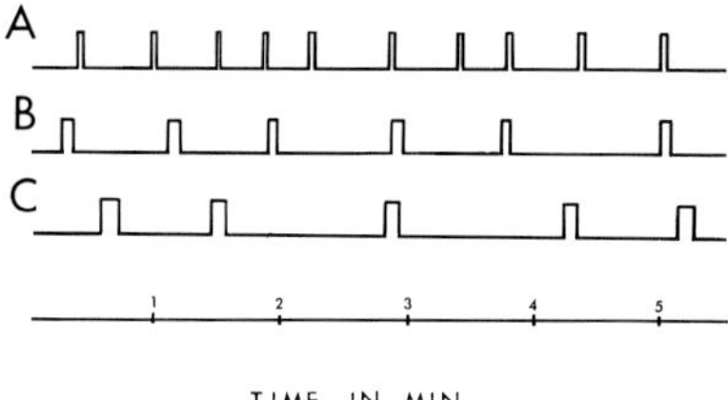

FIG. 9. Sample recordings of bouts of antennal movements of a single embryo of *D. pulex* at three progressively later stages. (A) 32 hours old, 1 hour before shedding the second egg membrane. (B) 36 hours old. (C) 40 hours old, 5 hours before the molt terminating embryonic development.

summer the embryos hatch out of their two egg membranes and immediately molt to become pelagic Stage I larvae (Fig. 11). Recently Pandian (1970a, 1970b) has studied the ecophysiology of embryos of *H. americanus* and the related *H. gammarus*. The usual 10–12 month development time can be shortened by at least 5 months by maintaining the "berried" female at 22°C (Barlow & Ridgeway, 1971).

The maternal assistance given the embryos in hatching is dramatic, and has been described by Templeman (1937) and Davis (1964) and was observed by the author. After ventilating her eggs throughout their development by

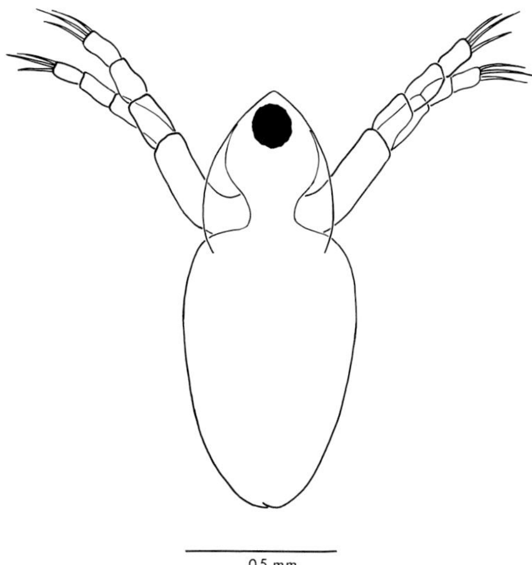

FIG. 10. *Daphnia pulex*, simplified drawing of an embryo 3–4 hours before the molt that terminates its embryonic development. It can swim now with its raised antennas, and its eye spots have fused completely.

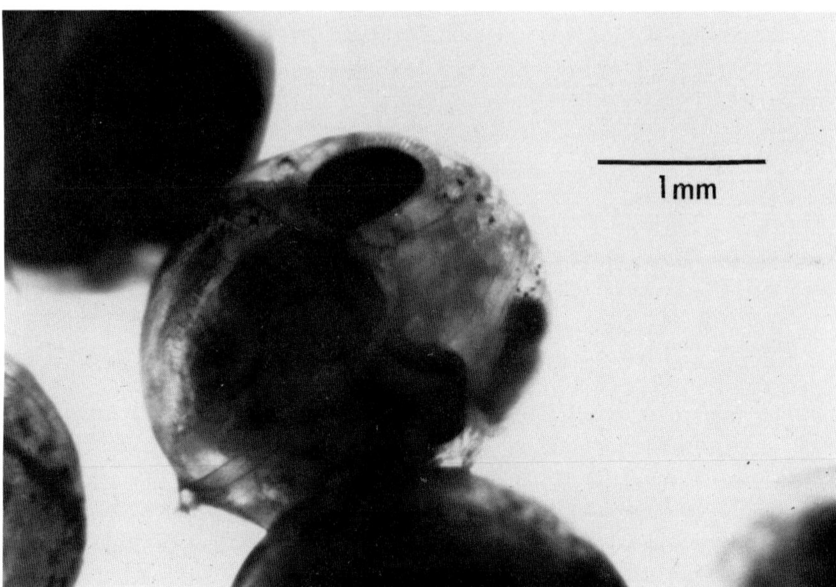

FIG. 11. *H. americanus* embryo 1 week before hatching, still motionless but for its heartbeat. The outer encapsulating membrane has been removed for a better view of the embryo.

flapping her swimmerets rhythmically back and forth, the female proceeds to shake her full-term embryos out of their surrounding membranes. She stands high on the tips of her walking legs and beats her swimmerets with such violence that the egg membranes rupture and the fully developed embryos molt into Stage I larvae. Embryos isolated from the female and left unagitated rarely succeed in hatching or molting by themselves.

Fortnightly visual observations were made by the author on embryos developing at 10° C, starting when the embryos were approximately 4 months old and continuing until they hatched 5 months later. Observations on embryos, both in their egg membranes and experimentally removed from them, showed them to be surprisingly inactive. Apart from the rhythmic gut contractions and the strong heartbeat which appeared to cause passive movements of the appendages and telson with each beat, the embryos showed no signs of the apparently spontaneous movements characteristic of the other crustaceans described. Perhaps more prolonged observations will yet reveal some.

Even during daily 1-hour observation periods during the last 2 weeks of their development, embryos still exhibited no other apparently spontaneous movements. When stimulated severely by hair tip, however, the embryos responded by abruptly bending their abdomens ventrally, in a manner similar to the avoidance reflex of the Stage I larvae. Gentle stimulation of the kind used on other crustacean embryos was without effect.

IV. Discussion

Considerable variation exists in the organization of embryonic movements of these six species of crustaceans. Spontaneous movements appear to precede reflex ones in the mysids and caprellid, but appear virtually nonexistent, but for heart and gut movements, in the lobster. Moreover, the mysid and cladoceran embryos begin to move relatively much sooner than those of the caprellid and lobster. On the other hand, most, though not all, of the embryonic movements of the six crustacean species appear to develop progressively into activities of the free-living stages. Examples are the heartbeats of all the species, antennal movements of the water flea, *D. pulex*, abdominal twitches of the mysids, and the curling movements of the skeleton shrimp, *C. unica*. Exceptions are the local twitches of the integument and the appendage tips of all the species in which they occur. However, progressive embryonic elaboration does not always characterize the activities of the newly emerged juvenile; for example, the synchronous beat of the swimming legs of the mysids, and the swimming of Stage I lobster larvae. Molts during or terminating embryonic development make it difficult to distinguish origins of later movements with certainty. (Premature removal of membranes that would soon be molted has not yet been accomplished without harming the embryos within.) The observations on the elaboration of embryonic movements are, moreover, purely visual ones, and other means of recording and analyzing the movements would be invaluable. (See Oppenheim, this volume, for a discussion of the origins of postembryonic movements in vertebrates.)

A number of the embryonic movements of the six species occur rhythmically or nearly rhythmically, a feature characteristic of certain vertebrate embryos (Corner, 1964; Decker, 1967; Hamburger, 1963). Such rhythmic organization of embryonic movements implies a possibly spontaneous neural basis to them, and since spontaneous neural activity occurs in adult locusts (Wilson, 1961) and postembryonic crickets (Bentley & Hoy, 1970), it should not be unexpected in crustacean embryos. This apparent separation of spontaneous movements from reflex responses in the crustacean embryos must, however, remain speculative until there is sufficiently detailed analysis of the neuromuscular development of the embryos in question.

Spontaneous neural activity may perhaps insure movement and hence normal morphological development in otherwise unstimulating conditions, and may thus prevent abnormalities. (Hamburger discusses spontaneous motility in vertebrate embryos in the first article in this section.) Passive and reflex movements of the embryo may also provide it with further stimulation, complementing or even substituting for the spontaneous movements. Certainly the crustacean embryos are subjected to extensive passive movements as a result of the maternal ventilation activities, and those embryos developing in crowded brood chambers may provide each other with a potentially continuous source of stimulation as well. Prolonged experimental anesthesia of these embryos, similar to that of amphibian embryos (Fromme, 1941) and chick embryos (Drachman & Sokoloff, 1966), has not been successful. Until it is, nothing can be said on whether embryonic movements insure normal mascular and joint development in crustaceans, as they do in other forms.

Embryonic stimulation in any organism may be adaptive in more than one way. For instance, Gottlieb (1966) has shown that perinatal stimulation of ducklings may influence their postnatal responses. An interesting aspect of the development of certain of the crustacean embryos is the number of membranes they shed or molt before becoming juveniles, and movements by these embryos appear to be involved in rupturing or removing one or more of these membranes. Davis (1968) emphasizes the dominant role that water absorption and hence growth of the embryo plays in the bursting of the egg membranes in particular, though he also recognizes that embryonic motility is sometimes involved as well. Perhaps embryonic motility is increasingly important to the embryo in shedding its membranes successfully, the more membranes it has to shed. Visual observations suggest that such a correlation exists. Such embryos, however, are also those apparently undergoing a greater degree of suppressed larval-like development, and their motility may instead be the remnants of ancestral larval movements and of no immediate value to the embryos now. Until the critical experiments of anesthesia and membrane removal, along with detailed analyses of the various neuromuscular systems, are completed, the functions of crustacean embryonic behavior remain unknown.

V. Summary

The extent of embryonic behavior varies considerably among six selected species of crustaceans (three mysids: *N. americana*, *M. relicta*, *P. flexuosus*; the cladoceran *D. pulex*; the caprellid amphipod *C. unica*; and the decapod lobster *H. americanus*) and may be correlated with the number of embryonic membranes shed or molted by the different embryos.

From purely visual observations, many of the embryonic movements of the six species appear to be spontaneous and rhythmically organized. Of these, some appear to develop progressively into activities of the post-embryonic organisms. Others, however, appear unrelated to later activities, and certain later activities may have no embryonic precursors.

The embryos of all six species are continually jostled by the ventilation movements of their mothers. All respond to tactile stimulation while still embryos, though the timing of onset of reflex responsiveness, again as determined solely by visual observation, also appears to vary considerably. Sources of stimulation other than possible spontaneous neural activity are therefore likely.

The motility of the crustacean embryos may perhaps insure normal development, and may assist in the shedding of the embryonic membranes, but these are merely speculative possibilities in the absence of experimental evidence.

References

Barlow, J., & Ridgeway, G. J. Polymorphisms of esterase isozymes in the American lobster (*Homarus americanus*). *Journal of the Fisheries Research Board of Canada*, 1971, **28**, 15–21.

Beeton, A. M. The vertical migration of *Mysis relicta* in Lakes Huron and Michigan. *Journal of the Fisheries Research Board of Canada*, 1960, **17**, 517–539.

Bentley, D. R., & Hoy, R. R. Post-embryonic development of adult motor patterns in crickets: a neural analysis. *Science*, 1970, **170**, 1409–1411.

Berrill, M. The embryonic behavior of the mysid shrimp, *Mysis relicta. Canadian Journal of Zoology*, 1969, **47**, 1217–1221.

Berrill, M. The embryonic behavior of *Caprella unica* (Crustacea: Amphipoda). *Canadian Journal of Zoology*, 1971, **49**, 499–504.

Berrill, M. The embryonic development of the avoidance reflex of *Neomysis americana* and *Praunus flexuosus* (Crustacea: Mysidacea). *Animal Behavior*, 1971, **19**, 707–713.

Berrill, M., & Henderson, C. The embryonic development of the swimming behavior of *Daphnia pulex*. (Crustacea: Cladocera). *Canadian Journal of Zoology*, 1972, **50**, 969–973.

Corner, M. A. Rhythmicity in the early swimming of anuran larvae. *Journal of Embryology and Experimental Morphology*, 1964, **12**, 665–671.

Davis, C. C. Osmotic hatching in the eggs of some fresh-water copepods. *Biological Bulletin*, 1959, **116**, 15–29.

Davis, C. C. A study of the hatching process in aquatic invertebrates. XIII. Event of eclosion in the American lobster, *Homarus americanus. American Midland Naturalist*, 1964, **72**, 203–210.

Davis, C. C. A study of the hatching process of aquatic invertebrates, XXII. Multiple membrane shedding in *Mysidium columbiae* (Crustacea: Mysidacea). *Bulletin of Marine Science*, 1966, **16**, 124–131.

Davis, C. C. Mechanisims of hatching in aquatic invertebrate eggs. *Annual Review of Oceanography and Marine Biology*, 1968, **6**, 325–376.

Decker, J. D. Motility of the turtle embryo, *Chelyda serpentina. Science*, 1967, **157**, 952–954.

Drachman, D. B., & Sokoloff, L. The role of movement in embryonic joint development. *Developmental Biology*, 1966, **14**, 401–420.

Esslova, M. Embryonalni vyvoj parthenogenetickych vajicek perloocky *Daphnia pulex*. *Vestnik Ceskoslovenske Spolecnosti Zoologicke*, 1959, **23**, 80–88.

Fromme, A. An experimental study of the factors of maturation and practice in the behavioral development of the embryo of the frog, *Rana pipiens*. *Genetic Psychology Monographs*, 1941, **24**, 219–256.

Gottlieb, G. Species identification by avian neonates: Contributory effect of prenatal auditory stimulation. *Animal Behavior*, 1966, **14**, 282–290.

Hamburger, V. Some aspects of the embryology of behavior. *Quarterly Review of Biology*, 1963, **38**, 342–365.

Herman, S. Vertical migrations of the opossum shrimp, *Neomysis americana*. *Limnology and Oceanography*, 1963, **8**, 228–238.

Herrick, F. H. Natural history of the American lobster. *Bulletin of the U.S. Fisheries Bureau*, 1911, **29**, 147–408.

Korschelt, E. Ontogenie der Decapoden. *Bronn's Klassen*, 1944, **5**, 671–861.

Lasenby, D. The ecology of *Mysis relicta* in an arctic and a temperate lake. Unpublished doctoral dissertation, University of Toronto, 1971.

Le Roux, M. L. Recherches sur la sexualité des Gammariens. *Bulletin Biologique de la France et de la Belgique, Supplement*, 1933, **16**, 1–138.

McCain, J. C. The Caprellidae of the western North Atlantic. *U.S. National Museum, Bulletin*, 1968, No. 278, 1–147.

Manton, S. M. On the embryology of a mysid crustacean, *Hemimysis lamornae*. *Philosophical Transactions of the Royal Society of London, Series B*, 1928, **216**, 363–463.

Nair, K. B. Reproduction, oogenesis and development in *Mesopodopsis orientalis*. *Proceedings of the Indian Academy of Sciences, Section B*, 1939, **9**, 175–223.

Obreshkove, V., & Fraser, A. W. Growth and differentiation of *Daphnia magna* eggs *in vitro*. *Biological Bulletin*, 1940, **78**, 428–436.

Orkand, P. M., & Kravitz, E. A. Localization of the sites of γ-aminobutyric acid (GABA) uptake in lobster nerve-muscle preparations. *Journal of Cellular Biology*, 1971, **49**, 75–89.

Pandian, T. J. Ecophysiological studies on the developing eggs and embryos of the European lobster *Homarus gammarus*. *Marine Biology*, 1970, **5**, 153–167.(a)

Pandian, T. J. Yolk utilization and hatching time in the Canadian lobster *Homarus americanus*. *Marine Biology*, 1970, **7**, 249–254.(b)

Templeman, W. Egg laying and hatching postures of the American lobster (*Homarus americanus*). *Journal of the Biological Board of Canada*, 1937, **3**, 339–342.

Wilson, D. M. The central nervous control of flight in a locust. *Journal of Experimental Biology*, 1961, **38**, 471–490.

Wotzel, F. Zur Entwicklung des Sommereies von *Daphnia pulex*. *Zoologische Jahrbucher*, 1937, **63**, 455–470.

Zucker, R. S., Kennedy, D., & Selverston, A. I. Neuronal circuit mediating escape responses in crayfish. *Science*, 1971, **173**, 645–649.

Section 3

HATCHING: HORMONAL, PHYSIOLOGICAL, AND BEHAVIORAL ASPECTS

INTRODUCTION

In the case of species which hatch out of eggs, such as birds, the "perinatal" changes do not seem quite as monumental as metamorphosis in amphibians, or as dramatic as birth in mammals. In birds, for example, significant behavioral changes are known to begin to occur several days *before* hatching. In this section, Dr. Oppenheim describes the hatching movements of several avian species in close detail, and he reviews for us that which is known about many vertebrate and invertebrate species which take their egress from an egg or egglike structure.

Despite their sensitivity to sensory stimulation and their high degree of active movement, embryos and fetuses are commonly regarded as being in more of a "sleeping" than a "waking" condition. The transition from a sleep-like embryonic state to an intermittently awake or alert state around the time of hatching in chicks is described in electrophysiological and behavioral detail by Drs. Corner, Bakhuis, and van Wingerden. Although problems of definition and measurement seem to be particularly acute in this area of investigation, the importance of posture (particularly involving the neck musculature) in mediating the change from one state to the other seems very clear in neonatal chicks. Since this postural aspect is also involved in inducing the phenomenon of tonic immobility ("animal hypnosis"), described in the literature of comparative psychology, one wonders whether Dr. Corner and his colleagues may be dealing with the precursor to that puzzling phenomenon.

PREHATCHING AND HATCHING BEHAVIOR: A COMPARATIVE AND PHYSIOLOGICAL CONSIDERATION

RONALD W. OPPENHEIM

Neuroembryology Laboratory
Division of Research
North Carolina Department of Mental Health
Raleigh, North Carolina

> I have nevertheless, many a time caused chickens to be hatched which had pecked their shell for the first time but half an hour, or even a quarter of an hour before: nor did I do this for the good of the chicken, but barely to satisfy the curiosity of the people who deserved to be indulged, and who having seen how eggs were artificially sat in my ovens, were desirous of having the pleasure to see a chicken come out of his shell.
>
> *M. de Réaumur, 1750*

I. Introduction

A. Historical Review

The embryos of all egg-laying animals are faced with a common biological problem at the termination of their incubation period; they must somehow escape from the shell or egg capsule. Although initially this problem may appear relatively simple, the least bit of serious thought given to the matter should prove otherwise. For example, the time of hatching must occur with some precision in order to insure the readiness of the embryo for a "free-living" existence. The assurance that hatching does occur at the appropriate time means, of course, that the multitude of mechanisms involved in the preparation and execution of this event, be they behavioral, hormonal, neural, etc., must also be carefully attuned, one with another. Looked at in this way, hatching can be seen to be a truly remarkable event encompassing many important embryological and general biological problems, as well as being interesting in and of itself.

The major portion of this article will deal with hatching and its mechanisms in birds, although hatching in other forms, ranging from monotremes to insects, will also be discussed. Such a restriction in emphasis was necessitated primarily by the lack of information available for nonavian egg-laying animals, and secondarily by the acquired predilection of the author. It is hoped that at the very least the reader will acquire sufficient knowledge from this article to enable him to detect the inaccuracy in Fig. 1.

While the opportunity to observe the hatching embryos of any egg-laying animal in nature has theoretically existed since the appearance of modern man, an event thought to have taken place about 300,000 years ago during the Pleistocene era (Buettner-Jansch, 1966), a more likely opportunity for such observations occurred with the advent of techniques for artificial incubation of avian eggs. Landauer (1967), in an excellent historical treatment of this field, describes such techniques as being used quite successfully

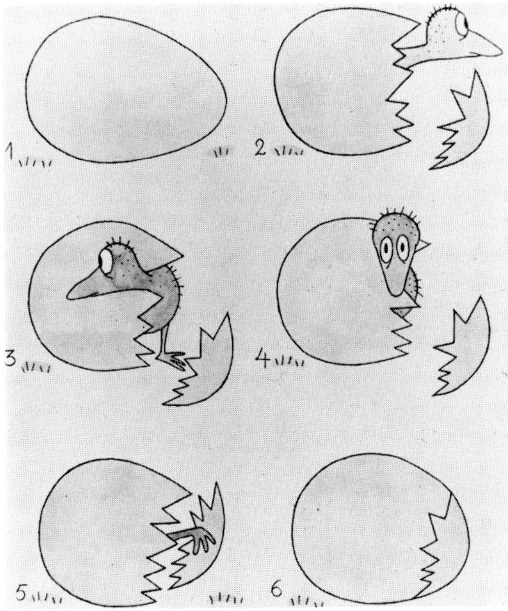

FIG. 1. Cartoon depicting hatching in an unidentifiable species of bird. Copyright by Talon Division of Textron.

by the Egyptians and Chinese in early antiquity, and the continuation of similar (and other) techniques down to the present day by many other peoples. Thus, it is surprising to learn that, in spite of these opportunities, no written description of hatching in birds exists prior to the eighteenth century. Whether this situation is due to a lack of interest, or simply to the failure of such descriptions to survive to the present time, cannot, of course, be determined.

Until relatively recent times, the observations and experiments of de Réaumur (1750) represented the most comprehensive and penetrating examination of hatching behavior in birds. A few of the discoveries of de Réaumur that have stood the test of time are: (a) the relationship between yolk-sac retraction and emergence from the shell; (b) the fact that the shell is broken by thrusts of the head and beak in a backward or uplifting direction and not by "pecking"; (c) the observation that the entire embryo rotates within the shell during hatching and that this rotation is dependent upon leg movements; and (d) a description of the movements involved in membrane penetration. Other observations, experiments, and speculations of de Réaumur concerning hatching are discussed below in Sections II and III. Although brief attention has been paid to certain specific problems related to hatching in the fowl since the time of de Réaumur (i.e., pipping and its mechanisms, Preyer, 1885; Sacc, 1847; von Baer, 1828), it was not until the twentieth century that hatching behavior in birds once again became the subject of serious experimental and descriptive investigation. Most of these later studies are extensively discussed below.

B. Definition of Terms

1. *Beak- or bill-clapping*: A rapid opening and closing of the mandibles, usually of low amplitude, and occurring in bursts.
2. *Climax*: The actual hatching process, including the rotation of the embryo and the cracking of the shell around the circumference.
3. *Egg*: A general term referring to the shell and its contents, including the membranes and the embryo.
4. *Egg burster*: A special set of spines, teeth, or ridges on the head and/or body of insects which aid in rupturing the shell at hatching.
5. *Egg capsule or envelope*: The noncalcareous egg of fish, amphibians, and many invertebrates.
6. *Egg caruncle*: A toothlike structure formed from the horny epidermal thickening at the tip of the snout (or beak) of turtles, crocodiles, birds, and monotremes. The monotremes also have a true egg tooth, though it is probably vestigial (Bellairs, 1970). The egg caruncle, like the egg tooth, is apparently used to help break the membranes and shell during pipping and hatching (Fig. 18).
7. *Egg membranes*: Refers to the vitelline and other membranes (gelatinous, chorion, etc.) found in forms not having a calcareous shell (e.g., fish).
8. *Eggshell*: The hard or leathery calcareous shell found primarily in reptiles, birds, and monotremes.
9. *Egg tooth*: A true tooth, of mesodermal origin, that is attached to the premaxillary bone in the midline and erupts from the gum shortly before hatching. A true egg tooth is found only in lizard and snake embryos, who use it to tear open the egg. It is lost soon after hatching (Fig. 22).
10. *Emergence*: The actual escape from the shell when the bird embryo pushes the shell cap off, unbends the head from its position on the chest and under the right wing, and comes completely out of the shell.
11. *Gaping or yawning*: A slow, but high amplitude (wide), opening of the beak or mouth; the beak usually remains open for a short time. Usually occurs once, rather than in bursts.
12. *Hatching behavior*: Behavior during the final stage of hatching when the embryo begins to finally break out of the shell (also see *Climax*).
13. *Hatching position*: The characteristic position of the embryo, that is assumed one or several days before hatching in most birds (Figs. 8, 9, 14, 17, and 19).

14. *Membrane penetration* (*MP*): The penetration of the beak into the air-space through the combined chorioallantoic (C-A) and inner shell membranes in birds (Fig. 10).

15. *Pipping*: The first cracking of the shell (Fig. 12).

16. *Prehatching behavior*: This refers to the various behavior patterns that are preparatory to hatching and which occur prior to the final stage (climax) when the embryo actually begins to emerge from the shell.

17. *Pretucking*: This term refers to the period of time prior to the occurrence of the first Type III, rotatory-tucking movements in the behavioral repertoire of the avian embryo.

18. *Startle*: A jerky, spasmodic movement that usually involves the entire body (see *Type II* activity, below).

19. *Tarsal joints*: The articulation between tibiotarsus and tarsometatarsus bones of the leg (Fig. 8).

20. *Tucking movements*: The sequence of movements by which the right side of the head is tucked under the right wing (Figs. 2–6).

21. *Type I overt somatic motor activity*: Irregular, convulsivelike movements, jerky in nature, that generally do not appear to have any coordination between the various parts of the embryo.

22. *Type II overt somatic motor activity*: Vigorous and rapid myoclonic-like jerks usually involving the entire body and having a high amplitude. (These movements are the same as startles.)

23. *Type III overt somatic motor activity*: Also referred to as rotatory or tucking movements. This type of activity is highly coordinated and more toniclike than Types I or II. During tucking it is these movements that serve to lift the head out of the yolk and to position the head under the right wing. It is also movements of this type that are responsible for pipping and for the circumferential rotation of bird embryos during climax.

24. *Wing flutter*: A rapid up-and-down movement of both wings, simultaneously, at the shoulder joint.

C. *Prehatching and Hatching Behavior in the Domestic Fowl,* Gallus domesticus

Although the prehatching and hatching behavior of the domestic fowl has been repeatedly described recently (Corner & Bot, 1967; Hamburger, 1968; Hamburger & Oppenheim, 1967; Kovach, 1970; Oppenheim, 1970, 1972b), the present account, while somewhat abbreviated, is nevertheless important for later comparison with other species in this article. The domestic chick is used as the "standard" simply because hatching behavior has been more fully described in the chicken than in any other egg-laying animal.

By about the sixteenth to the seventeenth day of incubation, the chick embryo has assumed a position similar to that seen in Plate I, Figs 2a and b. The attainment of the pretucking position is due primarily to passive gravitational factors acting on the embryo and other egg contents. Kuo (1932a) has described these passive changes in rather complete detail. The embryo's head is partially buried in the yolk in the lower half of the egg (with reference to the vertical axis) at this time (Plate I, Fig. 2). Wriggling movements are also thought to play a minor role in the attainment of this position. The extent to which these wriggling movements of the embryo are directed toward attaining a particular orientation (e.g., head toward the large end of the egg) is not known. However, Byerly and Olsen (1931) have reported a tendency for fowl embryos to orient toward an area of the shell where apparently the most gaseous exchange can take place. For example, eggs positioned with the large end up, but with paraffin coated on the shell over the airspace, result in a drastic increase in malposition II, head in small end, and malposition IV, head under right wing but away from airspace. Since the eggs were apparently treated at the beginning of incubation, it is not clear from the report by Byerly and Olsen whether this postural modification occurs during early (1–15 days), or prehatching and hatching stages. This experiment certainly deserves to be repeated with the addition of paraffin coating the shell on about day 15 or 16, just after the position of the embryo becomes fixed relative to the longitudinal axis of the egg. In this way it could be determined whether the embryo actively shifts its head away from a portion of the shell which has impaired gaseous exchange (or toward a part of the shell that has enhanced gaseous exchange).

Prior to day 16 of incubation, the embryo's overt somatic motor activity consists almost exclusively of jerky, convulsivelike activity (Types I and II). Although, according to one report, the coordinated head movement component (thrusts) of the Type III activity can occasionally be seen as early as day 14 or 15 (Kaspar, 1964), it is not until sometime on day 16 or 17 of incubation that the Types I and II activity is frequently interspersed with episodes of Type III activity. During these episodes the Types I and II activities are almost entirely absent, while the Type III movements appear to occur periodically about every 20 seconds (Hamburger & Oppenheim, 1967). At this time the Type III movements are called tucking movements, since over a long period of time (10–24 hours) they will result in the embryo getting its head tucked under the right wing (Plate I, Figs. 3–6). Although the Type III movements appear sporadically from the time of their inception at 16–17 days until climax, they represent only a small percentage of the *total* embryonic activity during this time (Hamburger & Oppenheim, 1967; Kovach, 1970; Oppenheim, 1972b).

In spite of the infrequent occurrence of the Type III activity *after* tucking,

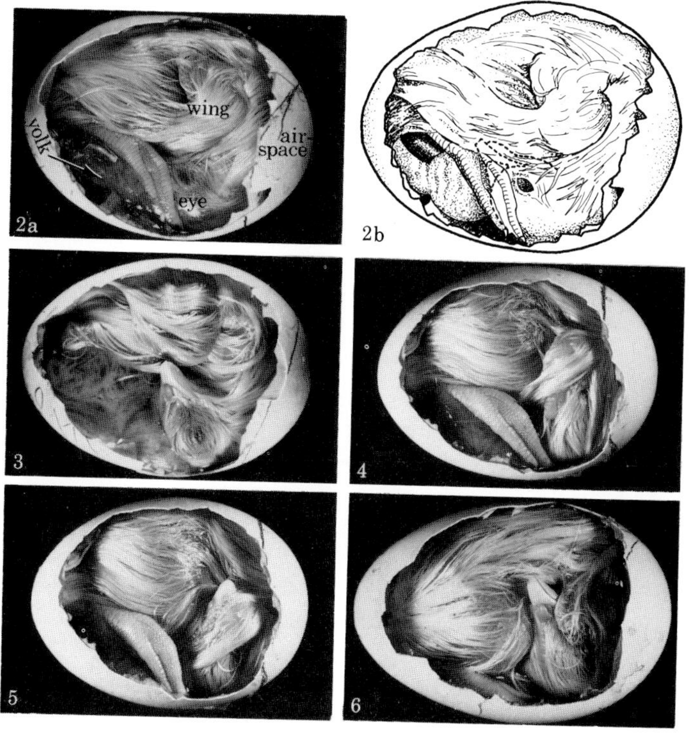

Plate I

FIG. 2a. The pretucking position of the domestic fowl embryo at 17 days of incubation.

FIG. 2b. Drawing of the pretucking position illustrating the position of the head and beak in the yolk.

FIG. 3. The first stage in the tucking process in which the head and beak are lifted out of the yolk.

FIG. 4. The second stage in the tucking process with the wing just slightly covering the beak.

FIG. 5. A later stage of tucking with the head tucked further under the wing.

FIG. 6. An embryo close to the culmination of tucking.

these movements are responsible for the embryo's changing position somewhat, such that soon the beak becomes pushed up into the membranes separating the embryo from the airspace (draping, Plate II, Fig. 7). It is at about this time that the embryo first begins true lung respiration (Kuo &

Shen, 1937); prior to this time all gaseous exchange is mediated by the highly vascularized allantoic membrane encompassing the major portion of the shell. At this same time beak-clapping becomes a more prominent part of the total behavioral repertoire. Changes in the rate of beak-clapping, as well as changes in other behavioral parameters, at this stage, are presented in Table I. It is interesting that at about the time of draping, when the embryo begins to be fairly tightly compacted in the shell (Plate II, Figs. 8 and 9), the frequence of embryonic movement begins to decrease. The question of whether this situation reflects a cause and effect relationship is discussed below.

Due to the increased uptake of calcium by diffusion from the shell to the albumen and yolk sac, and then to the embryo, there is a sharp *decrease* in shell strength during the latter part of incubation (Vanderstoep & Richards, 1970). This process probably aids in the pipping and later cracking of the shell, since artificial strengthening of the shell results in a lowered percentage of both pipping and hatching (Fisher, 1958).

Soon after draping, on day 18 or 19 of incubation, the chicken embryo penetrates its beak through the membranes separating it from the airspace (Plate II, Fig. 10). The specific movements that accomplish membrane penetration (MP) vary considerably such that beak-clapping, slight head jerks (Type I activity), startles (Type II activity), and coordinated head thrusts (Type III activity) all may cause the beak to penetrate the membranes (Hamburger & Oppenheim, 1967; Oppenheim, 1970).

Between day 17 and day 19 of incubation the musculus complexus ("hatching muscle"), located in the dorsal cervical region (Plate III, Fig. 11), enlarges rapidly due to an increase of lymph within the muscle; it then decreases in size until after hatching (Pohlman, 1919). The precise role played by this muscle in pipping and later cracking of the shell is still a controversial issue. However, about 7–12 hours after membrane penetration,* when this muscle is close to its maximum size, pipping occurs (see

*The interval between MP and pipping in the domestic fowl was estimated by Hamburger and Oppenheim (1967) to be about 20–24 hours. We now know that this is incorrect due to the windowing techniques used at that time. For it is now known that an open window in the egg produces a delay in pipping time (Visschedijk, 1968b). Recently, this interval was again determined by the author in the following manner: half of the embryos ($n = 20$) had a small hole (1 cm^2) made in the center of the airspace on day 17. This opening was then sealed with Parafilm. Beginning on day 18 0 hour the Parafilm was removed once every hour, for 1–2 minutes, around the clock to determine precisely when MP occurred. After MP, the Parafilm remained sealed and the eggs were again checked every hour for pipping. The mean MP–pip interval for these eggs was 8.0 hours. In the remaining eggs ($n = 20$), the shell was left intact, and the eggs were candled hourly to determine when MP occurred. The MP–pip interval for these embryos was 11.8 hours. Thus, the mean of these two groups ($\overline{X} = 10$ hours, $n = 40$) probably gives the best estimate of the true MP–pip interval.

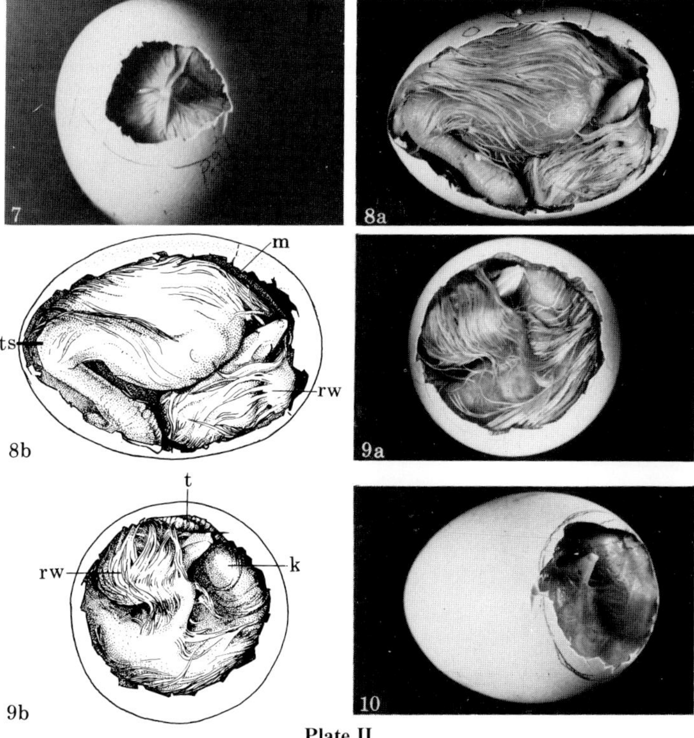

Plate II

FIG. 7. Removal of shell over the airspace shows the beak of the embryo pushed up into the airspace membranes (draping).

FIG. 8a. An embryo in the hatching position at about the time of pipping.

FIG. 8b. Drawing of the hatching position from Fig. 8a. m, Membrane; rw, right wing; ts, tarsal joint.

FIG. 9a. An embryo in the hatching position as seen after removal of the shell over the airspace. The arrow shows the direction of the head and beak movement during pipping and climax.

FIG. 9b. A drawing of an embryo in the hatching position. (From Fig. 9a.) k, Knee; rw, right wing; t, toes.

FIG. 10. Membrane penetration. Beak now extends freely into the airspace.

Table I). Pipping is invariably accomplished by a strong back-thrust or lifting movement of the head and beak (Type III activity), and is represented first by a cracked region at the large end of the egg; it is only later that pieces of shell are actually broken away, making a small (.5 cm^2) hole (Plate III,

TABLE I
OCCURRENCE, MEAN FREQUENCY, AND SD (IN PARENTHESES) OF VARIOUS EMBRYONIC

Species	Incubation period (days)	n	Pretuck I Age (day and hour)	Pretuck I Movement (per minute)	n	Pretuck II Age (day and hour)	Pretuck II Movement (per minute)	n
Precocial:								
Domestic chick	20–21 [b]	101	16-14 (7.1)	14.0 (4.2)	64	17-23 (13.9)	9.2 (3.0)	48
Mallard duck	25–26	—	—	—	—	—	—	18
Peking duck	26–27	16	20-12 (10.2)	22.8 (7.1)	20	22-9 (9.5)	11.6 (3.8)	14
Turkey	26–28	5	22-22	13.2	5	23-22	8.7	5
Bobwhite quail	23–24	20	17-9 (7.8)	17.7 (8.1)	19	19-10 (10.5)	11.6 (3.1)	10
Japanese quail	16–17	—	—	—	23	14-16 (3.0)	19.4 (5.6)	19
Semiprecocial:								
Gull	23–24	14	17-7 (6.1)	20.3 (7.4)	12	19-1 (4.2)	14.2 (4.8)	11
Altricial:								
Pigeon	17-18	4	14-7	11.9	24	14-23 (10.5)	12.9 (6.3)	19
Cardinal	12–13	—	—	—	—	—	—	—
Wren	12–13	—	—	—	—	—	—	—
Catbird	12–13	—	—	—	—	—	—	—

[a]The *tucked-draped or MP* category includes embryos which completed any one of these stages. Draping refers to the protrusion of the beak into the air-space membranes, and MP refers to when the beak actually penetrates these membranes. Post-pip refers to different times after the first pip-crack. In the gulls the *post-pip* stage refers to varying times after the first pip-crack occurred in the *air-space* region.

Fig. 12). Beginning about the time of pipping, there is an increase in the rate of bill-clapping (Table I). The significance of this increase is unknown, as is the functional significance of beak-clapping in general; it may be that beak-clapping, which begins at about 11–12 days of incubation in the chicken embryo, is related to the ingestion of albumen that is first observed at about this same time in the embryo (Foote, 1969; Kuo, 1932c). Beak-clapping also often occurs in rather stereotyped bursts (Corner & Bot, 1967).

Activities of Precocial and Altricial Species of Birds at Specific Prehatching Stages[a]

Tucked-draped or MP						Post-pip				
Age (day and hour)	Movement (per minute)	Head thrust (per minute)	Beak-clap (per minute)	Respiration (per minute)	n	Age (day and hour)	Movement (per minute)	Head thrust (per minute)	Beak-clap (per minute)	Respiration (per minute)
18-16	4.6	.3	11.0	11.7	27	19-14	4.2	.8	16.9	44.7
(10.7)	(2.6)	(.2)	(7.1)	(5.2)		(14.5)	(1.6)	(.4)	(11.8)	(9.7)
23-15	7.5	.2	14.0	30.0	16	25-10	4.2	.8	19.1	58.0
(12.7)	(3.2)	(.2)	(9.6)	(10.3)		(11.7)	(1.9)	(.6)	(16.3)	(17.1)
24-2	4.3	0.4	13.0	26.5	7	26-12	4.7	2.3	34.7	53.5
(13.9)	(1.7)	(.1)	(6.2)	(12.8)		(9.9)	(2.1)	(1.3)	(19.9)	(16.2)
25-0	3.2	.6	15.9	44.0	8	25-22	3.2	1.5	14.4	59.8
						(12.7)	(2.0)	(.5)	(9.9)	(17.3)
21-0	5.0	.4	16.7	21.5	9	22-20	2.5	.2	22.4	74.0
(6.2)	(1.2)	(.2)	(10.9)	(11.7)		(7.3)	(.9)	(.1)	(10.1)	(13.0)
17-17	7.0	.4	10.4	55.7	—	—	—	—	—	—
	(2.1)	(.3)	(5.6)	(11.9)						
20-19	7.5	.2	15.9	16.7	14	22-12	4.0	.3	12.5	25.0
(5.3)	(2.7)	(.1)	(11.2)	(7.6)		(4.5)	(2.7)	(.1)	(7.7)	(10.3)
15-16	6.5	.3	6.8	15.2	18	16-20	7.4	.6	9.8	24.0
(7.2)	(2.8)	(.2)	(4.3)	(7.1)		(7.3)	(3.6)	(.4)	(4.2)	(8.3)
—	—	—	—	—	3	11-12	6.5	.0	6.5	18.3
—	—	—	—	—	3	12-12	4.0	.3	7.4	24.0
—	—	—	—	—	3	12-0	4.6	.6	13.9	19.0

[b]Age is determined by calling an embryo 1-day old only after the first 24 hours of incubation have elapsed. Thus, an egg incubated for 30 hours is 1 day, 6 hours old (1-6).

Lung respiration exhibits a rise in both rate and regularity between the time of MP and postpipping (Tables I and III). The level of all somatic motor activity (except beak-clapping) remains relatively low after pipping and generally no further cracks are made in the shell from the time of pipping until climax, a period averaging 20–24 hours in the white leghorn fowl embryo (Table II). During this time Types I and II activities predominate. However, quite suddenly these types of activity become very infrequent, or disappear entirely, and are replaced by coordinated and periodically occurring back thrusts of the head and beak (Type III). This rather abrupt transi-

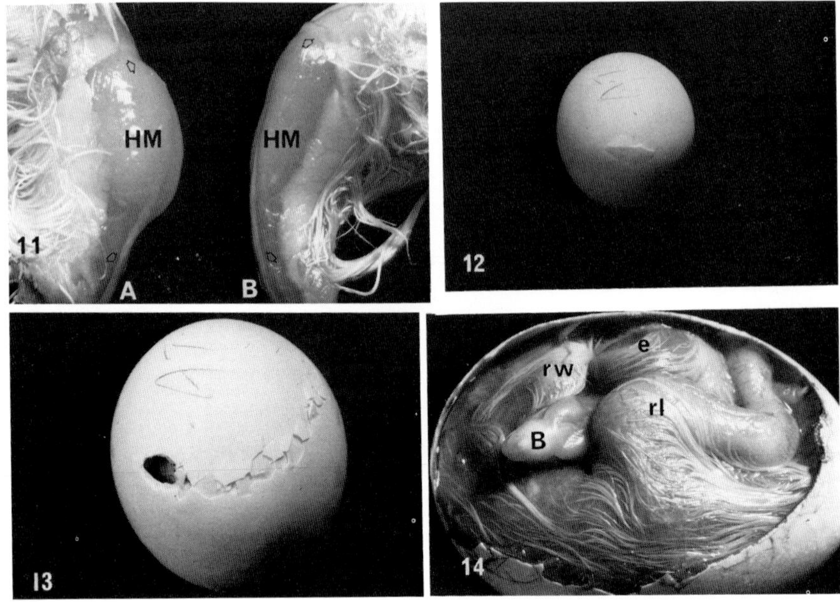

Plate III

FIG. 11. Hatching muscle of a domestic fowl embryo within 30 minutes after pipping (A) and about 48 hours earlier (B). HM, Hatching muscle.

FIG. 12. The first pipcrack in a domestic fowl egg.

FIG. 13. Climax rotation in a counterclockwise direction as seen from outside the egg looking at the blunt pole. Pipping occurred at the hole.

FIG. 14. Hatching position of the pigeon embryo at 16 days incubation. B, Beak; e, eye; rl, right leg; rw, right wing.

tion represents the onset of climax behavior in the domestic fowl, which has been described in greater or lesser detail by numerous authors (Balaban & Hill, 1969; Breed, 1911; Corner & Bakhuis, 1969; de Réaumur, 1750; Hamburger & Oppenheim, 1967; Kovach, 1970; Kuo, 1932; Spalding, 1875). Within a short time (10–15 minutes) strong coordinated struggling movements of the trunk, and specific flexions and extensions of the legs are added to the Type III climax head movements. It is at this time that the embryo begins to visibly rotate counterclockwise within the shell and to crack the shell around the circumference (Plate III, Fig. 13). At the initiation of climax, the yolk sac is usually entirely withdrawn into the embryo and the *chorio-*

allantoic respiration has almost ceased to function (Kuo, 1932a; Visschedijk, 1968a); the rate of *lung* respiration reaches a peak during climax (Table II). The coordinated Type III rotatory or climax movements occur regularly about once every 10–12 seconds, and are interspersed by periods of relative quiescence. Beak-clapping and vocalization also often accompany each climax movement sequence. Climax lasts on the average 49 minutes in the domestic fowl embryo (Table II), during which time the embryo rotates approximately two thirds of the way counterclockwise around the shell, pushes the shell cap off, and emerges (see the article by Corner, Bakhuis, and van Wingerten, this volume, for a more detailed discussion of emergence). Bock and Hikida (1969) have claimed that in the domestic fowl "the birds make only one opening in the shell with the egg tooth and break the rest of the shell with thrusts from their feet and wings [p. 102]." The above description, as well as filmed records, demonstrate that such a contention is clearly incorrect; the egg tooth (or beak) is involved in *both* pipping and climax via head thrust movements and Type III movements.

D. *Problems and Questions*

Numerous problems or questions can be asked concerning almost every aspect of the prehatching and hatching behavior described above. Several of these questions are delineated below.

1. Is prehatching and hatching behavior, as described for the domestic fowl, similar in all *birds*?
2. Are there any similarities in this behavior between birds and other egg-laying animals?
3. What are the mechanisms responsible for the onset and maintenance of coordinated Type III activity from day 16–17 to climax?
4. Does the intra- or extraegg *sensory* environment play more than a permissive role in prehatching or hatching behavior?
5. Do hormones have an active and direct role in the timing, manifestation, or maintenance of hatching behavior?
6. How are the modifications in gaseous exchange and the related metabolic changes during the final few days of incubation related to prehatching and hatching behavior?
7. What are the mechanisms responsible for the sudden onset and maintenance of climax behavior?
8. Are clearly identifiable precursors of posthatching behavior seen during prehatching and hatching stages?

The objective of the remaining portion of this article will be to attempt to carefully and critically present and examine most of the available evidence dealing with each of the above questions.

TABLE II

MEAN FREQUENCY, DURATIONS, AND INTERVALS OF VARIOUS PREHATCHING AND

Species	Age at climax (day and hour)	Interval between pipping and climax (hours and minutes)	SD	*n*	Respi-ration rate during climax (per minute)	SD	*n*	Duration of climax intervals (seconds)
Precocial:								
Domestic chick	21-3	20-8[b]	12-5	23	82.5	9.9	5	11.6
Mallard duck	26-7	31-11	8-45	12	92.9	13.6	9	11.1
Peking duck	27-9	33-47	9-20	5	80.8	—	4	12.6
Turkey	26-16	19.26	9-1	12	80.5	9.6	8	13.6
Bobwhite quail	24-12	40-54	7-22	19	79.5	10.4	11	6.7
Semiprecocial:								
Gull	24-5	40-57[c]	6-43	6	41.5	6.3	5	20.9
Altricial:								
Pigeon	18-3	33-38	7-32	8	34.8	5.7	11	11.4
Cardinal	12-1	—	—	—	31.0	—	3	13.3
Wren	13-3	15-26	—	4	—	—	—	12.5
Catbird	12-20	20-50	—	3	34.0	—	1	12.2

[a]The E and I refer to number of embryos and number of intervals actually recorded. Duration of climax intervals refers to the time between each climax movement.

[b]We recently have collected data on this *pip-climax interval* in a broiler strain of chicken (*Vancress arboracre*; $n = 137$) and have found a mean interval of 11.5 hours. Thus, there are clearly breed-specific differences in this and other hatching events.

II. Comparative Aspects of Prehatching and Hatching Behavior

A. Altricial and Precocial Birds

While intuitively it might at first seem that all birds should use the same behavior and mechanisms for escape from the shell, a consideration of the developmental differences among birds at the time of hatching (Nice, 1962) should, if not cause consternation, at least caution one to maintain an open mind on the matter. Due to the almost exclusive use of the precocial domestic fowl in studies of avian embryology, there is very little information available for altricial embryos. It is known that altricial embryos at the time of hatching have still not completed much of their development (Daniel, 1957; Nice, 1962). This developmental lag is clearly evident in the nervous system, where the state of neurogenesis of an altricial form at hatching has been compared

HATCHING (CLIMAX) EVENTS IN ALTRICIAL AND PRECOCIAL SPECIES OF BIRDS[a]

	n		Duration of climax			Rota-tion (in de-grees)			Head thrust		
SD	E	I	(minutes)	SD	*n*		SD	*n*	(per minute)	SD	*n*
7.7	5	407	49.0	18.2	8	250	57	8	6.6	2.3	5
8.5	10	505	49.2	32.4	11	273	72	5	7.5	1.7	10
8.5	4	164	60.4	27.9	5	260	60	5	5.4	3.6	5
8.3	13	1021	44.7	21.3	12	288	99.5	11	7.0	4.1	13
2.5	16	876	34.9	15.6	20	528	249	23	11.3	1.2	16
13.2	6	291	60.1	52.1	7	65	26.5	6	4.8	4.2	6
5.7	13	579	48.6	25.3	9	320	74.5	11	6.6	3.4	13
5.2	3	107	7.8	—	3	280	—	3	7.0	—	3
9.2	3	232	29.0	—	3	240	—	3	8.3	—	3
3.5	2	62	6.3	—	2	—	—	—	4.2	—	2

[c]Since, as was mentioned in the text, the gull often pips below the airspace and *prior* to tucking, the interval between pipping and climax was calculated from the time that the first pipcrack was made in the *airspace region*, making the data from the gull more comparable to the other species.

to that of a 12-day domestic fowl embryo (Sutter, 1950; Tuge, Kanayama, & Chang, 1960). There are also marked differences in the timing and sequence of myelination between atricial and precocial forms (Anokhin, 1964; Portmann, 1947; Schifferli, 1948).

Once again de Réaumur (1750) appears to be the first to have left a written record of his thoughts and observations on the comparative aspects of hatching. For example, he discusses the different problems faced by various birds, due to differences in shell thickness, the size, and shape of the beak, etc., and he actually observed that climax movements of a duck embryo were similar to those of the domestic fowl.

With the exception of a recent study of the prehatching and hatching behavior of altricial and precocial birds by the present author (Oppenheim, 1972b), all of the reports on species other than the domestic fowl are merely brief descriptions of those rather select aspects of hatching that can be observed without disturbing or opening the shell. Nevertheless these reports are often informative, and consequently I have chosen to review them below within the context of the various stages of prehatching and hatching behavior.

1. Pretucking

Except for the species presented in Table I, I have been unable to find any direct information on the exact position (posture) or the behavior of other embryos during this stage. However, since the species seen in Table I represent a rather diverse sample of the class Aves, and since their pretucking posture conforms closely to that depicted in Plate I, Fig. 2 for the domestic fowl, it seems reasonable to suppose that all birds assume a similar pretucking posture before the onset of prehatching and hatching behavior. This possibility appears all the more likely when it is realized that the assumption of this posture is dependent primarily upon passive gravitational adjustments within the egg; and that such adjustments are related to the yolk sac, the weight of the embryo, egg position, turning of the eggs, etc., factors that are all rather similar in most birds (Asmundson, 1938; Bolotnikov, Kamenskii, Afanaseva, & Yakonenko, 1970; Drent, 1967; Kuo, 1932a; Lind, 1961; Oppenheim, 1970). Since those embryos for which quantitative data are available (Table I) exhibit a higher rate of *overall* somatic motor activity at the pretucking stage than at any subsequent stage, it would also be interesting to know whether *all* birds are similar in this regard.

2. Tucking

With one exception, all the avian embryos for which adequate information is available appear to tuck their head under the right wing (Plate I, Fig. 3–6; Plate III, Fig. 14; Plate V, Fig. 17). In addition to those species noted in Table I, the following birds also appear to follow this rule: ostrich (Sauer & Sauer, 1966), pheasant and partridge (Asmundson, 1938), black-tailed godwit (Lind, 1961), coot (Steinmetz, 1930), black-headed gull (Kirkman, 1931), arctic tern, rhea, starling, merganser duck, Barrows goldeneye duck (Oppenheim, personal observations), herring gull and eider duck (Drent, 1967), and heron (Portmann, 1950). The exception mentioned above is the Megapode, *Alectura lathami*. According to Baltin (1969), these birds remain in the pretuck position until just shortly before emergence (see Plate I, Fig. 2).

Minor differences have been observed in the movements used to accomplish tucking between pigeons and gulls, on the one hand, and the domestic fowl, on the other (Oppenheim, 1972b). Although the pigeon is the only altricial species for which such information is available, it is possible that other altricial forms might also exhibit similar differences. However, because the end result is the same in all the above species, i.e., tucking the head under the right wing, and since Types I and II movements *alone* could not accomplish tucking, the description of tucking behavior (Type III) presented above for the domestic fowl is probably basically correct for all other avian embryos that are found to tuck the head under the right wing.

3. Membrane Penetration (MP)

In the domestic fowl, and also among other avian species, MP generally occurs prior to pipping, but after tucking and draping. However, there are individual exceptions to this, even in the fowl (El-Ibiary, Shaffner, & Godfrey, 1966; Freeman, 1964; Oppenheim, personal observations), in which occasionally a crack occurs in the shell *without* the embryo's beak having penetrated into the airspace. This, of course, happens routinely in most fowl eggs that are in malposition II (head in the small end of egg). The occurrence of pipping, *before* membrane penetration, in embryos that are in a *normal* position, however, is the rule rather than the exception in some species: pigeon (Oppenheim, 1972b), herring gull (Drent, 1967), black-tailed godwit (Lind, 1961), coot (Steinmetz, 1932), laughing gull (Oppenheim, 1972b). A discussion of this phenomenon is presented in Section II, A, 4, below.

The actual act of membrane penetration has been observed in only a few embryos representing four species: domestic fowl, turkey, bobwhite quail, and pigeon (Oppenheim, 1972b; Hamburger & Oppenheim, 1967). Yet, these observations are all in agreement that over a long period of time the membranes are first worn thin over the tip of the beak until, at one point in time, almost any kind of a head or beak movement serves to finally penetrate the beak into the airspace. At first only the tip of the beak is exposed in the airspace, but later, as a result of slight shifts in the embryo's posture, the entire beak down to the nares becomes exposed. In all the species examined so far, respiratory movements begin just prior to, or after, membrane penetration. Just how much gaseous exchange is being accomplished by the lungs at this time is not entirely clear, since these initial respiratory movements are rather sporadic, irregular, and low in rate (Kuo & Shen, 1937; Oppenheim, 1972b; Vince & Salter, 1967). Beak-clapping occurs in all precocial and altricial species I have observed at this time (Table II). *Prior* to draping or MP the beak can often only be clearly observed after cutting of the C–A membrane and manipulation of the embryo. For this reason no attempt was made to observe beak-clapping any earlier in many of the species in Table 2. Beak-clapping rate tended to be consistently lower in all the altricial forms.

Vocalizations, either spontaneous or evoked, first occur in the embryos of most precocial species sometime between MP and pipping (Collias, 1952; Goethe, 1955; Gottlieb & Vandenbergh, 1968; Lind, 1961; Oppenheim, 1972b). Altricial embryos are somewhat retarded in this regard. I have never heard vocalization, nor have I been able artificially to evoke it, prior to the final climax stage of hatching in altricial forms, although the intensity may be so low before this that I simply missed any such vocalizations (Oppenheim, 1972b).

Another prominent sound that is produced by avian embryos is clicking. In some species such clicking may be heard soon after MP (Kear, 1967), while in others (chick, quail) clicking may not be heard until 12–15 hours before emergence (Vince, 1969). Margaret Vince (1969, 1970; also see Vince, this volume) has reviewed her own work and that of others regarding the function and mechanisms of clicking in birds, thus alleviating the need for an extensive treatment here (but see Section III, A, 4 of this chapter). Suffice it to say that from her work with quail, it is now clear that clicking, by accelerating or, under special experimental conditions, delaying development through the communication of sensory information between eggs, serves to synchronize the time of hatching, such that all embryos in a clutch hatch within a short time of one another. It is interesting that the hatching time of a species that apparently does not normally synchronize its hatching (domestic fowl) can be modified by exposure to artificial clicking (Vince, Green, & Chinn, 1970). Clicking seems to be a special type of respiration that is intimately associated with the function of the glottis (McCoshen & Thompson 1968a, 1968b).

4. Pipping

There is practically no data available on the duration of the interval between MP and pipping. Vince (1970) has estimated this interval to be about 10 hours in the bobwhite quail. The data for the domestic fowl appears to be conflicting. For example, Vince (1970) cites an unpublished dissertation of Freeman (1964) as evidence that the MP–pip interval is 8 hours in the fowl. However, in an earlier published report, Freeman (1962) stated that this same interval was only 3–4 hours in the fowl. On the other hand, El-Ibiary *et al.* (1966) have reported an approximate MP–pip interval of 5–6 hours for the fowl, and McCoshen and Thompson (1968b) find a similar interval (4–6 hours). Such inconsistencies may result from the use of different strains of fowl (it is known that the pip-hatch interval can vary as much as 10 hours between strains [Oppenheim, 1972b]), and from the reliance of the above authors on *indirect* measures of MP, such as onset of recordable respiration, vocalization, and oxygen uptake. The mean value of 10.0 hours obtained by the present author (see footnote to p. 170) is based on a more direct measure of MP and thus probably represents a rather more reliable estimate of the MP–pip interval for the strain of fowl used (white leghorn).

Many authors, including the present one (see Hamburger & Oppenheim, 1967), who have dealt with embryonic behavior or hatching in birds, have either stated or have given the impression that sometime in the past, someone has claimed that the domestic fowl embryo pecks its way out of the shell, the implication being that the movements used for getting out of the shell (pipping and climax) are the same as those used for obtaining food after hatching. In the preparation of the present article I have made a rather careful attempt to locate this villain, but to no avail! The closest I have come to locating such a claim are in the following quotes. First from Helmholtz, as cited in Spalding (1875): "The young chicken very soon pecks at grains of corn, but it pecked while it was still in the shell. . . . [p. 507]," and then from Windle and Barcroft (1938): "A brief rapid series of ineffective *pecking* movements like those used to pip the shell sometimes occurred [p. 686]." These statements, which, incidentally, are partially correct in that certain muscular components of the pecking act undoubtedly occur in the egg (see Breed, 1911; Kaspar, 1964; Kuo, 1932b), clearly make no explicit claim for the chick pecking its way out of the egg.

Pipping in the domestic fowl was first correctly attributed to lifting or backthrust movements of the head and beak, away from the breast and toward the shell, by de Réaumur (1750) and later confirmed by other authors (Breed, 1911; Rennie, 1833; Spalding, 1875). Subsequently, I have observed that many other species use similar movements for pipping (Oppenheim, 1972b). However, as mentioned above, in two species examined in that study, pipping (first crack made in shell) occurred *prior* to membrane penetration, and in at least one of these two (laughing gull), pipping frequently occurred during the tucking stage. Thus, here is at least one case of pipping occurring before the completion of tucking. Although the movements involved in "early" pipping of this kind have never been directly observed it is likely that, since tucking and pipping movements are both of the Type III variety, the movements responsible for early pipping in the gull are probably not so different from those used by the domestic fowl. A more important question, however, is whether the stimuli or mechanisms underlying pipping differ between gulls (as well as other species exhibiting the same "early" pipping phenomenon) and the domestic fowl. This problem is dealt with below in Section III.

Although it is not entirely clear whether it is the musculus complexus ("hatching muscle"), or the other dorsal cervical muscles that are responsible for pipping (see below), it is likely that the muscles used during pipping and hatching are different from those used for pecking, since the directionality of movement is different in the two cases (Kaspar, 1964). Thus, the recent suggestion that breaking out of the egg in the pigeon represents a "pecking experience" probably has no factual foundation (Wortis, 1969).

After the first crack has been made in the shell it is common, among those species that pip after MP (e.g., the fowl), for no further cracks to be made in the shell until the final climax stage of hatching ensues (Oppenheim, 1972b). This first crack generally becomes enlarged, however, and often a hole is made at this site. In those species that commonly pip *prior* to MP (e.g., the gull), the situation is somewhat different. The following description is taken from Drent (1967) for the herring gull, but the situation is apparently similar in the black-headed gull (Kirkman, 1931), the godwit (Lind, 1961), the laughing gull (Oppenheim, 1972b), and the ringed plover (Laven, 1940). Approximately 64 hours prior to emergence the first crack appears at a point 75° to the right of vertical, as one views the blunt pole of the egg (Plate IV, Fig. 15). Then, during the next 35 hours, a series of such cracks extend from the first breaking point obliquely upward or counterclockwise. After an average of 60° has been traversed, a hole appears (Plate IV. Fig. 16). Gradually the area of breakage is extended and reaches an average of 115°, at which time the embryo is able to push the shell cap off and emerge. I have suggested (Oppenheim, 1972b) that the primary reason the gull and other species noted above pip the shell in this way, and not like the domestic fowl, might be due to the differences in both the dimensions of the beak and the strength of the shell. The gull's long narrow beak, coupled with its relatively thin shell (Fisher, 1966), makes it likely that the early series of cracks result from tucking movements and from head thrusts which occur during the change in position of the head and trunk after tucking. The path traversed by the cracks (Fig. 16) lends support to this notion. Thus, if it were possible to lengthen and narrow the beak of the domestic fowl, and to weaken its shell, it might also exhibit a series of cracks in the shell similar to the gull. It is also possible that species differences in the size of the embryo relative to the size of the shell at this stage may play a role in this phenomenon (see Corner *et al.*, this volume).

In all birds that have been examined, the first crack in the shell (pipping) almost always occurs on the side of the egg that is uppermost in the nest or incubator (Drent, 1967; Kovach, 1968, 1970; Oppenheim, 1970, 1972b; Poulsen, 1953), and this crack is generally located more toward the blunt or airspace end of the egg. The possible mechanisms underlying this phenomenon are discussed in detail below (Section III, A).

Although the size of the so-called hatching muscle (musculus complexus) is drastically increased around the time of pipping in most, if not all birds (Fisher, 1966; Oppenheim, 1972b), the precise function of this muscle, and the factors responsible for its increase in size during this time, have not yet

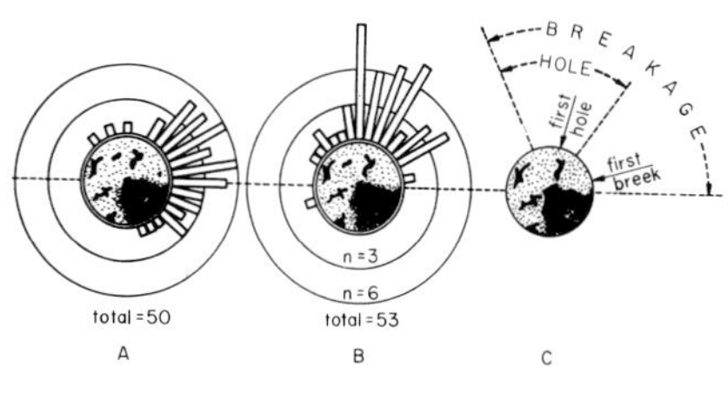

15

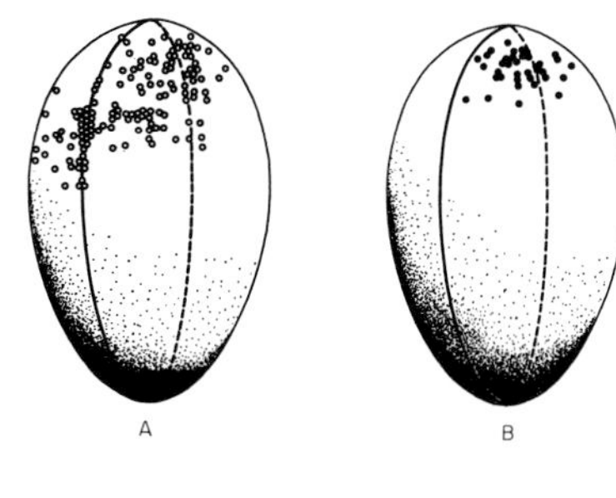

16

Plate IV

FIG. 15. Events during hatching of the herring gull with respect to the position of the egg in the water dish. The egg is seen from the blunt pole and the horizon indicated by the broken line. (A) first break, (B) first hole, (C) summary for 53 eggs showing means from A and B, and in addition the mean arc of breakage (115°) and span of the hole (60°) attained before the cap is broken off and hatching accomplished. From Fig. 20, Drent (1967).

FIG. 16. Measurements made on 36 herring gull eggs during hatching. (A) shows the progress of cracking in relation to the first crack made, plotted along the solid line. Each dot represents an observation of the position of the billtip. The broken line shows that portion of the egg which, on the average, lies uppermost in the nest. (B) shows the site of the first hole, each dot representing the measurement for one egg. From Fig. 21, Drent (1967).

been unequivocally determined (see below). The actual movements that result in pipping have been observed in a few domestic fowl, peking duck, laughing gull, pigeon, and quail embryos (Brooks & Garret, 1970; Hamburger & Oppenheim, 1967; Oppenheim, 1970, 1972b). However, in the absence of direct electromyographic recordings during this event, it is impossible to know whether it is the hatching muscle or the other dorsal cervical muscles that are actually contracting during this time.

While the frequency of general somatic movements (Types I, II) remains at about the same level during the pipping stage, beak-clapping and respiration rate have been found to increase during this time (Kuo & Shen, 1937; Oppenheim, 1972b; Vince & Salter, 1967; Vince, 1972). Head-thrusting (or head-lifting) movements, which are a component of Type III activity, also exhibit a rise in frequency during this time in all species that have been examined (Hamburger & Oppenheim, 1967; Oppenheim, 1970, 1972b; see Table II).

The final hatching position, which is generally attained by the time of pipping, is similar in altricial and precocial embyros, although slight variations exist due to the relative size of different body parts (Plate II, Fig. 8, 9; Plate III, Fig. 14; Plate V, Fig. 17; Plate VI, Fig. 19). The aforementioned difference in beak-clapping rate between precocial and altricial species was also found to persist during and after the pipping stage (Table II).

The form or character of beak-clapping has been found to be rather different between the chick, quail, turkey, and altricial species on the one hand, and duck and gull embryos on the other. In the latter, the movements consisted of very fast opening and closing, with the mandibles usually opening only slightly; whereas the other species did not beak clap as rapidly, and

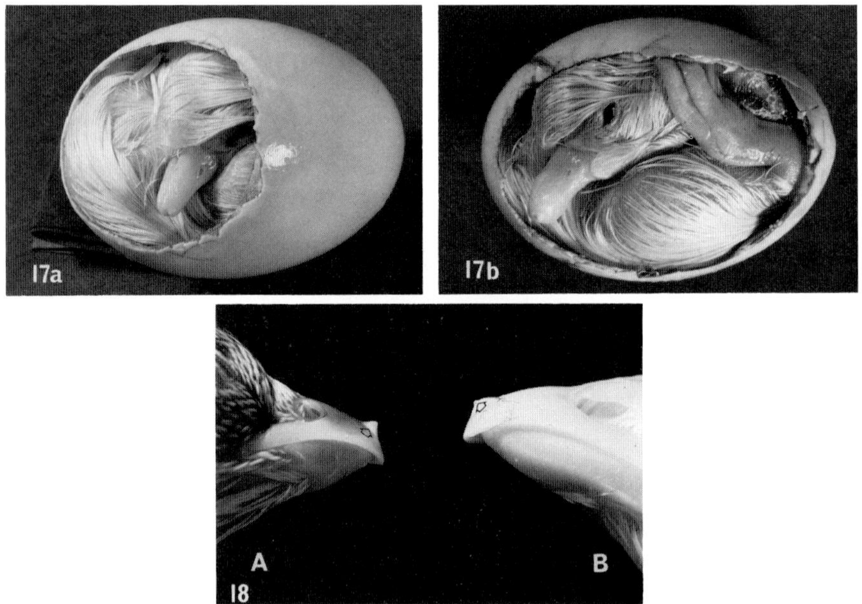

Plate V

FIG. 17a. Hatching position of duck embryo as seen from the airspace.

FIG. 17b. Hatching position of duck embryo as seen from the side of the egg.

FIG. 18. Egg caruncle ("tooth") of quail (A) and domestic fowl (B) showing the sharper and more prominent caruncle of the quail.

the beak often opened somewhat wider. Furthermore, the altricial forms tended to have shorter and less rapid bouts of beak-clapping. Gaping is exceedingly rare in all species examined, and we are reasonably certain that no differences exist in the frequency of gaping between altricial and precocial embryos. This is in agreement with Kaspar's (1964) observations.

Finally, with the exception of the observation that a small percentage of domestic fowl embryos with a genetically determined short upper beak (and thus lacking an egg caruncle) are able to pip and hatch (Landauer, 1967), there appears to be no *experimental* evidence for or against the role of the avian egg caruncle ("egg tooth") in pipping, or in the other phases of prehatching and hatching behavior (i.e., MP, climax).

5. Climax and Emergence

With only one or two exceptions (the megapode, and possibly the rhea, both of which are discussed below), all birds, precocial and altricial, that have been examined so far, exhibit an interval between the time of pipping and climax or emergence ("pip–hatch interval"). However, the time between these events is rather variable both between and within species. A partial list of the available data on this interval can be found in Vince (1970), and in Table II. In addition to the information found in those two sources the following, perhaps less reliable information would appear to exhaust the available data for this interval. Gross in Bent (1932) reports that the pip–hatch interval in the greater prairie chicken is about 24 hours; Sauer and Sauer (1966), in a study of the African ostrich, state that "the hatching effort may drag on for hours or even days [p. 69]." Although it is not entirely clear whether the Sauers refer here to climax, or to the period between pipping and climax, they earlier mention "the first crack in the shell [p. 66]," which might be taken to mean pipping. Whitman (1919) observed that the pigeon pips about 24 hours prior to hatching; Laven (1940) reports a period of 48–72 hours between the first crack and emergence in the ringed plover; Allen (in Bent, 1923) observed that *one* black duck merged pipping with climax, such that there was no interval between the two. This occasionally happens in individuals of other species, e.g., quail (Abbott & Craig, 1960), domestic fowl (Hamburger & Oppenheim, 1967). This fact, plus data available on the pip–hatch interval for other species of the subfamily Anatinae (Fisher, 1966; Table I), makes it likely that this was an aberrant individual observed by Allen and not representative of black ducks. And finally, a recent report by Wayre (1966) has indicated that *Rhea americana* may normally merge pipping with hatching. He found that of four eggs checked at 1900 hours only one had pipped, but by 2400 hours all were hatched.

Although in many species the "pip–hatch" interval is a period during which no *new* cracks are made in the shell (although pieces of shell may be broken away at the original site of pipping during this time), other species continue making a series of cracks, that usually proceed in an orderly manner across the shell (Figs. 15 and 16). This latter situation appears to be most commong among the order Charadriiformes, as it has been observed in the herring gull (Drent, 1967), the laughing gull (Oppenheim, 1972b), the godwit (Lind, 1961), and the black-headed gull (Kirkman, 1931). It has also been observed in the altricial pigeon (Oppenheim, 1972b). It has been suggested (Vince, 1970) that those species that make an extensive series of cracks or pips may not have a period of relative motor quiescence between pipping and hatching, as has been documented for other embryos (Oppenheim, 1972b). The implication here is apparently that these embryos are in a more or less continuous climax-like state from the first pip to emergence. This seems rather unlikely for the following reasons: (1) observational recordings of somatic motility of Types I, II, and III from laughing gull embryos during this period demonstrate that they do have a period of motor quiescence, or a lowered rate of activity, compared to earlier stages (Table I); (2) although there is a low rate of occurrence of Type III movements during this time (especially head thrusts), these same recordings from the gull did not indicate a *continuous* state of climax. As was suggested above, the presence of a series of cracks in the shell of these species prior to climax appears to result not from any marked differences in their behavior during this time, but rather from the differences in the dimensions of the beak and the strength of the shell. Furthermore, although there appears to be little, if any, rotation during the climax stage in these species, this does not mean that the motor activity (Type III movements) during this time is somehow different. Because of the prior weakening of the shell (from the extensive series of cracks) the shell cap is pushed off before very much rotation can be manifested. Artificially strengthening the shell with tape, however, clearly shows the capability of the laughing gull embryo to rotate within the shell during climax (Oppenheim, 1972b).

Although there appear to be no major changes in the rate or frequency of overall somatic motility between pipping and the onset of climax in those species that have been observed (Tables I and II), in the absence of either continuous recordings or a series of sequential recordings during this time, such a statement may be misleading. For example, it has been found that about 16 hours before emergence (and after pipping) the peking duck embryo begins to exhibit a higher frequency of uplifting, or backward-directed head movements (Oppenheim, 1970). Other species also exhibit a higher frequency of similar head movements prior to climax (Tables I and II). Thus, by directing attention to *specific* types of movements, important changes in behavior might be found to occur during this period (e.g., see Balaban & Hill, 1969).

Respiration rate, which shows a rather steady rise toward the end of incubation (Table I; Kuo & Shen, 1937; Romijn, 1948), increases significantly during climax in all species, except the bobwhite quail (Table II; Vince, 1969). There is also an increase in the occurrence of both eye-opening and vocalization during climax and emergence (Balaban & Hill, 1969; own observations), but no apparent change occurs in the rate of beak-clapping.

The transition from the preclimax (or postpipping) to the climax stage occurs rather suddenly in many species (Oppenhein, 1972b). Except for the occasional appearance of coordinated Type III movements prior to climax, this transition represents a modification in overt motor behavior from a prominence of jerky, convulsivelike movements (Types I and II), to an almost exclusive incidence of stereotyped and coordinated Type III movements. Outwardly the onset of climax is manifested by a continuation of the cracks in the shell. Such a transition was seen in all those species presented in Table II. There were some individuals in many of the species that entered the climax stage, only to later return to either the preclimax type of activity (Types I and II), or to short periods (> 1 minute) of complete quiescence (except for breathing and beak-clapping). This was especially striking in the laughing gull embryos. In some of these gull embryos such alternations between the two stages occurred a number of times before emergence. However, with this exception, most individuals of the various precocial and altricial species only terminated the climax stage by emergence from the shell.

The actual motor pattern utilized during climax, while classified as a Type III activity, nevertheless differs in certain respects from movements seen prior to this time. Climax movements have been described in detail for the chick and duck (Hamburger & Oppenheim, 1967; Kovach, 1970; Oppenheim, 1970) and appear to be similar for all the other species presented in Table II (but see p. 191). The sequence begins with a slight withdrawal of the head and beak away from the shell, followed by a specific pattern of extension and flexion of the two legs at the tarsal joints together with a powerful thrust of the head and beak backward toward the shell. After 10–20 minutes a powerful rotatory trunk movement ("wriggling") is added to the repertoire, and it is at this time that the embryo first begins to visibly rotate within the shell (see Hamburger & Oppenheim, 1967 for a more detailed description of these events). Vigorous steppinglike movements at the tarsal joints also accompany these rotatory trunk movements. When viewed from the blunt end of the shell, the rotation of the embryo is outwardly seen as a series of cracks leading around the circumference of the airspace in a counterclockwise direction. Each entire climax movement sequence lasts 1–3 seconds and is followed by a quiescent interval whose duration is highly stereotyped for each species (Table II). The amount of rotation that is necessary before the shell cap can be pushed off varied from 65° in the gull (starting from the point of the first pip crack made in the airspace), to 528° in the bobwhite quail (also seen Johnson, 1969), with 279° being the interspecies average (see Table II). The extensive rotation by the quail is undoubtedly related to the difficulty in tearing the thicker shell *membranes* (Romanoff & Romanoff, 1949; Vince, 1970), since the *shell* becomes cracked during the first 360° rotation. The quail's egg caruncle seems better constructed to handle the thicker shell membranes (see Fig. 18). The time necessary to complete climax and emergence also varied substantially between species, from 6 minutes for the catbird to 60 minutes for the gull and peking duck (Table II). We have no data on the exact time at which the yolk sac becomes *completely* withdrawn into the body, or when the ductus arteriosi become occluded, shifting all blood flow through the lungs, in the various species. However, in the domestic fowl (Hamburger & Oppenheim, 1967; Kovach, 1970), these events occur around the time of climax onset.

The degree to which species of birds other than those discussed above follow this same pattern of hatching (climax) is difficult to assess from the nature of some of the reports in the literature. However, based on the available information, it appears that climax behavior, similar to that described above, is also found in the following species: greater prairie chicken (Gross, in Bent, 1932), dove (Craig, 1912), black duck (Allen, in Bent, 1923), broad-breasted bronze turkey and ring-necked pheasant (Abbot & Craig, 1960), Japanese quail (own observations), partridge (Asmundson, 1938), godwit (Lind, 1961), coot (Steinmetz, 1930, 1932), and the black-headed gull (Kirkman, 1931).

Perhaps even more interesting for the present discussion would be the discovery of a species in which hatching is accomplished *differently* from that described above. Although de Réaumur (1750) briefly and tangentially discussed the possibility that hatching might be different in species such as the ostrich with its exceedingly thick shell, it was really Wallace Craig in 1912 and Kirkman (1931) who first discussed this possibility seriously. Indeed, there is documentation (some of which is rather anecdotal) that significant differences may exist in the hatching behavior of the ostrich and other species, as well.

Hudson (1895) describes the hatching of the American jacana as follows: ". . . while I was looking closely at one of the eggs lying on the palm of my hand, all at once the cracked shell parted, and at the same moment the young bird leaped from my hand and fell into the water . . . it swam rapidly to a second small mound [p. 112]." It is impossible to know, from this description, whether the "cracked shell" was only pipped or whether a longer series of cracks were already present. This "exploding" from the shell has also been described by Moore (1912) for embryos of the least sandpiper. He states that ". . . one egg hatched before my eyes. It broke open violently, as if by explosion, the two sections shooting to opposite sides of the nest and the new youngster burst violently into the world [p. 218]." Even though these two observations may only indicate that at the time of emergence (and at the end of a typical climax stage) the embryos of these species are extremely precocious, they certainly deserve to be followed up by a more systematic study.

In the past it has been supposed that the thick shell of the ostrich (and other Ratites) posed problems that might require a unique means of hatching. Unfortunately, there is practically no information available on hatching in these birds. However, the position of the ostrich embryo just prior to hatching is similar to that seen in other birds (Beebe, 1906; Fig. 19); and Sauer and Sauer (1966) have reported that as in the domestic fowl, the first crack in the shell (pipping) is made by "head thrusts." They described a "small window" being made in the shell over the beak by such head thrusts (see Fig. 20 and 21). The major behavior pattern used to complete hatching in the ostrich is described as a "stretching of the body and kicking downward." Such movements result in a fragmentation of the entire shell. Rarely does an ostrich embryo emerge without breaking the shell into many pieces. The authors make no mention of whether rotation occurs during hatching. From personal observations of *Rhea americana* eggs close to the end of incubation I have found that the shells, in addition to being thick, are quite brittle such that it would be difficult for the embryo to chip away a clean-cut intact shell cap, as happens in most other birds. In other words, even if the behavior of the ostrich during climax were identical to the domestic fowl, the shell cap would still most likely be broken into many fragments at the time of emergence.

Two members of the sandpiper family, the woodcock and the willet, have been reported to use a rather unique means of breaking the shell during emergence (Wetherbee & Bartlett, 1962). After pipping a rather large hole in the shell, the embryos brace their beak in the opening and with "large convulsive movements" rip the shell open along the *longitudinal* axis, rather than chipping and opening the shell around the circumference of the airspace, as in most other birds. Long spinal processes on the cervical and thoracic vertebra, and an egg carunclelike structure on the *lower* beak apparently aid in this longitudinal tearing of the shell. Although from the brief description of this behavior by these authors, it is not possible to decide whether hatching behavior in these two species is truly different, it is quite likely that the convulsive movements these authors mention are typical climax movements, which, because of the tearing of the shell by the specialized spinal vertebral processes, are never able to manifest themselves in the typical counterclockwise rotation. Applying tape to the shell to prevent the longitudinal tearing might settle this question.

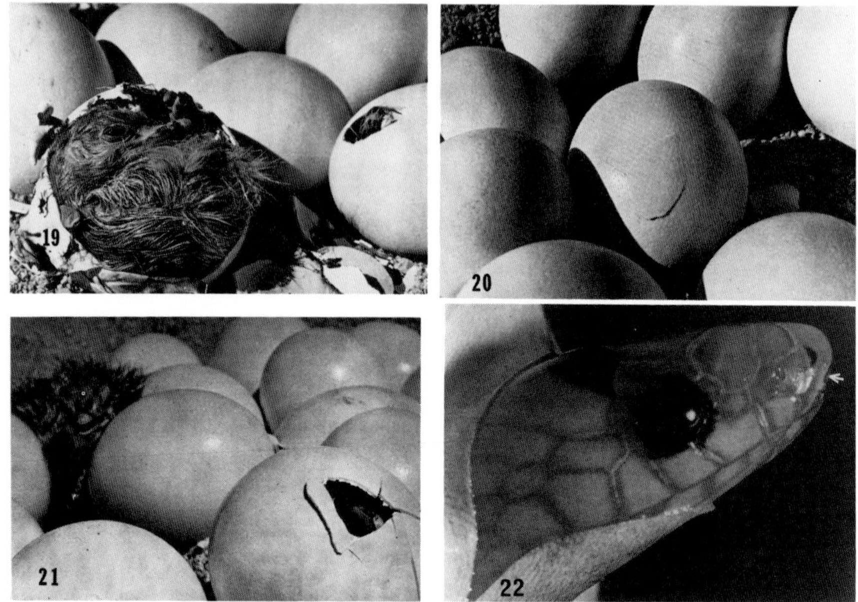

Plate VI

FIG. 19. Apparent hatching position of the African ostrich. From Fig. 27, Sauer and Sauer (1966).

FIG. 20. First pip crack in shell of African ostrich. From Fig. 24, Sauer & Sauer (1966).

FIG. 21. Hole made at site of pipping of African ostrich. Note beak in hole. From Fig. 26, Sauer & Sauer (1966).

FIG. 22. Snake embryo with head protruding from the shell showing the triangular egg tooth (arrow). From Time-Life, Inc. *The Reptiles*.

The apparent peculiarities in the hatching behavior of the several species discussed above, although in need of more systematic verification, can, in my opinion, be more easily related to either the degree of precocity of these birds (Nice, 1962), or to morphological differences, and thus would not appear to represent a serious divergence from the hatching behavior seen in the majority of birds. Therefore, the question still remains as to whether any living avian species differs *markedly* in its method of hatching or emerging from the shell. From brief reports concerning hatching in the megapodes (Bertling, 1904; Frith, 1962), it has been conjectured that these birds might, indeed, present striking differences in their hatching behavior. However, until the recent interesting report of Baltin (1969), the extent of this difference was not at all clear. The following description is a condensation from Baltin.

Baltin studied captive individuals of one species of megapode, the brush turkey, *Alectura lathami*. These birds are found on the eastern coast of Australia, and are primarily characterized by the reptilianlike habit of laying their eggs in holes in the ground; the eggs are then covered by large mounds of decaying vegetation which produces natural heat for incubation. The young, after hatching, must make their way to the surface unaided. Generally 25–30 eggs are laid and these are positioned vertically, with the large end facing upward. The eggs are not turned during the 45- to 55-day incubation period nor are they in contact with each other. When incubated artificially they require a temperature of 33–34°C and a relative humidity of almost 100%. Apparently, any exposure of light during incubation interferes with hatching. Unlike the eggs of all other birds, the egg of the brush turkey has a *movable* airspace that always assumes a position at the uppermost part of the egg. This is due to the presence of the airspace *under*, rather than between, the two shell membranes, as in other bird eggs. The airspace is also quite small (1 cm^2). As late as a few hours before hatching (emergence) the head of the embryo is between the legs and still points toward the small end of the shell, similar to the *pretuck* position of other bird embryos (Plate I, Fig. 2).

A day or so prior to hatching, movements and tremors can be clearly seen to occur within the egg. Later the embryo attempts to break the shell by bracing its back against the shell and making forceful movements with the legs and wings. The head is still buried between the legs at this time. After many of these forceful movements, cracks begin to appear at numerous points on the shell. The legs break out first, and only after a large amount of shell is broken away does the chick lift the head and fully emerge. The hatching process was timed in two cases and took 60 and 48 minutes respectively. The hatching muscle remains relatively small in this species, and while an "egg tooth" or caruncle is present at the midpoint of incubation (day 20–21), it is subsequently lost prior to hatching. Breathing apparently does not begin until the final hatching process is underway. During hatching the entire shell is broken into very small pieces (1 cm^2). Bertling (1904) has reported that the shell of *Alectura* is rather thin, which may account for such extensive breakage.

Thus, although the description above is incomplete, it seems likely that not only is the prehatching and hatching behavior of *Alectura* (and probably other megapodes, as well) markedly different from that of any other bird, but because of this behavioral difference, the underlying mechanisms are also probably different. It would be well to keep this in mind when reading the sections below on mechanisms of hatching, since most of the studies discussed there have used the domestic fowl embryo.

B. Hatching in Other Vertebrates and Invertebrates

As was obvious from the above discussion, the systematic investigation of hatching behavior in birds has been, for the most part, neglected. Yet, if possible, the situation for other egg-laying animals is even worse. I have been able to find only a few studies whose primary concern was with hatching behavior and/or its mechanisms in fish, amphibians, reptiles, or the egg-laying monotremes. The situation in invertebrates is only slightly better. In most cases investigators have been content to simply state that "the embryo hatched or emerged." Thus, in a review such as the present one, one has the choice of either simply ignoring all other nonavian animals, or of trying to compare hatching in birds with other forms in the face of frustratingly brief, incomplete, and often anecdotallike accounts of hatching in these nonavian egg-laying animals. Because a consideration of the few apparent similarities that do exist in the hatching of birds and other forms may serve to generate some interest—and perhaps even an investigation or two—into this problem, I have chosen the latter course.

1. Fish

There is good evidence that in some, if not all, fish specialized glands of endodermal origin secrete a proteolytic hatching enzyme which breaks down or weakens the egg capsule membranes just prior to emergence (Ishida, 1944). In the fresh water killifish, *Oryzias latipes* (Medaka), studied by Ishida, respiratory movements are apparently important in rupturing the hatching glands so that secretion of the hatching enzyme occurs. Although active somatic movements of the embryo are also said to be important for actual emergence, Ishida (1944) does not describe the type of movements used by this fish during hatching, nor whether anesthetized embryos can hatch (see below).

Tracy (1926) found that as hatching approaches in the toadfish, *Opsanus tau*, the egg membranes become weaker, while the movements of the embryo are increasing in vigor. Strong side-to-side coils or swimming movements eventually result in a tear being made in the membranes. Usually it is the tail which breaks out first, and only later, as a result of more vigorous movements, do the head and trunk emerge. The coils or swimminglike movements that result in emergence are no different than the movements seen just prior to, or after, emergence. It is apparently the weakening of the egg membranes that allows these movements to be effective in hatching. It is interesting that at the time the toadfish hatches, it is not capable of responding to any kind of external sensory stimulation.

Gideiri (1966) has briefly described the hatching of a number of teleost fish. *Salmo trutta* (trout), *Trachinus vipera* (lesser weever), and *Motella cimbria* (rockling) embryos are all reported to break through the egg membranes by rapid bursts of coillike or swimming movements. These bursts occurred at intervals of 0.5–18 minutes until a tear was made in the membranes. The final exit from the egg was brought about by "violent" movements of the body. Hatching in *Salmo solar* was rather similar to this description except that after the initial tear was made in the membrane, the final emergence was a gradual process which laster 30 minutes to 3 hours, with no violent movements.

A quiescent period, just prior to hatching, has been reported by Kyûshin (1968) for *Hemitripterus villosus* embryos (a cottid fish), in which the vigorous twitching movements seen earlier almost cease entirely.

Thus, hatching in fish differs markedly from birds, as might be expected of an anamniote with such striking morphological differences, and which also lacks a hard calcareous shell. However, the most interesting point about hatching in fish is that, unlike birds, they apparently do not utilize any new or special kinds of behavior for escaping from the egg capsule. This strongly suggests that, with the exception of the hatching enzymes, they do not develop any special physiological or behavioral mechanisms associated with, or needed for, hatching.

A note of caution should perhaps be added here concerning the relative role of active somatic movements *versus* the hatching enzymes in the hatching of fish. Wintrebert, as cited by Needham (1931), has found that active movements are *not* necessary for hatching in several fish, including the rainbow trout, *Salmo irideus*, and the carp, *Carassius auratus*. When placed in an anesthetic at the time of hatching these embryos emerge (hatch) as well as unanesthetized ones. Thus, the hatching enzyme alone is sufficient for hatching in these fish. The extent to which these findings may apply to other fish, including those discussed above, can only be determined analytically and not by observation alone. For instance, Armstrong (1936) found that both movement and the hatching enzyme were necessary for hatching to occur in *Fundulus heteroclitus*.

2. Amphibians

While there are only a few useful reports of hatching in amphibians the situation appears, for the most part, to be similar to that just described for fish.

Shortly before hatching, in many species of Anura, a series of unicellular glands appear on the snout of the embryo or larva. These glands function in producing a secretion which "digests" the egg capsule (Noble, 1926), thereby greatly weakening it. Although it is often stated or assumed that the larvae then use active somatic movements to complete the hatching process (Bragg, 1940; Fernandez-Marcinowski, 1921), experiments referred to above by Wintrebert seriously question this assumption, at least for certain species. Both the salamander, *Ambystoma tigrinum*, and the toad, *Alytes obstetricans* are able to hatch when anesthetized, although they tend to get stuck in the jelly surrounding the eggs.

In the bufonid genus *Eleutherodactylus*, the eggs are laid on land, and development is entirely enbryonic, i.e., there is no larval or tadpole stage (the young hatch out as tiny frogs). These forms have a spinelike projection on the tip of the upper jaw (egg tooth?) which apparently aids in slitting open the rather tough egg capsule. Although no detailed descriptions of the actual hatching process are available, Hughs (1966), while observing embyronic behavior in *Eleutherodactylus martinicensis*, has reported that, as in bird embryos, the frequency of active somatic movements of all kinds decreased in these embryos as hatching approached, but that just prior to hatching *spontaneous* "stepping" or ambulatory movements became more frequent than at prior stages. *Reflexogenic* "repeated kicking" movements are also prominent at the time of hatching. Lutz (1944) has reported that at least in some individuals of this genus the "egg tooth" is actually used to break the egg capsule. Because of its rather unique mode of development, making it more similar to bird embryos than other amphibians, a detailed description of hatching in *Eleutherodactylus* would be most desirable.

3. Reptiles

The well-known phylogenetic relationship of birds and reptiles (especially crocodiles, see Walker, 1972), suggests that a careful comparison of the methods and mechanisms of emerging from the egg by reptiles might reveal interesting and useful similarities. Again, however, the lack of information necessitates that, at the present time, any such comparison be rather superficial.

Lizards and snakes develop a true *egg tooth* that is attached to the premaxillary bone in the midline and which curves forward in front of the snout (Plate VI, Fig. 22). Its edge is flattened and quite sharp. However, the crocodilian and turtles (like birds) have an *egg caruncle* that is formed from a thickening of epidermis at the tip of the snout (Plate VII, Fig. 24). Both of these structures are shed soon after hatching (Bellairs, 1970).

Hill and DeBeer (1950) have observed that the python snake uses its egg tooth to slit the egg shell by slashing movements of the head. Pope (1946) has described the hatching of snakes as follows: "When the young snake is ready to emerge it makes many slits in the shell with the egg-tooth and then pokes its snout through the area thus weakened. It does not crawl out immediately but lingers, thrusting first its snout and then its head in and out many times.... There is much variation in the length of time that elapses between the first slit of the egg-tooth and final desertion of the egg [p. 73]." An African rock python, observed by Pope, took 3 days from the first slit to emergence; in other snakes this interval has been found to vary from hours to 2 days (see Smith, 1954). The interval between first cutting a hole in the shell and later emergence appears rather similar to the pip–hatch interval of birds (see Plate VII, Fig. 23). Apparently some snakes do not use the egg tooth for tearing the shell, but rely more upon generalized struggling (Smith, 1954).

Goldstein (cited in Hill & DeBeer, 1950), Smith (1954), and Hughes, Bryant, and Bellairs (1967) have all described snout thrusting movements by the embryo of the lizard, *Lacerta vivipara*, during hatching. Smith, however, has questioned the use of the egg tooth in this species for tearing open the shell. The most detailed description of hatching in any reptile has been provided by Smith (in Hill & DeBeer, 1950) for the limbless lizard, *Anguis fragilis* (slowworm). Smith actually observed the entire hatching process in four separate embryos, and because of the transparency of the shell membranes, he was able to see the embryo and its movements quite clearly. The head was repeatedly thrust forward until finally the membrane was pierced and the young one crawled out head first. There were no side-to-side movements of the head, nor was there any struggling of the body. The entire hatching process took only a minute or two. Smith is of the opinion that the egg tooth is used more as a piercing organ in lizards, instead of for slashing as is the case with snakes.

I have been able to locate only three descriptions of hatching in either turtles or crocodilians (but see Plate VII, Fig. 24). Olexa (1969) briefly describes hatching in the musk turtle, *Sternotherus odoratus*, as follows: "... I have found one egg in a different position and on close inspection I could see a tiny bulge on the shell. A young turtle was about to break through the shell with its egg-tooth. It worked so hard that the whole of the egg was shaking, and soon a small triangular hole was seen to which the young pressed its eye.... At first it tried to enlarge the hole only be drilling movements of its head but later, when the aperture became larger, it also used its front feet [p. 28]." Preliminary personal observations of hatching in snapping turtles, after the first hole appears in the shell, agrees with this report. They seem to "crawl" out of the egg by ambulatory movements, but the initial opening in the shell and membranes have not actually been observed.

Deraniyagala (1939) has described hatching in two species of crocodiles, *Crocodylus palustris kimbula* and *Oopholis porosus.* In the former a hole or crack was made in the shell prior to emergence. In one specimen the interval between the first crack and emergence was 5 hours. In *Crocodylus*, the

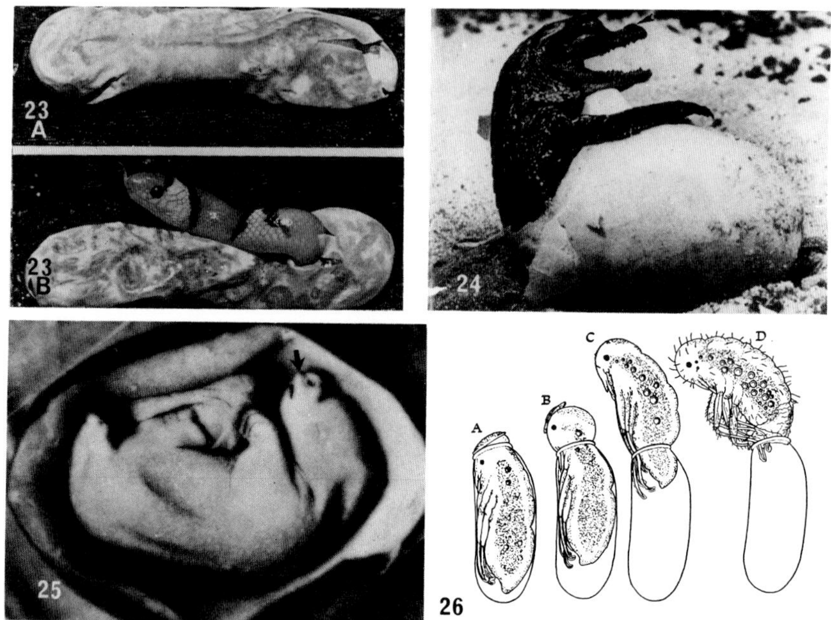

Plate VII

FIG. 23. Hatching in a snake embryo showing the first slit in the egg (A), and then after a long interval emergence from the egg (B). From Time-Life, Inc. *The Reptiles.*

FIG. 24. A hatching crocodile partially emerging from one end of the shell. Note egg caruncle (arrow). From Time-Life, Inc., *The Reptiles.*

FIG. 25. A monotreme embryo inside the shell showing the prominent egg caruncle (arrow). From Hill & DeBeer (1950).

FIG. 26. Hatching of an invertebrate, *Cimex*, from the egg. (A) Cap of the egg being forced off; gut and abdomen showing waves of peristalsis toward the head; a few air bubbles in the gut; (B) active peristalsis continues; head distended and bulging from the egg; (C) larva, almost free from the egg has started to swallow air vigorously; (D) embryonic cuticle has split and slipped backward, allowing spines to stand erect; tracheae have filled with air; larva swallowing air. From Wigglesworth (1965).

hatching position is basically a sitting posture with the head and snout flexed down upon the chest. The first sign of hatching was a crack at one end of the shell. The snout soon emerged from this opening, although it was another 2 hours before actual emergence. Both species were found to emit loud vocalizations during the entire hatching process. The major behavior pattern used for hatching in these species is snout-thrusting.

4. MONOTREMES

Although it may seem odd to include a mammalian form in a discussion of hatching behavior, anyone remotely familiar with the duck-billed platypus or the spiny anteater, will realize that among the numerous other peculiarities of these two monotremes, the habit of egg laying is perhaps the strangest.

Like marsupials, the monotremes have an abdominal pouch within which the young are maintained until weaning. However, unlike marsupials, these animals first lay a single egg within the pouch where it is subsequently incubated until hatching, a period lasting 10–11 days in the spiny anteater or echidna, *Tachyglossus aculeatus* (Griffiths, McIntosh, & Coles, 1969). After hatching, the young attach themselves to a nipple within the pouch. At the time of hatching, the monotreme embryo possesses a true egg tooth, which is probably vestigial, and a larger, but apparently functional, egg caruncle (Hill & DeBeer, 1950; see Plate VII, Fig. 25). Although Griffiths (personal communication) has not yet been able to observe the actual movements that the monotreme embryo uses for making the first cut in the shell, he has found that after this first opening is made, emergence occurs by "the hatchling struggling with a continuous writhing motion of the head and forelimbs. . . ." The description of these movements seems rather similar to the activity of the newborn marsupial when making its way from the birth canal to the pouch (Hartman, 1920).

To briefly summarize this necessarily incomplete picture of hatching in vertebrates, the following tentative conclusions may be drawn: (a) Many avian and reptilian species appear to use a specialized type of head movement (thrusting of one kind or another) for making an initial opening in the shell or egg capsule. (b) In most birds, a similar type of specialized movement is also involved in the final emergence from the shell, while in reptiles and other forms a more generalized struggling, often involving ambulatory or locomotorylike movements, results in final emergence. (c) Many avian and reptilian forms have an interval between the time the first opening is made in the shell and later emergence, which may last from hours to days. (d) Numerous vertebrate embryos have an egg tooth or egg caruncle on the snout or beak which apparently aids in escaping from the shell, although such a function has never been experimentally demonstrated. (e) The situation in most fish and amphibians is not entirely clear, since it has been shown in a few cases that secretion of a hatching enzyme, which breaks down the egg capsule, is often sufficient for emergence without any active somatic movements on the part of the embryo.

5. Invertebrates

The inclusion of invertebrates in the present discussion of hatching behavior is motivated not so much by the possibility that more insight will be gained into the vertebrate hatching story, but rather by a desire to ascertain how other, quite distantly related animals, solve a similar biological problem, namely, escaping from the egg.

Of course, compared to birds or reptiles the problem is not quite the same for many invertebrates, since they have only embryonic membranes to contend with, rather than an actual calcareous shell. Most invertebrates have a chorionic membrane, vitelline membranes, and the serosal cuticle. Some forms also lay down an embryonic cuticle before hatching (Wigglesworth, 1965).

As was found to be the case with vertebrates, adequate descriptions of hatching behavior are practically nonexistent for invertebrate forms. Sikes and Wigglesworth (1931) and Davis (1968) have ably reviewed most of the available literature in this area (also see Berrill, this volume).

Some invertebrates hatch by a passive process in which large amounts of fluid are taken into the egg chamber by osmotic means causing the membranes to expand and burst (e.g., *Turbellaria*).

In others, though the process is still primarily passive, the hatching mechanisms are slightly different. The embryo may actively swallow the amniotic fluid or air within the egg again producing an expansion of the egg chamber which serves to rupture the egg membranes (Patay, 1941).

Another interesting example of passive hatching occurs in the American lobster *Homarus americanus.* The mother appears to assist hatching by shaking the embryos out of their egg capsule by violent and rhythmic contractions of her swimmerets (Berrill, this volume).

Cephalopodes possess a larval gland (organ of Hoyle) that secretes an enzyme which digests the shell or egg membranes. Although these embryos exhibit increased activity around hatching time (Choe, 1966), Wintrebert (in Needham, 1931) has shown, by anesthesia experiments, that such activity is not necessary for hatching to occur (but see von Orelli, 1959).

Many invertebrates possess egg bursters or hatching spines, which are rather sharp cuticular objects that aid in breaking through the membranes (Van Emden, 1946). These spines or bursters, which are often located on the head, but may also be distributed on other portions of the body, can be used passively, such as during the swelling of the egg as mentioned above, or they may break the egg membranes as a result of active muscular contraction on the part of the embryo (see below).

In addition, many invertebrates comibine these various techniques in different ways (Plate VII, Fig. 26; Davis, 1968), while still others use entirely unique methods of hatching. For example, Fabre (as cited in Needham, 1931) has described hatching in the reduvius bug (order Hemiptera) in which, when ready to hatch, a membranous bag filled with gas (CO_2?) lifts off the lid of the egg and explodes, leaving an opening for the escape of the embryo!

For the present account the most interesting examples of hatching of invertebrates are those cases that involve active embryonic movements, presumably under neural control. Balfour-Brown (1913) has left one of the clearest descriptions of active hatching available for an invertebrate. He studied the life history of the water beetle (family Hydrophiloidea) which, shortly before hatching, begins exhibiting very specific rhythmic swallowing movements. Later other *total* body movements (nonswallowing) occur which cause the head to be pushed into the shell. Accompanying these body movements are vertical (up and down) movements of the head itself. Due to the presence of hatching spines on the head, these various movements scrape and eventually burst open the membranes. By slight "writhing" movements the embryo works its way out of the shell. Thus, here is a case in which very specific bodily movements are used for hatching by an invertebrate.

Some insects merely crawl around inside the egg so that the hatching spines, acting like a can opener, make a neat slit in the membranes; and still other invertebrates exhibit various kinds of generalized wriggling, stretching, and pulsatile movements for apparent use in hatching (Berrill, 1971; Davis, 1968; Sikes & Wigglesworth, 1931).

Judson (1963) has investigated the mechanisms underlying the active hatching of the mosquito, *Ades aegypti.* He reports that a reduction in ambient pO_2 initiates hatching by stimulating active movements of the embryo. Due to the pressure of the hatching spine on the head, these movements force the head into the shell, eventually rupturing the membranes. Judson suggests that the nervous system is involved in both stimulus reception (of pO_2), and actual mediation of the hatching movements.

In summary there appears to be considerably more variation in the methods of hatching of invertebrates than is the case among the vertebrates reviewed above. However, similar to vertebrates, the invertebrates have devised both passive and active means of escape from the egg; they have evolved special morphological structures (egg spines or bursters) to aid in hatching; and in some cases they appear to use specialized movements for breaking and escaping from the shell. These similarities, of course, in no way imply a direct phylogenetic relationship (homology), since even among closely related groups of *invertebrates*, mechanisms of hatching are markedly different and are not particularly useful as taxonomic aids (Davis, 1968). Such similarities or analogies are simply examples of convergent evolution.

III. The Possible Mechanisms Associated with the Various Stages of Prehatching and Hatching Behavior in Birds

A. Passive and Sensory Factors

1 Pretucking and Tucking

The factors that lead to the domestic fowl embryo attaining the characteristic pretucking position as seen in Plate I, in Fig. 2, have been extensively studied by Kuo (1932a) and are apparently in part due to active wriggling movements and in part to passive factors. It should be pointed out, however, that the critical experiment to prove that active movements are involved has not been carried out, namely, a long-term immobilization of the embryo by anesthesia or curarization *prior to* attainment of pretucking position. In any case, it does appear, from the observations of Kuo and others, that such factors as egg position, turning of the eggs, and gravitational shifts of the embryo and other egg contents are of primary importance for determining the pretucking position (Byerly & Olsen, 1933, 1936). Thus, by about 15 days of incubation, the position of the embryo in relation to the longitudinal axis of the shell becomes relatively constant as opposed to earlier stages (Byerly & Olsen, 1933; Drent, 1967; Kuo, 1932a), although *minor* changes in the embryo's position still occur between this time and the time when the final pretucking position is attained on day 16–17 (Kuo, 1932a; Oppenheim, personal observations).

As described earlier, beginning on day 16 or 17 the more jerky, convulsive-like activity of the fowl embryo (Types I and II) becomes sporadically interspersed with smooth, coordinated Type III movements. These Type III movements serve to lift the head, upper trunk, and right wing out of the yolk sac and to the embryo's right as depicted in Plate I, Fig. 3–6. Eventually these movements lead to the head being tucked under the right wing. Although a small number (3–5%) of domestic fowl embryos direct their tucking movements to the left side and eventually get the head under the left wing, the vast majority lift their head out of the yolk to the *right*, and also tuck under the right wing (Asmundson, 1938; Byerly & Olsen, 1936; Hamburger & Oppenheim, 1967).

The degree to which this stereotyped directional preference during tucking is due to sensory factors is not clear. It is interesting, however, that a preference for turning the head to the right has been documented for the duck embryo at both early stages (Gottlieb & Kuo, 1965) and during the pretucking period (Oppenheim, 1970), and that this preference has been shown to disappear *after* tucking (Oppenheim, 1970). Gottlieb and Kuo (1965) have suggested that the early right turning preference may be due to stimulation of the left side of the head by the yolk sac (since the embryo lies on its left side) causing the embryo to turn its head to the right. Some support for such a suggestion comes from a recent study by Hamburger and Narayanan (1969), who found that domestic fowl embryos that have had the head and beak regions deafferented by early bilateral removal of the primordia of the trigeminal ganglia subsequently fail to tuck or to hatch. Acceptance of this evidence assumes, of course, that there is a direct relationship between the early right turning preference which consists of jerky Type I and II movements, and the latter more coordinated Type III right turning movements. This has not been proven. Furthermore, severe brain damage also accompanied the deafferentation, making it impossible to relate the behavioral deficit to the loss of sensory input (Hamburger & Narayanan, 1969). This study is further discussed in Section III, D below.

On the other hand, both the initiation of the Type III tucking movements on day 17 and their directional preference might be due to vestibular input resulting from the posture of the embryo on day 17. Indeed, Decker (1970) has found that early *bilateral* removal of the anlage of the vestibular apparatus in the fowl embryo results in the complete absence of Type III tucking movements. While these results might seem to suggest that tucking is mediated by vestibular input, such a conclusion, in the light of the following evidence, is probably premature: (a) Decker reports that there was a massive secondary cellular degeneration in the CNS after such a bilateral vestibular operation, which could well have been responsible for the behavioral results. (b) *Unilateral* removal of either the right or left vestibular apparatus did not interfere with the embryo's attaining a normal tucked position toward the right side. (c) Only very slight secondary CNS degeneration occurred after a *unilateral* vestibular operation. (d) Kovach (1970) was unable to elicit a righting response of any kind in normal chick embryos until well after the tucking stage. This evidence makes it rather unlikely that stimulation of the vestibular system initiates or guides the Type III tucking movements. In fact, according to Decker (1970) and Kovach (1970), and contrary to the early report of Visintini and Levi-Montalcini (1939), the vestibular system may not be functional (at least with regard to righting) until just a few hours prior to hatching.

Except for experiments dealing with the role of particular parts of the brain in the mediation of Type III tucking movements, discussed in Section III, D, below, I have been unable to ferret out any other information dealing with the mechanisms associated with tucking.

2. Membrane Penetration and Pipping

Once the embryo assumes the typical hatching position after tucking, with the beak pushed tautly up into the airspace membranes (draping), it is prepared for membrane penetration. Any sufficiently strong movement of the head or beak, at this time, results in tearing or penetration of the membranes. Since both the time of MP and the movements utilized for MP are quite variable, it seems unlikely that any *specific* behavioral mechanism is utilized for this event. It is interesting, however, that the frequency of head movement increases significantly during this time and up until pipping (Hamburger & Oppenheim, 1967). The possibility that this increase may be related to gaseous metabolism is discussed in Section III, B, below.

Kuo (1932a) has made the claim that if tucking does not occur, i.e., the head or right face lies *over* rather than *under* the right wing, MP cannot occur. A recent experiment in which the right wing was surgically removed, however, has shown that MP *can* occur in the absence of tucking (Narayanan & Oppenheim, 1968).

The average interval between MP and pipping in the domestic fowl is about 10 hours and is determined primarily by gaseous exchange in the airspace (see Section III, B, below). The actual motor pattern used to effect pipping is a strong lifting or backward thrust of the head toward the shell, as indicated by the broken arrow in Plate II, Fig. 9a. This head movement is of the Type III variety and is often accompanied by coordinated trunk and leg movements, similar to those seen during climax (Hamburger & Oppenheim, 1967).

It was mentioned earlier in this article that the first pipcrack, in all species that have been examined, occurs more toward the blunt or airspace end of the shell, and that when eggs are positioned more or less horizontally (long axis parallel to the ground as is generally the case in the nest or hatching tray of the incubator), the first pip crack appears on the side of the shell facing uppermost (upper half of Fig. 27D). Although this upward pipping phenomenon has been described and studied rather extensively (Byerly & Olsen, 1937; Drent, 1967; Driver, Higgins, & Newman, 1968; Goethe, 1953; Hutt & Pilkey, 1934; Kovach, 1968, 1970; Lind, 1961; Oppenheim, 1970, 1972b; Poulson, 1953; Tinbergen, 1953; Waters, 1935), several questions about the mechanisms underlying this event are still not entirely settled.

In the avian egg the embryo generally assumes a position on the upper surface of the yolk during a major portion of the incubation period, and, as pointed out by Kuo (1932a), this position is largely determined by weight factors (gravity) within the egg. Lind (1961) and Drent (1967) have both demonstrated that shifting of the eggs in nature by the brooding parent tends to maintain the egg and embryo in this position. This probably has the function of bringing the embryo into closer proximity to the warm brood patch of the parent.

By day 16 or 17 of incubation in the domestic fowl, the passive gravitational factors described above leave the embryo in a position as seen in Plate

I, Fig. 2. Changing the position of th e egg by as much as 180° at this stage (Fig. 27) will result in a certain amount of passive gravitational readjustment of the embryo and egg contents. Whereas almost all eggs that remain in the natural position ("up") pip at the *uppermost point* on the shell (arrows, Figs. 27 and 28), eggs that have been rotated 180° ("down") at this stage do not show this tendency, although a substantial number of them do pip somewhere on the *upper half surface* of the shell (Fig. 28A–D). And when such a manipulation is done at stages later than pretucking or tucking few, if any, "down" positioned eggs pip at the uppermost *point*, and only a small percentage even manage to pip in the upper half surface of the shell (Fig. 28E).

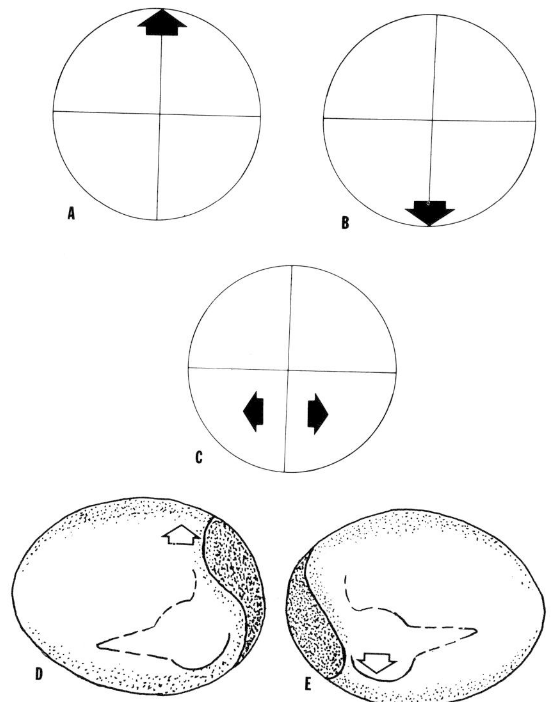

FIG. 27. Schematic illustration of the position of domestic fowl embryos within the shell as determined from X-ray photographs. (A) Viewing the egg from the blunt pole the arrow indicates the approximate position of the beak at the time of draping when the egg is allowed to assume a position determined passively by gravity ("up"); (B) position of the beak when the egg is rotated 180° from A ("down") at the time of draping; (C) position of the head-beak prior to tucking (day 17), when the egg is "up" (apparent left arrow) or rotated 180° ("down"; apparent right arrow); (D) schematic outline of head and beak as drawn from X-ray photos when egg is positioned "up" prior to tucking. Note the asymmetry of the air-space (darkened area); (E) similar to D but now the egg has been rotated 180° ("down").

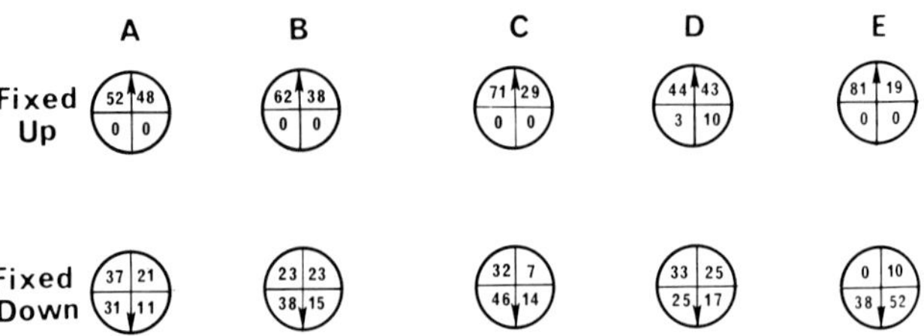

FIG. 28. The percentage of domestic fowl embryos pipping in the various quadrants of the shell when the egg is positioned horizontally (parallel to the ground), and either fixed in the natural position assumed in the nest or incubator ("up"), or when the egg is rotated and fixed 180° from this position ("down"). The eggs are viewed from the blunt or airspace pole. All eggs used in A-D contained embryos that were in the pretucking or tucking stages. (A) Kovach, 1968; (B) Drent, 1967; (C) Oppenheim, 1970; (D) Byerly and Olsen, 1937; (E) Oppenheim, unpublished experiment. Of 16 eggs that were opened and examined from the batch of eggs used in E, 75% were either tucked or draped, whereas 25% had recently penetrated the membranes. There were 24 eggs in each of the groups ("up-down") in (E). It should be pointed out that although the embryos from the Byerly and Olsen study (D) were said to be 19 days old, they were apparently delayed in development since they had not yet completed tucking. Furthermore, although it is not included in this figure, almost all eggs positioned "up", pip not just in the upper half of the shell, but also very close to the *uppermost point*, as indicated by the arrows; whereas while many of the positioned "down" eggs pip in the upper half of the shell, very few of these embryos manage to pip at this uppermost point (arrows). There is also a tendency for a majority of pipping to occur on the right side of the shell (apparent left in the figure) in both the "down" and "up" groups. This is due to the tendency of tucking movements to be preferentially to the embryos right (see text).

These findings can be explained by the action of two primary mechanisms: (a) as the embryo gets older and becomes more tightly compacted in the egg, passive gravitational readjustment is less likely to occur; (b) the prominence of Type III tucking movements during tucking and their gradual decrease after this time. On day, 17, the domestic fowl embryo exhibits rather massive tucking movements that lift the head and beak out of the yolk, and toward the upper half of the egg (Plate I, Figs. 3–6). Whether the egg is positioned normally ("up") or rotated 180° ("down") at this stage, these movements continue. In the "up" position virtually all embryos will manage by such tucking movements to obtain a position with the head in the upper portion of the shell where they will eventually pip; whereas of the eggs positioned "down," less than half pip on the upper part of the shell. This difference is due to the fact that the "down" positioned embryos, although still making tucking movements to the right, must shift their head and break from the lower right quadrant to the upper left quadrant of Fig. 27C in order to pip in the upper half of the shell, whereas the normally positioned "up" group need only lift their head and beak from the lower to the upper left quadrant. Once tucking and draping are accomplished, turning the eggs 180° "down" results in almost all pipping occurring on the lower part of the shell (Fig. 28E). In summary, then, eggs normally positioned according to gravity either in nature or in the incubator become pipped in the upper half of the shell as a result primarily of Type III tucking movements that are preferentially to the right.

There is some disagreement in the literature concerning the interpretation of the pip-up phenomenon discussed above. Kovach (1968), while originally agreeing that an active behavioral process (tucking?) was involved, has more recently modified his view. Kovach (1970) is now of the opinion that gravitational shifting of the embryo and egg contents (aided only indirectly by the embryo's movements) are primarily responsible for the embryo attaining a position, as seen in Plate I, Fig. 6. However, as the present account has attempted to show, in the absence of tucking movements, the embryo remains in the position seen in Plate I, Fig. 2. It is impossible for pipping to occur in the upper half of the shell from this position. Kovach (1968) appears to maintain that even up to the time of draping or membrane penetration, a majority of embryos rotated 180° from the natural egg position can *passively slide* around inside the shell and reattain the normal position. The results depicted in Fig. 28E clearly show this not to be the case (also see Drent, 1967). However, even if a substantial number of these embryos were able to passively slide or shift position within the shell in order to pip up after such a manipulation, it is difficult to see what relevance these results would have for the natural situation where the parents, by their egg-shifting movements, ensure that an egg never deviates from the normal up position for more than a few minutes (Drent, 1967). Thus, the necessity for such extensive gravitational shifting by the embryo, as proposed by Kovach, apparently never occurs in nature.

The mechanism by which the embryo gets its head into the upper portion of the shell subsequent to pipping has been attributed by Byerly and Olsen (1937) to a sensory guided gravitational response. However, since both Decker (1970) and Kovach (1970) have shown that vestibular, and other postural righting reflexes do not appear in the domestic fowl embryo until long after tucking, this seems a rather unlikely possibility. It may be that other kinds of tactile or proprioceptive stimulation resulting from the position of the embryo (Plate I, Fig. 2) are responsible for the preferential lifting of the head out of the yolk and to the right, or alternatively, the tucking movements could be initiated by an endogenous motor mechanism (see below).

What selective advantage, if any, there may be in pipping in the upper half of the shell is not clear, although Kovach (1968) has reported more deaths in eggs that pip on the lower shell surface. This may be related to an impaired respiration in those embryos that pip on the lower shell.

3. Post Pipping and Climax

After pipping there are virtually no modifications in the embryo's tightly cramped position within the shell. It was suggested earlier in this article that perhaps the cramped position of the embryo at this time is directly related, via inhibitory mechanisms, to the decrease in somatic motility that occurs in most species (see Tables I and II). Kuo (1932b) has reported that removing the entire shell and shell membranes from around the embryo at this time has no effect on the embryo's behavior or on hatching time. However, Kuo presents no data on this point and in light of the evidence discussed below (and in Section III, C), it seems unlikely that even Kuo's observation on the hatching time of these shell-less embryos is correct.

Numerous studies are available in the poultry science literature which show a clear-cut decrement in the hatchability of embryos that are in abnormal positions (e.g., Asmundson, 1938). It seems likely, however, that these findings are not so much due to the altered sensory input arising from the abnormal position, but rather can be attributed to interference with the onset and normal development of lung respiration, since many of these abnormal positions impede membrane penetration and pipping (Asmundson, 1938; Byerly & Olsen, 1931, 1936).

The behavior and hatching time of embryos pulled partially out of the shell, just after membrane penetration, has been studied by Balaban and Hill (1969), although these authors present no comparable data on the embryo's behavior *prior* to being pulled out. However, they do find certain reliable behavioral changes that may have their counterpart in embryos in intact eggs. For example, they found an increase in the frequency of head lifting and vocalization during what they term stage 3, which they compare to the climax stage of hatching as defined here. Such an increase also occurs in the intact embryo (personal observations). It is curious, however, that embryos in the Balaban and Hill (1969) study remained in stage 3 (or climax) for more than 12 hours before emergence, whereas in normal intact eggs, climax lasts for only about 50 minutes. Apparently excess drying plus the small opening in the shell (2.5 cm^2) of eggs in their study prevented the embryos from escaping, even though they were already partially pulled out. Perhaps of more interest is the report by Balaban and Hill that embryos partially pulled out of the shell hatched considerably earlier than unopened controls. Similar results have been found with duck embryos (Gottlieb, 1971), and this observation is discussed more thoroughly in Section III, B, below.

Few behavioral differences have been found in embryos during prehatching and hatching stages after the experimental modification of the normal sensory environment within the shell (Helfenstein & Narayanan, 1969; Narayanan & Oppenheim, 1968; Oppenheim, 1970). For example,

Helfenstein and Narayanan (1969) surgically removed both legs in early embryonic stages and subsequently observed that, in spite of not being as restricted in the shell as control embryos, these embryos exhibited typical climax behavior. These findings would appear to cast some doubt on the statement of Kovach (1968) that, "... the essential stimulative factors for hatching were provided by the particular cramped position of the body inside the egg ... [p. 284]." However, it would be even more informative to have detailed comparisons of the behavior of embryos that are in a normal "cramped" position *versus* "uncramped" embryos that had been partially pulled out of the shell. Such a study is currently being conducted in our laboratory by Hinda Levin, and some of the preliminary findings are reported and discussed below. Although this study included making behavioral comparisons at stages ranging from *draping* to *climax*, only the data for embryos treated within 2 hours after pipping (or later) are complete enough to be presented at this time.

Each experimental egg was opened over the airspace on the day it was to be observed by removing most of the shell over the airspace and then coating the inner shell membrane with warm Vaseline (petroleum jelly), according to Kuo's (1932a) observation technique. The opening was then sealed tightly with nonporous Parafilm and the egg was returned to the observation box (temperature and humidity controlled) for 30 minutes. A yolked control egg was prepared at the same time and in the same way. At the end of the 30-minute acclimation period, a 5-minute recording (*open*) was made of various specific motor patterns with the use of a keyboard attached to an event recorder. At the end of this recording the head, neck, and upper trunk region of the experimental embryo were pulled out of the shell (Plate VIII, Fig. 30), and the embryo was placed on a bed of tissue in the observation box; the control embryo remained in its normal, intact position within the shell. After another 30-minute acclimation period, a second 5-minute recording (*out 1*) was made of both control and experimental embryos. Two hours later a third recording (*out 2*) was taken. Only embryos that went on to hatch are included in the final analysis. Motion picture records were also made of select embryos. The data in Fig. 29 includes only those groups that were manipulated and recorded either 2 hours, or 12+ hours after pipping. Each data point in Fig. 29 is the summary of 10 individual embryos.

The pulled-out embryos hatched (i.e., completely escaped from the shell) earlier than the controls by 7.8 hours (SD = 6.1 hours, $p = .0003$) confirming the earlier report of Balaban and Hill (1969). With few exceptions there were no behavioral differences either between the pulled-out and yolked controls or between the three separate recordings from the pulled-out group (*open*, *out 1*, *out 2*). The latter comparisons are the most germane concerning the question originally posed, namely, does the same embryo behave dif-

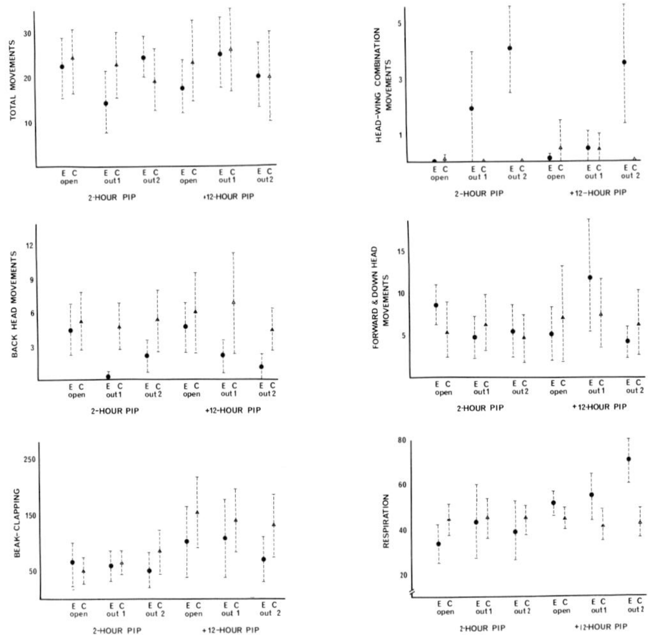

FIG. 29. Means and 95% confident limits for various movements for pulled-out (E) and control (C) embryos in one group treated 2 hours after pipping, and in another group treated more than 12 hours after pipping. The respiration rate is per minute, and all other measures are for 5 minutes.

ferently before and after being partially pulled out of the egg? Perhaps the most interesting category is that of *total movements.* Here all movements of the embryo that could be seen (except beak-clapping) were recorded, so that this single category is more inclusive than could be attained by the addition of the other separate categories in Fig. 29. Although these are preliminary results, the fact that there are no reliable differences in *total* movements when an embryo is pulled out of the shell *versus* when it is in a cramped or restricted hatching position within the shell, indicate that the suggestion made earlier in this article that the normal decrease in the frequency of movement after tucking and draping may have been due to inhibition from the cramped position attained at that time is probably not correct.

In summary, it may tentatively be concluded that, for the most part, the frequency and gross character of the movements seen between pipping and the onset of climax are not closely associated with the cramped posture of the embryo within the shell. This is not to say that loss of all sensory input (tactile and proprioceptive) by deafferentation would not modify these movements; certainly the fine tonic, phasic, and amplitude characteristics of muscular activity would most likely be changed by such an operation (in fact some evidence for such an effect is reported in the article by Corner *et al.*, this volume). But it seems unlikely, in light of the above experiments, that those aspects of behavior that have been examined (i.e., general patterns of motor activity) can be related to sensory input in any direct fashion.

But what about sensory factors during climax? Both Kovach (1968) and Corner and Bakhuis (1969) have briefly reported that by freeing the embryo from the shell *during* climax, the typical coordinated and rhythmical climax movements cease. In fact, Corner and Bakhuis (1969) report that stretching of the neck is even sufficient to bring about a cessation of climax behavior. On the other hand, Kirkman (1931) reports that the black-headed gull continues periodic struggling movements (that may or may not be similar to climax, see Corner *et al.*, this volume), for as long as 45 minutes after emergence. We have also observed and recorded on film, in the course of studying the partially pulled-out embryos, that when climax ensues there is a clear periodic struggling, including head raising, wing fluttering, and kicking that lasts until the embryo manages to completely escape from the shell. Thus, while perhaps differing in detail from Kovach (1968) and Corner and Bakhuis (1969), we would agree that certainly the specific motor pattern seen during the climax stage of an embryo in the normal hatching posture is markedly different from that of a *partially pulled-out* embryo (compare Plate II, Fig. 8, 9, and Plate VIII, Fig. 30). What is similar in the two cases is the sudden onset of climax, the periodic struggling, and the more alert overt behavior of the embryo at this time.

The above observations all suggest that something about the normal hatching posture of the embryo within the shell during climax may be responsible for the characteristic motor pattern seen during that time, as well as for the typical counterclockwise rotation and cracking of the shell. The most likely candidate for this would seem to be a postural or righting reflex. Indeed, Pohlman (1919) and Steinmetz (1930), as well as Kaspar (1964), and Kovach (1970) more recently, have all suggested that the climax movements are the manifestation of postural or righting reflexes resulting from the specific position of the embryo at this time. Corner (1968) has also suggested that proprioceptive input from the neck may be important in maintaining climax. This is an attractive suggestion, and for the most part, the available evidence would seem to support it. For example, Kovach (1970) has demonstrated that by 19 days 18 hours, the chick embryo consistently exhibits a postural reflex. That is, when held on its back with its neck twisted to the right or left, the legs of the embryo assume a characteristic posture (see Figs. 5 and 7 in Kovach, 1970). However, as has been clearly shown by Kleitman and Koppányi (1926), these reflexes are present in chickens even after the bilateral destruction of both labyrinths! Thus, the reflexes demonstrated by Kovach (1970) are probably tonic neck reflexes and do not involve the vestibular system. This conclusion is supported by the very interesting report of Decker (1970) that chick embryos with a *unilateral* labyrinth removal were able to complete hatching in a normal manner, including counterclockwise rotation, even though they were *unable to right themselves* when placed on their back, until 5 hours after hatching (compared to within 1 hour after hatching for the controls). Thus, since these unilateral birds lacked a righting response and yet managed to hatch normally, it is difficult to see how a *vestibular* mediated righting reflex could be involved in climax. At the present time it seems more likely that *other* postural reflexes (e.g., tonic neck reflex) may be involved in climax. In this regard it is interesting that it is at about the time that the hatching muscle reaches its maximum size (at pipping) that Kovach (1970) was consistently able to elicit a postural reflex in the chick (19 day 18 hours). The increased sensory input that might be thought to result from additional pressure on the neck proprioceptor organs, via the enlarged hatching muscle, may be involved here. Although the fact that the size of the hatching muscle tends to *decrease* after pipping suggests some caution in applying such a mechanism to the climax stage, it does appear likely that the specific motor characteristics of climax (counterclockwise rotation, the stereotyped interval between climax movements, etc.) are somehow mediated by postural reflexes of a proprioceptive nature.

The fact that the embryos of different species of birds vary considerably in the extent to which they rotate during climax, before finally emerging from the shell, raises another interesting question; namely, is the degree of rotation controlled by endogenous factors that program a certain amount of rotation for each species, or is it related to extrinsic factors such as shell or membrane strength (e.g., quail, gull), the extent of shell breakage before climax (e.g., gulls), and the size or shape of the beak or egg caruncle? Although we have not systematically examined these various factors, we have observed that strengthening the shell of the laughing gull by the application of tape thus making it more difficult for the embryo to break out of the shell, results in considerably more rotation than normal. Provine (1971) has examined this question more systematically in the domestic fowl. He found the mean amount of rotation in control eggs ($n = 80$) to be 243° (SD = 56°), while embryos with the shell cap taped down ($n = 80$) rotated 440° (SD = 164°, $t = 10.05, p < .001$). Corner and Bakhuis (1969) have also found that the rhythmic climax movements may continue for many hours if the embryo is kept in the shell. It is clear from these results that the amount and duration of climax rotation is controlled by extrinsic mechanical factors and is in no way regulated by an intrinsic species-specific program.

One final point to be discussed concerns the "turning-on" of the climax stage. Although Kovach (1968) has suggested that the cramped position of the embryo is responsible for initiating climax, the fact that embryos are cramped in the shell for 20–40 hours prior to emergence, while climax does not begin until about an hour prior to emergence—plus the fact that pulled-out embryos show the same sudden "turning-on" of climax, even though they are no longer cramped in the shell—makes such a suggestion rather unlikely. The "turning-on" of climax is apparently related to some neural or hormonal mechanism, as discussed in Section III, C and D, below.

4. Clicking and the Synchronization of Hatching Time

This topic has been extensively reviewed recently by Vince (1969, 1970, and this volume), who, by her careful and detailed studies of synchronization in quail embryos, has greatly increased our understanding of this phenomenon. It has been demonstrated, for example, that the clicking (and other) sounds produced by Japanese and bobwhite quail embryos *in nature* can selectively advance the development of other embryos with which they are in contact, thereby bringing about a closer synchronization of hatching time than if the eggs (embryos) were not in contact with one another. Thus, under natural conditions, a large clutch of eggs that has been laid at 24-hour intervals over many days manages to hatch within an hour or two of one another. Such a phenomenon is of obvious survival value if the newly hatched birds make their exodus from the nest very soon after hatching, as is thought to occur with quail. Bobwhite quail apparently leave the nest about 4–5 hours after hatching (Stoddard, 1931), as compared to more than 16 hours for the mallard duck (Bjärvall, 1967). One possible complicating factor related to the survival value of synchronization is the recently reported fact that quail and chicks that hatch about 24 hours earlier than controls, as a result of being accelerated, show slight by significant deficits in motor behavior (Vince & Chinn, 1971). They first begin to walk about 1 hour later than controls. If quail chicks do in fact leave the nest soon after hatching, as suggested above, then a difference of 1 hour in the ability to walk could mean that an accelerated chick is more likely to be left behind. Yet in terms of survival value this may not be a meaningful factor, since in nature quail eggs are probably seldom advanced or delayed in hatching by as much as 24 hours.

Of great interest for the present discussion is the mechanism involved in this synchronization. It is already known that the eggs must be in contact with one another for synchronization to occur, but it is not yet clear whether the stimulation and synchronization between eggs is mainly mediated by the auditory system or the cutaneous tactile receptors. It has been shown (Vince, 1966), that stimulation from either a loudspeaker, or a vibrator (in contact with the egg) have a similar synchronizing effect. Regardless of the specific stimulus, however, it is clear that the synchronization of hatching time in quail embryos involves some kind of sensory mediated communication.

Other questions concern the precise way in which the clicking affects an embryo. Does it selectively increase or decrease Type III hatching movement? Does it affect the metabolic rate or the secretion of certain hormones? Does it selectively stimulate a certain part of the nervous system that may be involved in hatching? Although these questions presently have no answers, it is known that the onset of lung respiration can be advanced by stimulation in the Japanese quail (Vince *et al.*, 1971; Vince, this volume).

It is also known that sounds of different *frequencies* appear to selectively mediate the retarding or accelerating effects of synchronization (Vince, 1970). Low frequency (nonclicking) sounds produced by less advanced embryos retard the development of more advanced embryos (before the latter are clicking). Once the more advanced embryos begin clicking (higher frequency), however, their clicking tends to accelerate the less advanced embryos. Of the clicking sounds themselves, high *rates* of clicking (90–400 per second retard, while rates between $1\frac{1}{4}$–80 per second tend to accelerate hatching time. Thus, there seems to be a certain degree of sensory or perceptual specificity in the mechanisms at work here.

While clicking does not affect pipping time or the click–hatch interval in the bobwhite, it can advance pipping and MP in the Japanese quail. Also the interval between lung respiration and hatching is markedly changed by contact with advanced or retarded eggs (Vince & Cheng, 1970).

Recently Vince, Green and Chin (1970) have shown that hatching time in the *domestic fowl* can be accelerated by exposure to clicks at a rate previously shown to be effective with quail (three clicks per second). They also found that the variation in hatching time was markedly reduced by such stimulation.

In attempting to confirm or reject an unsupported statement by Taylor, Kreutziger, and Abercrombie (1971) that there is an inverse relationship between the time of pipping and the duration of the pip–hatch interval, we have also found an interesting phenomenon that lends further support to the idea that some sort of communication occurs between domestic fowl embryos.

A group of 200 eggs that were in contact with one another were divided into early *versus* late pippers (means = 19 day 4 hours *versus* 19 days 16 hours, respectively). Pipping time and hatching time were determined by hourly checks of all eggs around the clock. A Spearman rank order correlation coefficient (Rho_s) was then performed on these data and a significant inverse relationship was found between the time of pipping and the pip–hatch interval ($Rho_s = .59$, $p < .01$). The mean pip–hatch interval for the early pip group was 22 hours, and for the late pip group 15 hours. There was also a significant difference in hatching time (means = 20 day 2 hours for early, and 20 day 8 hours for the late pippers, Mann-Whitney Test, $p < .00003$, two-tailed). Hatching success was similar in both groups. These results suggest that something resulting from the contact of chicken eggs may be acting between pipping and hatching to reduce the variability in hatching time. Similar findings have been reported by Lind (1961) for the godwit and in the American coot and certain ducks by Fisher (1966). If this conclusion is correct then eggs kept in physical contact with one another should exhibit less variability in hatching time than separated eggs.

We have recently tested this by placing two groups of white leghorn chicken eggs into a common incubator. In one group, 30 eggs were in tight physical contact with one another, and in the other group 33 eggs were physically separated from one another by 1–2 inches. All eggs were checked for pipping and hatching time by hourly checks around the clock. The eggs were either separated or placed in contact at 17 days 0 hour of incubation. Although there were no reliable differences between the groups regarding the *mean* hatching time, the *variation* in hatching time was reliably different. The contact group had a SD of 3.5 hours, while the separated group had a SD of 6.1 hours ($p < .05$, F-test, one-tailed)*. These results are remarkably similar to the findings of McChosen and Thompson (1968b) with chickens, and when considered along with the findings of Vince *et al.* (1970), they strongly suggest that domestic chicken embryos may also synchronize their hatching as a result of interegg communication, though this synchronization is certainly not as tightly controlled as in quail. Furthermore, because the results of a Spearman rank order test on the *separated* eggs, from the experiment above, yielded a nonsignificant correlation ($\text{Rho}_s = .10$) between the time of pipping and the pip–hatch interval; whereas, as reported above, a similar test on eggs *in contact* yielded a significant correlation, it appears that the length of the pip–hatch interval of chicken eggs in physical contact can be altered, presumably by the communication of sounds between the eggs.

In light of all these results concerning the effects of external stimulation on hatching time, and because of the known effects of the thyroid hormone upon respiration, metabolism, and hatching time in the domestic fowl (see Section III, C, below), it would be most interesting to determine whether

*A similar experiment has recently been conducted by Gilbert Gottlieb (personal communication) with Peking ducks. As in the experiment reported above, there was not a significant difference in the *average* time of hatching between the eggs in contact with one another (mean = 26.8 days, $n = 134$) *versus* those not in contact (mean = 27.0 days, $n = 84$), but the *variation* in hatching time (SD) between the two groups was reliably different: contact SD = 11.8 hours, noncontact SD = 17.6 hours ($p < .01$, F-test). Therefore, ducks would also seem to possess at least a limited capacity for synchronizing their hatching time.

all these "synchronizing" effects, in both quail and fowl, result from the sensory input (clicks, etc.) causing an increase or decrease in the production of thyroxine. For example, the sensory input might stimulate or inhibit the secretion of thyroid-stimulating hormone (TSH) from the hypophysis via the hypothalamic-pituitary system which in turn would affect endogenous thyroxine levels.

B. Gaseous or Respiratory Factors

1. Tucking and Membrane Penetration

Windle and Barcroft (1938) report that nonrhythmic respiratory movements can be elicited artificially by destruction of part of the allantois as early as day 10 in the domestic fowl embryo. However, since it is exceedingly difficult to differentiate such movements from the spontaneous activity normally going on at that time, the more conservative report of Kuo and Shen (1937) that true respiratory movements can not be artificially elicited until 15–16 days is probably more correct. Windle and Barcroft (1938), Kuo and Shen (1937), and Corner and Bot (1967) are all in agreement that under *normal* conditions pulmonary respiration does not begin until shortly before membrane penetration at 18 or 19 days.

It is clear from the experiments of Windle and Barcroft (1938), Windle, Scharpenberg, and Steele (1938), and Dawes and Simkiss (1969), for the domestic fowl, and of Windle and Nelson (1938) for the domestic duck, that under normal conditions the initiation of pulmonary respiration is determined by an accumulation of excess carbon dioxide in the blood. This finding has also been confirmed more recently by Freeman (1962), who has demonstrated a rise in oxygen consumption a few hours prior to pipping, and apparently at the time when pulmonary respiration begins. It is important to point out, however, that the increase in CO_2 that is responsible for initiating pulmonary respiration need not involve air exchange in the airspace, since Visschedijk (1968b) failed to find an advanced onset of pulmonary respiration after inhibition of gaseous exchange through the airspace.

Both Windle and Barcroft (1938) and Freeman (1962) have suggested that the anoxia resulting from oxygen want in the airspace may be responsible for stimulating the muscular activity associated with pipping (see below). It also seems possible that a lack of O_2 or an excess of CO_2 *prior to* the initiation of pulmonary respiration may be involved in the onset of the Type III tucking movements that become so prominent at about that time. On the other hand, Vince *et al.* (1971) have suggested that the high blood pCO_2 that apparently initiates pulmonary respiration may itself be produced by the increased frequency of coordinated Type III tucking movements that occur just prior to this time (Hamburger & Oppenheim, 1967).

By continuously measuring the oxygen consumption of single eggs, Freeman (1962) has been able to provide rather complete information on this parameter, as well as its possible relationship to prehatching and hatching behavior. He finds that there is constant uptake of O_2 until just a few hours prior to pipping, at which time it increases. The point at which the increase occurs is related to the onset of pulmonary respiration. This increased O_2 consumption continues for several more hours until pipping occurs, and then gradually falls off to another plateau. About 2 hours before hatching, the uptake of O_2 rapidly increases again and continues to rise until 1–2 hours after hatching. The latter rise, just prior to hatching, was first observed and recorded by Preyer (1885).

For the most part changes in the pO_2 and pCO_2 of the blood of the domestic fowl embryo follow the pattern discussed above (Freeman & Misson, 1970). Furthermore, the increase in the *rate* of respiration (as hatching approaches) that has been demonstrated repeatedly, can also be related to these changing patterns of O_2 consumption (Oppenheim, 1972b; Kuo & Shen, 1937; McCoshen & Thompson, 1968b; Tremor & Rogallo, 1970; Windle *et al.*, 1938). It is interesting that Tremor and Rogallo (1970), using a sophisticated recording device that did not require the opening of the egg, found respiration rates similar to those reported by the other authors who were using somewhat less automated techniques.

2. Pipping and Climax

Once pulmonary respiration has begun, less and less of the gaseous exchange of the embryo is mediated through the chorioallantoic membrane, and more and more by the lungs (Romijn & Roos, 1938). In a series of three articles, Visschedijk (1968a, 1968b, 1968c) has examined a number of questions related to this transition. For example, he has experimentally verified the suggestion mentioned above that an excess of CO_2 in both the airspace and blood stimulates pipping. By modifying the gaseous exchange across the airspace, the time of pipping could be selectively held back or advanced by 6–7 hours. A high CO_2 concentration was about twice as effective as a comparable decrease in O_2 for stimulating pipping. It is possible that the increased frequency of head movement that occurs between MP and pipping in the domestic fowl (Hamburger & Oppenheim, 1967) is also related to these gaseous changes which control pipping time.

Experimental modifications in gaseous exchange through the airspace are only effective *after* the normal initiation of lung respiration, and even then they do not advance the time of hatching (Visschedijk, 1968b). Thus, although there is a clear rise in oxygen consumption about 2 hours prior to hatching (Freeman, 1962), a sharp rise in the rate of respiration during climax, and in spite of the fact that respiration via the allantois must be reduced to about 15% of the total before climax ensues, the "turning-on" of climax is apparently not directly related to gaseous metabolism. Freeman (1962) has postulated that a hormone is responsible for the onset of climax (see Section III, C, below). It has also been suggested by Taylor *et al.* (1971) that changes in the gaseous metabolism of domestic fowl embryos may be related to both pipping and hatching time via the adrenal gland by an increased or decreased retention of glycogen stores. The suggestions are discussed more completely in Section III, C, below.

To summarize then, it appears quite likely that the onset of pulmonary respiration is initiated by an accumulation of CO_2 in the embryo's blood. Once respiration begins, shortly before membrane penetration, it increases in intensity, regularity, and rate. After membrane penetration another increase occurs in the concentration of CO_2 in the airspace and blood, which leads to the stimulation of pipping. Although there are further changes in gaseous metabolism after pipping, they do not appear to be causally related to the *onset* of the final climax stage of hatching. The precise way in which these changes in gaseous metabolism are related to prehatching behavior (membrane penetration and pipping) is not known.

C. *Hormonal Factors*

1. Metabolic Effects

Although Freeman (1962) was apparently the first to suggest that a hormone might be responsible for initiating the final climax stage of hatching he presented no data at that time to support this contention. Because most, if not all, of the hormones produced by the adult chicken are found already in the embryo just prior to hatching (Betz, 1971), the suggestion of a causal relationship between hormones and hatching is certainly reasonable.*

In an unpublished dissertation, Freeman (1964) reported injecting the following substances onto the airspace membranes of chicken eggs at 18.5 days of incubation, prior to the onset of pulmonary respiration: thyroxine (T4), triiodothyronine (T3), 2-thiouracil, and thyroid-stimulating hormone (TSH). Both T3 and T4 (1 μg) advanced the onset of pulmonary respiration by an average of .6 hour, and hatching occurred 4.6 hours earlier in both the T3 and T4 groups.

*An interesting example of the role of hormones in a hatchinglike behavior (eclosion) in the silk moth (*Hyalophora cecropia*) has recently been reported (Truman & Sokolove, 1972). These authors have shown that the entire sequence of behavior patterns leading to emergence, which lasts about 1–2 hours, can be "activated and read off" by a hormonal trigger without benefit of any sensory feedback during the 1-hour process.

Injections of either 2 or 4 mg of 2-thiouracil, an inhibitor of thyroid secretion, delayed hatching by 10.7 and 17.3 hours, respectively.

Injections of 1 or 2 IU of TSH advanced hatching by 3.5 and 4.0 hours compared to controls. The simultaneous injection of 2 mg thiouracil and 2 μg thyroxine resulted in no apparent differences in hatching time.

These experiments, as well as several other studies (Adams & Bull, 1949; Balaban & Hill, 1971; Brandstetter, Watterson, & Veneziano, 1962; Grossowicz, 1946; Romanoff & Laufer, 1956; Sinha, Ringer, Coleman, & Zindel, 1959; Watterson, Zimmerman, & Johnson, 1964; Zamenhof & van Marthens, 1971) have all implicated the thyroid gland in certain aspects of hatching. However, none of these studies have provided conclusive evidence that the final climax stage of hatching is stimulated hormonally, as implied by Vince (1970). Thyroxine and thiouracil both act on general metabolic activity (Romijn, Fung, & Lokhorst, 1952). Embryos injected with *thiouracil* at prehatching stages (at least as early as 10–12 days of incubation), for example, not only hatch later, but also develop more slowly (Sinha *et al.*, 1959); and thyroxine injected as early as day 4 produces an increase in metabolic rate, and general development that lasts until the end of incubation (Beyer, 1952; Portet, 1960). A more direct test of the Freeman hypothesis might be to increase the level of thyroxine between pipping and the onset of climax and then determine whether the hatching time is advanced. This experiment has recently been done in the author's laboratory and is reported below.

Balaban and Hill (1971), in the most complete study to date dealing with thyroxine and hatching, have reported an advance in the hatching time of embryos injected with thyroxine any time between 17 days and the time of membrane penetration (19 days). Data collected in the present author's laboratory have confirmed these results for 17-, 18-, and 19-day embryos (Table III).

Our own data (Table III) are also in agreement with the findings of Balaban and Hill (1971) in that injection of a thyroid inhibitor (either thiourea or 2-thiouracil) at 18 days incubation, and *prior* to lung respiration, does not induce a delay in hatching time. Since injection of an inhibitor of thyroid secretion *before* 18 days (even at 16 or 17 days) *does* delay hatching, it would seem that sufficient thyoxine is already present in the blood stream by 18 days to insure normal development and hatching.

Perhaps one of the most interesting findings of the Balaban and Hill (1971) study concerns the complex relationship between thyroid function, metabolism, yolk-sac retraction, and hatching. They report that around the time of membrane penetration or pipping the yolk-sac of control embryos begins to be retracted at a rate that is significantly faster than at earlier stages (approximately 8 times faster). Embryos injected with thiourea at 17 days of

TABLE III

PIPPING AND HATCHING TIME OF CHICK EMBRYOS INJECTED WITH VARIOUS

Injection treatment	Dose per egg[a]	Day 17		
		n	pip age	Hatch age
Progesterone	100 μg/.05 ml	20	19 days, 18 hours	20 days, 5 hours
Control	.05 ml Ethylene glycol	15	19 days, 8 hours	20 days, 9 hours
Hydrocortisone	1 mg/.05 ml	—	—	—
Control	.05 ml Dioxane	—	—	—
Estradiol	25 μg/.05 ml	14	19 days, 7 hours	20 days, 8 hours
Control	.05 ml Ethylene glycol	17	19 days, 9 hours	20 days, 7 hours
Cortisone acetate	1.25 mg/.05 ml	22	19 days, 5 hours	20 days, 4 hours
Control	.05 ml Dioxane	24	19 days, 5 hours	20 days, 4 hours
Thyroxine	25 μg/.1 ml	23	18 days, 21 hours	19 days, 17 hours
Control	.1 ml 100% ETOH	37	19 days, 12 hours[c]	20 days, 5 hours[c]
Thiouracil	2 mg/.2 ml	—	—	—
Control	.2 ml Distilled water	—	—	—
Prolactin (LSH)	25 μg/.1 ml	20	19 days, 12 hours	20 days, 6 hours
Control	.1 ml LSH dilutent	20	19 days, 8 hours	20 days, 2 hours
Corticosterone	1 mg/.025 ml	24	19 days, 9 hours	20 days, 7 hours
Control	.025 ml Dioxane	27	19 days, 9 hours	20 days, 4 hours
Noradrenaline	25 μg/.05 ml	—	—	—
Control	.05 ml Distilled water	—	—	—
2,4-Dintrophenol	.1 mg/.2 ml	—	—	—
Control	.2 ml Distilled water	—	—	—
2,4-Dinitrophenol	.2 mg/.2 ml	26	19 days, 16 hours	28 days, 8 hours
Control	.2 ml Distilled water	17	19 days, 17 hours	20 days, 10 hours

[a]In all cases half of each group were injected onto the airspace membranes and half into the small end of the egg.

[b]The 19-day thyroxine group was injected within 1.5 hours after MP by pulling the head out

incubation failed to exhibit this change in the rate of yolk-sac retraction, and subsequently only a few hatched, in spite of the fact that they had begun to breathe at the same rate as controls and even though they subsequently penetrated the membranes (MP). It seems likely, in view of the observation of Kuo (1932a) that contractions of the abdominal musculature serve to draw in the yolk sac, that the increase in the rate of respiratory movements is closely related to the increased rate of yolk-sac retraction. Indeed, Kuo (1932a) has reported that respiratory movements aid in yolk-sac withdrawal. Although cortisone acetate has been reported to advance yolk-sac retraction (but not MP or pipping) when injected at 16–17 days of incubation (Moog & Richardson, 1955), we have been unable to confirm these findings (Table III).

HORMONES ON DAYS 17, 18, OR 19

Day 18			Day 19[b]		
n	Pip age	Hatch age	*n*	Pip age	Hatch age
33	19 days, 12 hours	20 days, 8 hours	21	19 days, 19 hours	20 days, 2 hours
23	19 days, 13 hours	20 days, 8 hours	20	19 days, 17 hours	20 days, 3 hours
18	19 days, 4 hours	19 days, 22 hours	—	—	—
15	19 days, 4 hours	20 days, 0 hours	—	—	—
18	19 days, 10 hours	20 days, 4 hours	18	19 days, 14 hours	20 days, 2 hours
13	19 days, 14 hours	20 days, 7 hours	20	19 days, 17 hours	20 days, 3 hours
—	—	—	—	—	—
—	—	—	—	—	—
33	18 days, 21 hours	19 days, 19 hours	43	—	20 days, 11 hours
27	19 days, 10 hours[c]	20 days, 9 hours[c]	38	—	20 days, 18 hours[d]
22	19 days, 9 hours	20 days, 8 hours	—	—	—
18	19 days, 12 hours	20 days, 7 hours	—	—	—
34	19 days, 10 hours	20 days, 6 hours	—	—	—
29	19 days, 10 hours	20 days, 5 hours	—	—	—
—	—	—	—	—	—
—	—	—	—	—	—
—	—	—	19	19 days, 19 hours	20 days, 17 hours
—	—	—	18	19 days, 21 hours	20 days, 15 hours
23	19 days, 16 hours	20 days, 8 hours	22	19 days, 18 hours	20 days, 8 hours
24	11 days, 14 hours	20 days, 6 hours	21	19 days, 19 hours	20 days, 5 hours
30	19 days, 14 hours	20 days, 10 hours	—	—	—
10	19 days, 14 hours	20 days, 7 hours	—	—	—

of the shell and injecting into the neck muscles. The noradrenaline group was also injected by the same method but within .5 hours after *pipping*.

[c] $p < .001$ Mann-Whitney Test, two-tailed.

[d] $p < .0003$.

The results discussed so far all suggest that thyroxine injected into the egg on days 17, 18, or 19 affects hatching time by its general metabolic effect (oxygen consumption, respiration rate, yolk-sac retraction). There are at least two bits of information in the literature, however, which indicate that the thyroxine hatching effect may also be somewhat more specific. Both Balaban and Hill (1971), and Rogler, Parker, Andrews, and Carrick (1959) have reported that some thyroid-deficient embryos manage to assume a typical hatching position, initiate breathing, MP, and in some cases reported by Rogler *et al.* (1959), are even able to pip the shell. Since few, if any, thyroid-deficient embryos go on to hatch these observations suggest that, behaviorally, it may be the final climax stage of hatching that is primarily modified by the thyroid deficiency.

With these observations in mind we have recently carried out the following experiment. Pipping time was determined by hourly checks of each egg. At the time of pipping the shell over the airspace was removed and this opening resealed with Parafilm. Approximately 5 hours after pipping ($\pm$ 30 minutes), an injection of either thyroxine or solute (control) was made into the neck or wing muscles. The opening in the egg was resealed, the egg was placed back into the incubator, and hatching time was determined by hourly inspections. The results of this experiment can be seen in Table V (thyroxine A). The time of hatching was not reliably altered by this treatment. Although this finding suggests that thyroxine is not acting to modify those hatching events that are occurring during the final 10–20 hours prior to emergence, the possibility that these negative results are due to some methodological artifact must be considered. The technique of injecting substances intramuscularly into the intact embryo, *in ovo*, while desirable, has at least one major disadvantage, namely, that it is difficult to get adequate access to the embryo without damaging membranes, etc. Thus, in the present situation we could not always be sure that the injected substances were getting into the muscle, nor could we rule out the possibility that the injections were made so superficially that leakage occurred. For these reasons the embryo was pulled partially out of the shell for injection (Plate VIII, Fig. 30) in the following experiments.

Twenty-five control and 25 experimental embryos (thyroxine B) were prepared by making an opening in the airspace and pulling the head and upper trunk out through this opening. This manipulation was done at 5 hours postpipping. The embryos were then injected with either thyroxine (25 μg/. 1 ml) or a control (. 1 ml) solution into the dorsal neck muscles, and hatching time was later determined. Furthermore, in order to determine whether a larger dose might be effective, another similarly treated group was injected with 40 μg of thyroxine (Group C). The results of these experiments are summarized in Table V. Once again there were no significantly reliable differences in the hatching time of either group (25 or 40 μg). Because thyroxine injected on *early* day 19 (at MP), or before, does speed up hatching, whereas these latter results from embryos injected *late* on day 19 (5 hours postpipping) indicate no such acceleration there appears to be a critical time after which thyroxine no longer is effective. The reason for this changing responsivity is not clear at this time.

Another approach to determining whether the role of the thyroid in hatching is primarily a nonspecific metabolic one, or whether it has a more specific neural or behavioral role (as has been well documented for certain aspects of metamorphosis in amphibians; see the recent reviews by Kollros 1968, and

Hughes, Volume 2 of this series) would be to modify metabolism without modifying or mimicking thyroid function. The pharmacological agent 2,4-dinitrophenol (DNP) apparently has this effect (Cutting, Mehrtens, & Tainter, 1933; Sollmann, 1957). This drug is reported to accelerate metabolism within a few hours after injection, without any of the other typical actions of thryoxine. For example, it does not stimulate amphibian metamorphosis nor does it alleviate hypothyroidism, except for raising the low basal metabolic rate (Cutting & Tainter, 1933). The effects of a single does lasts for 1–2 days.

We have recently tested the effects of DNP on both the metabolism and hatching time of domestic fowl embryos. The results are presented in Tables III, IV, and V. Injection of about 3 mg/kg of DNP (a dose known to raise the metabolic rate of humans by 50%) on either 18 or 19 days of incubation, had no effect on the time of pipping, the time of hatching, or the metabolic rate, as measured by respiration rate and yolk-sac retraction. Even after a second dose of DNP was injected intramuscularly no effect on these parameters was produced (Table IV).

In a second experiment, similar in most respects to that just described, except that twice the original dose of dinitrophenol was administered (6 mg/kg), negative results were once again obtained (see Table III). We detected no reliable differences in either the time of pipping or hatching between control embryos and embryos injected with DNP on either day 17 or 18 of incubation. Another group of embryos was injected at precisely 5 hours after pipping, by removing the shell over the airspace, and injecting directly into the neck or wing muscles (see above). Respiration rate was recorded a few minutes prior to injection and again 2 hours after injection. Hatching time was also determined. These results are presented in Table V. No reliable differences were detected in any of the parameters measured in this experiment.

Because of the possible unreliability of the *in ovo* intramuscular injection technique used in this study (see above), this experiment was repeated with the following modification: 5 hours after pipping, the head and upper trunk of the embryo were pulled out of the shell through an opening made in the airspace, and the embryo was injected with either DNP (.2 mg/.2 ml) or control solution (.2 ml) into the dorsal neck muscles ($n = 26$ in each group). Approximately 2 hours (± 15 minutes after injection the respiration rate of all embryos was recorded for 2 minutes. Hatching time was then determined by hourly inspections of all eggs. Respiration rate was found to be 64.4 per minute (SD = 17.4) for the dinitrophenol group, and 58.4 per minute (SD = 17.3) for controls. This is not a statistically reliable difference. There were also no significant differences in hatching time between the two groups (DNP = 20 day, 16 hours; control = 20 day, 16 hours). Taken together, all

TABLE IV

RESPIRATION RATE (PER MINUTE), YOLK-SAC WEIGHT, AND HATCHING TIME (DAY AND HOUR) OF CHICK EMBRYOS INJECTED WITH 2,4 DINITROPHENOL (.1 MG/.2 ML DISTILLED WATER)

Time of 1st injection	*n*	Respiration rate at 19 days, 17 hours	Respiration rate at 19 days, 18 hours	2nd Injection at 19 days, 19 hours	Respiration rate at 19 days, 22 hours	Yolk-sac weight in mg at 19 days, 24 hours	Age at hatch
Day 18							
Experimental	12	31.6	49.1		66.8	0.3	20 days 9 hours
SD		9.2	12.5		12.1	—	(6.3)
control	12	48.0	60.0		70.2	0.0	20 days, 2 hours
SD		27.1	14.6		11.5	—	—
Day 19							
Experimental	10	25.7	57.4		73.6	0.2	20 days, 9 hours
SD		13.3	12.7		12.4	—	(5.6)
Control	12	29.5	52.4		67.3	1.3	20 days, 8 hours
SD		17.1	16.7		10.4	—	(4.5)

NOTE: The second injection was made intramuscularly, whereas the first injection was made into the small end of the egg.

TABLE V

TIME OF PIPPING, HATCHING, AND RESPIRATION RATES (PER MINUTE) OF EMBRYOS INJECTED WITH THYROXINE OR DINITROPHENOL[a]

Substance injected	Dose	*n*	Pip age (SD)	Respiration rate, 5 hours post-pip	Respiration rate, 7 hours post-pip	Hatch age (SD)
Thyroxine A	25 μg/.025 ml	24	19 days, 13 hours (9.2)[b]	—	—	20 days, 14 hours (5.2)
Control	.025 ml	17	19 days, 12 hours (4.8)	—	—	20 days, 12 hours (4.8)
Thyroxine B	25 μg/.025 ml	25	19 days, 14 hours (7.1)	—	—	20 days, 15 hours (6.3)
Control	.025 ml	25	19 days, 12 hours (8.7)	—	—	20 days, 16 hours (5.6)
Thyroxine C	40 μg/.05 ml	20	20 days, 4 hours (7.8)	—	—	20 days, 16 hours (5.3)
Control	.05 ml	21	20 days, 5 hours (8.0)	—	—	20 days, 17 hours (5.5)
Dinitrophenol	.2 mg/.2 ml	27	19 days, 13 hours (12.5)	40.4 (14.6)	50.0 (13.7)	20 days, 14 hours (5.3)
Control	.2 ml	14	19 days, 18 hours (5.7)	38.8 (25.4)	50.4 (23.2)	20 days, 15 hours (4.9)

[a]All embryos used in these experiments were injected intramuscularly *for the first time* 5 hours after pipping.

[b]All SD values are in hours.

of these negative findings with dinitrophenol suggest that in the chick embryo this substance does not affect those overt metabolic events that are most closely associated with hatching, namely, respiration and yolk-sac retraction. Therefore, because of the failure of these experiments to modify metabolism in the chick embryo (as measured here), we are presently not able to draw any conclusions about the possible nonmetabolic role of the thyroid hormone in hatching behavior, the original goal of these experiments. Apparently, DNP acts differently in different species. Waterman (1939), for instance, failed to find any effect of DNP on either the metabolic or developmental rate of *Oryzias latipes* embryos, a freshwater killifish.

2. Hatching Muscle

It was discussed earlier that by the time of pipping the hatching muscle has increased tremendously in size primarily by an uptake of fluids (Plate III, Fig. 11), and that this enlargement of the hatching muscle has been related to cracking of the shell during pipping and climax, although the *specific* role of the hatching muscle in these processes (e.g., active or passive), has not been entirely clarified (Brooks & Garrett, 1970; Smail, 1964). Numerous studies are also available which demonstrate unique biochemical and other changes occurring in the hatching muscle around the time of pipping (George & Iype, 1963; Hsiao & Ungar, 1969; Klicka, Edstrom, & Ungar, 1969; Klicka & Kaspar, 1970; Ramachandran, Klicka, & Ungar, 1969; Rigdon, Ferguson, Trammel, Couch, & German, 1968).

In attempting to examine the mechanism underlying the increase in the size of the hatching muscle, Brooks and Ungar (1967) injected domestic fowl eggs with various steroid hormones and subsequently evaluated the size of the hatching muscle and hatching time. These authors report an increased uptake of fluid into the hatching muscle (and thus an increased size) after the injection of progesterone but not after estradiol or testosterone. The injections were made on day 18 of incubation. Furthermore, progesterone was reported to advance the time of hatching, while testosterone and corticosterone delayed hatching. From these results Brooks and Ungar suggested that progesterone serves as a "hatching hormone."

In view of the importance of this suggestion, and because of the small sample size and the rather imprecise methods used to determine hatching time by Brooks and Ungar (1967), we have recently attempted to replicate these results on a larger material.

Injections were made either into the airspace or the small end of the egg through small diameter holes (2 mm^2) made with a dental drill. Each experimental group always had a contemporary control group in order to rule out possible batch differences in pipping and hatching time. After injection, the eggs were positioned horizontally in large hatching trays within an incubator maintained at 99–101°F and 60–70% relative humidity. Each egg was checked at least once every hour around the clock from 18 day 0 hour, until hatching, in order to rather precisely determine pipping and hatching time. In addition, the hatching muscles of a group of the day 18-progesterone-injected embryos (and controls) were weighed between membrane penetration and pipping on day 19.

The results of these experiments, including data from the injection of other substances and hormones not investigated by Brooks and Ungar (1967), are presented in Table III. Except for thyroxine, the results of which were presented earlier, none of the substances, including progesterone, had any effect upon the time of either pipping or hatching. The data for the weight of the hatching muscle are as follows: Progesterone $\overline{X} = 597$ mg, SD $= 73$ ($n = 12$); Control $\overline{X} = 617$ mg, SD $= 61$ ($n = 12$). This is not a statistically reliable difference.

In view of these findings it seems exceedingly unlikely that progesterone, or any other adrenal steroid for that matter, serves as a "hatching hormone" as originally suggested by Brooks and Ungar (1967). This is not to say, however, that those unique biochemical events that occur in the hatching muscle are not under hormonal control. All that can be said at the present time is that the onset of hatching behavior (pipping and climax) is not influenced by either adrenal steroid hormones, or (with the exception of thyroxine) any other hormones that have been examined to date (Tables III, IV, and V).

Of the hormones that have not yet been examined in relation to hatching, perhaps the most likely candidate for future study is adrenaline. The measured concentration of adrenaline in the adrenal glands of domestic fowl embryos increases 5-fold between 17 days of incubation and hatching (Wasserman & Bernard, 1970), an interval encompassing the entire prehatching and hatching period. Adrenaline has also been repeatedly demonstrated to cause a rapid depletion of glycogen stores and an increase in blood glucose levels (Freeman, 1969; Gill, 1938; Leibson & Leibson, 1943). For example, Freeman (1969) has shown that injection of adrenaline into 19-day chick embryos induces a significant depletion of glycogen stores from the liver, while adrenergic blocking agents (e.g., propranolol) *reduced* the loss of glycogen from the liver.*

The increased energy requirements of the embryo during climax (both neural and otherwise) may be related to these findings regarding adrenaline. Adrenaline and the biogenic amine, dopamine, however, may both be more

*Preliminary results from experiments in our laboratory, however, have so far failed to detect any effect of noradrenaline upon hatching in the chick embryo (Table III).

directly involved in climax, since these agents are both associated with levels of behavioral arousal in young chicks (Key & Marley, 1962; Spooner & Winters, 1965). In light of the fact that the forebrain is important for the normal manifestation of hatching behavior (see Section III, D below) it is interesting that dopamine exhibits a marked rise in concentration in the forebrain of chick embryos between 17 days of incubation and hatching (Kobayashi & Eiduson, 1970). It has also been reported that reducing catecholamines in the brain of the chick embryo with reserpine can modify hatching time (Sparber & Shideman, 1968).

To summarize this section on hormones, it may be said that the only hormone that has experimentally been shown to affect hatching is thyroxine. Both the time of pipping and hatching can be significantly advanced or retarded by the modification of endogenous levels of thyroxine. The precise way in which these effects are mediated is not yet entirely clear. Certainly it has been well documented that events associated with metabolism are altered by experimentally modifying endogenous levels of thyroxine, and that similar metabolic changes normally occur as hatching approaches. However, the extent to which thyroxine influences hatching by acting on general metabolism (accelerated maturation) and the extent to which it acts more directly on specific neural and behavioral events associated with hatching requires further experimentation. It is clear, however, from the experiments reported and discussed here that by about 18 days of incubation, sufficient levels of endogenous thyroxine are available to the embryo for the normal initiation and timing of hatching. Preventing further production of thyroxine after this time does not modify hatching, although as we have seen, *raising* the base level of thyroxine by injection of exogenous sources can affect hatching time even when the injection is done only 20 hours before the final hatching climax.

D. *Neural Factors*

1. Tucking, Membrane Penetration, and Pipping

It has been repeatedly mentioned in this article that on about day 17 a new motor pattern is added to the repertoire of the domestic fowl embryo (i.e., Type III, movements). The question has arisen as to whether this new behavior pattern is related to the maturation of a specific morphological or physiological neural system at this time. If this is, in fact, what is occurring, and if this new system does not extend throughout a major portion of the CNS, then it should be possible, by selective destruction or isolation of parts of the CNS, to identify and study such a system.

As a first approach to this problem we have isolated the brain from part of the spinal cord by making a gap in the thoracic spinal cord at early (48 hours embryonic stages (Oppenheim & Narayanan, 1968). Ideally, such a gap should be made at the cervical-medullary border in order to completely isolate spinal from brain regions, but such an operation seriously interferes with the onset and maintenance of normal pulmonary respiration at late stages of incubation.

It was found that thoracic operations of this kind do not interfere with tucking, MP, pipping, or the onset of climax. However, whereas normally the climax movements result in a counterclockwise rotation and simultaneous cracking around the shell cap by the embryo, the thoracic embryos, although continuing to make climax movements, were never able to rotate or emerge on their own. A closer examination of the behavior of these embryos suggested that it was a deficit in leg movements that was primarily responsible for this. Although thoracic embryos did exhibit leg movements, these were primarily of the jerky Type I and II variety. It was very rare to observe the vigorous simultaneous or alternate stemming of the tarsal joints that is so typical of climax. Furthermore, the leg movements that did occur were out of phase with the climax movements occurring *above* the thoracic gap. Thus, due to these deficiencies in leg movements, the thoracic embryos were unable to rotate or hatch on their own. Because these thoracic embryos were able to exhibit alternating leg movements after hatching, it is clear that the basic motor mechanism for such movements is in the spinal cord. However, in order for these movements to be properly organized into the *total* climax pattern, the spinal mechanism must be functionally and anatomically connected to the brain. A similar conclusion has recently been reached by Corner *et al.* (this volume) after making cervical spinal transections in the chick, and a similar specific defect in rotation during climax was found by Helfenstein and Narayanan (1969) after early bilateral removal of the hindlimb buds of chick embryos. These experimental findings reinforce the observational evidence previously discussed that suggested an important role of the legs in climax rotation. However, it is not certain to what extent these results are generalizable to other birds. Johnson (1969) has recently reported briefly that the legs may not be necessary for climax rotation in the bobwhite quail. We have confirmed this suggestion in the following way. Both legs of bobwhite quail embryos were extended outside the shell through a small (3-mm diameter) hole at the pointed pole, and the egg was suspended such that the legs could not come into contact with any solid surface for leverage. By use of time-lapse cinematography we followed 12 such embryos for 30 hours from the time of pipping. Most of these embryos exhibited rotation and cracking of the shell cap. The median amount of rotation was about 90°, although some embryos rotated as much as 300°. It was clear from the film that, while the legs were not playing any role in rotation, they still exhibited the typical flexions and extensions that accompany each climax movement (see above). The fact that these embryos rotated less than controls and did not hatch is probably due to excessive drying within the shell through the opening made for extending the legs. Because of these rather surprising results with quail, we decided to repeat this experiment with the domestic chicken. Again 12 embryos were prepared with the legs completely extended out through a hole in the small end of the shell. Five control embryos had a hole made in the shell to control for dessication, but their legs were left in the normal position. All five control embryos rotated normally (200–300°) and hatched. The average amount of rotation of the experimental group was 7°. These results support the experiments reported above concerning the important role of the legs in climax rotation of the chicken. The quail, on the other hand, clearly depends more upon the activity of other body parts for such rotation.

In a study designed to examine the effects of deafferentation of the trigeminal area on the motility of the chick embryo, Hamburger and Narayanan (1969) have reported that their experimental embryos were never observed to exhibit the Type III movements, and subsequently failed to tuck or hatch. These embryos always remained in the typical pretucked position (Plate I, Fig. 2). While these results might suggest that sensory input to the facial region is important in evoking selective patterns of prehatching and hatching behavior (i.e., Type III movements), Hamburger and Narayanan (1969) point out that, because of inadvertent brain damage accompanying this operation, such a conclusion is premature. The cerebellum was absent or abnormal in 58% of their cases, and the mesencephalon was damaged in 65% of the cases. Therefore, the failure of these embryos to tuck or hatch may be due to the loss of brain regions that are indispensable for the occurrence of Type III behavior. An alternative explanation of these results might be that those neural mechanisms underlying prehatching and hatching behavior require a certain level of nonspecific tonic sensory input in order to be properly facilitated or triggered. At the present time it is not possible to decide between these several alternatives.

Further insight into the role of specific brain parts in prehatching behavior has been gained from the study of Decker (1970), discussed earlier. Briefly, Decker found that embryos lacking *both* otocysts exhibited severe degeneration of the vestibular nuclei in the medulla and that these embryos failed to exhibit Type III tucking movements, whereas embryos with a *unilateral* otocyst removal had only minor CNS degeneration, and these embryos were able to tuck and hatch normally. Decker concludes from these findings that it was not the loss of vestibular sensory input that prohibited Type III tucking movements, but rather that the secondary CNS degeneration produced a loss of the relationship between centers in the CNS that are apparently essential for the mediation of these Type III movements.

Although such an interpretation is plausible, the technical difficulties inherent in any attempt to separate *secondary* degenerative effects from a *primary* loss of functional sensory (vestibular) input, should serve to make one exceedingly cautious in drawing such a conclusion. To argue, as Decker does, that the reason the unilateral embryos were able to tuck and hatch was due to the absence of severe secondary CNS degeneration in these cases is also not entirely convincing. The intact otocyst may have been able to compensate functionally for its missing partner during development by modification in synaptic connections, etc. More reassuring support for the contention that *specific* vestibular input is not of primary importance in the mediation of Type III tucking movements comes from another observation of Decker (1970), namely, that the unilateral embryos that hatched did not exhibit a postural righting response until 5 hours *post-hatching*. This consideration, when added to the report of Kovach (1970) that it is not possible to consistently elicit a postural reflex in domestic chick embryos until *after* tucking, makes it unlikely that the vestibular system is involved in type III activity until shortly before climax ensues, thus providing strong support for Decker's (1970) original contention. Again, however, this is not to rule out the possibility that the role of the vestibular system in tucking (and later prehatching behavior) may be the more *nonspecific* one of contributing a certain level of tonic excitability to particular portions of the CNS.

In summary, then, information from the experiments by Hamburger and Narayanan (1969) and Decker (1970) suggests that either the hindbrain (medulla-cerebellum), the midbrain, the vestibular system, or some combination of these is involved in the Type III movements associated with tucking.

In a recent study attempting to elucidate the role of the midbrain and forebrain in prehatching and hatching behavior, we have been able to demonstrate that the midbrain is involved in the mediation of Type III tucking movements (Oppenheim, 1972a). Embryos with the midbrain completely ablated at early prefunctional stages (i.e., 40–45 hours incubation) were never observed to exhibit Type III movements, and they failed to tuck or hatch. On the other hand, if the cerebral hemispheres, or only the dorsal midbrain (primarily the optic lobes), were removed, the embryos did exhibit Type III movements, and subsequently managed to tuck and hatch.

Although this finding does not allow one to rule out the hindbrain or the vestibular apparatus as also being involved in tucking, it does make it clear that if these latter parts are involved they are part of a larger system which definitely includes the midbrain as well.

2. Climax

In the experiment just discussed, involving early embryonic removal of the midbrain or forebrain (Oppenheim, 1972a), it was found that climax behavior was entirely eliminated by forebrain removal (a substantial portion of the diencephalon was left intact in these forebrain cases). Domestic fowl embryos with the forebrain removed were able to assume a typical hatching position (Plate VIII, Fig. 31), and subsequently penetrated the membrane and made pipping movements (they were not able to actually break the shell, as the primordia of the upper beak was inadvertently removed during the

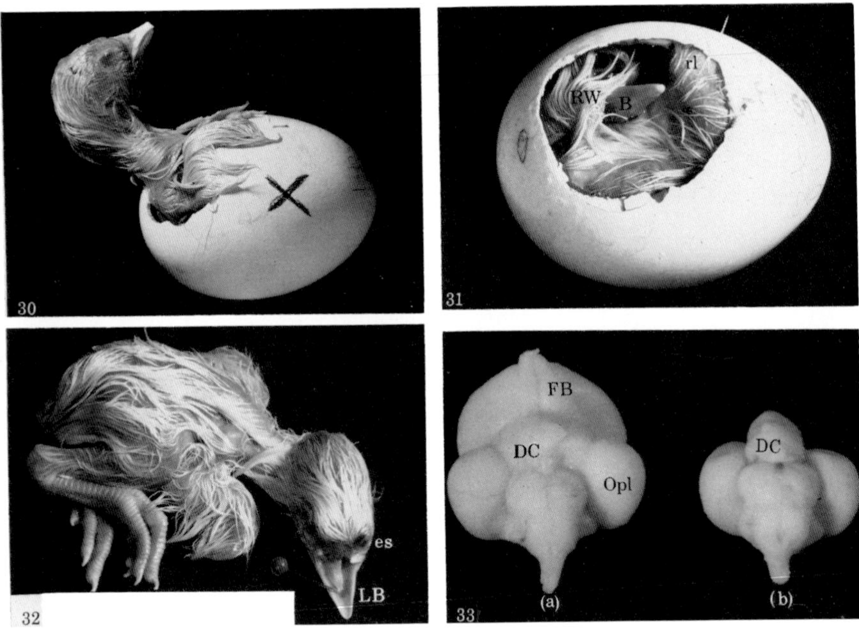

Plate VIII

Fig. 30. An 18- to 19-day embryo pulled partially out of the shell (see text for details).

Fig. 31. A domestic fowl embryo in which the forebrain was removed on day 2 of incubation, and which has now attained a rather typical and normal hatching position. B, Beak; rl, right leg; RW, right wing.

Fig. 32. A forebrain-less embryo after being pulled out of the shell on day 21 of incubation. Note the absence of upper beak and eyes. es, Eyespot; LB, lower beak.

Fig. 33. Ventral view of the brain of a control (A), and a "*total*" forebrainless embryo (B) at 16 days of incubation. DC, Diencephalon; FB, forebrain; Opl, optic lobe.

brain ablation; see Plate VIII, Fig. 32). These embryos were observed to initiate and maintain pulmonary respiration and to withdraw the yolk sac at the normal time; the hatching muscle also became normally enlarged at the appropriate stage. Yet, at the time when control embryos were beginning climax or emerging from the shell, the forebrainless embryos continued in the preclimax pattern of behavior (i.e., primarily Types I and II activity), for 1 to 3 more days, at which time they often died. Histological examination of the brains of these embryos showed that the entire cerebral hemispheres and part of the diencephalon had been successfully removed (Plate VIII, Fig. 33). In one group of embryos, in which the entire diencephalon and a small portion of the posterior telencephalon were spared (Fig. 34), climax and emergence were normal. Although the rather gross ablations necessitated by the technique of early embryonic microsurgery make it impossible to identify the precise areas within the forebrain that are indispensable for climax, it is at least now clear that in the future, one should begin to make more precise lesions in both the anterior diencephalon and the posterior and basal portions of the cerebral hemispheres.

As a result of making acute transections of the brain of fowl embryos at *late* stages of incubation (18–20 days), after pulling the head and neck out of

FIG. 34. Ventral view of the brain of a control (a), and a *partial* forebrainless embryo (b). FB, Forebrain; Hyps, hypophysis.

the shell, Corner and Bakhuis (1969; also see Corner *et al.*, this volume) have arrived at a similar conclusion regarding the role of the forebrain in climax behavior. They attempted to transect the brain in such a way as to isolate the forebrain from the midbrain. Although the behavior of the embryos in their study was not systematically (i.e., quantitatively) examined, Corner and Bakhuis did report that their operated embryos failed to show the typical behavior pattern characteristic of climax. Recent evidence of these authors (Corner *et al.*, this volume) tends to give more support to the *diencephalic* location of the "climax-trigger" mechanism. Corner *et al.* (this volume) also present evidence that certain aspects of the behavior controlled by this diencephalic mechanism may be involved in arousal, and sleep–wakefulness patterns in the newly hatched and adult chicken.

IV. The Problem of the Embryonic Origin of Posthatching Motor Patterns

Although a consideration of the relationship between embryonic and postnatal behavior is peripheral to the main subject matter of the present article, the fact that the chicken embryo does not begin to exhibit clear patterns of coordinated motor behavior until the onset of prehatching and hatching stages, and since such behavior is more similar to coordinated posthatching behavior than is the early embryonic (Types I and II) activity, such a consideration might prove fruitful in gaining a fuller understanding of the embryonic antecedents of posthatching behavior.

This question of what relationships exist between the embryonic and postnatal motor behavior of birds has been, and remains, an intriguing but elusive problem in developmental psychobiology. In amphibians and fish such a relationship has been clearly shown to exist, and many of the details of the underlying neuroanatomy are known (Coghill, 1929; Hughes & Prestige, 1967; Tracy, 1929). In these forms overt motor behavior during the embryonic period develops gradually, becoming ever more complex as it comes more and more to resemble the larval and adult behavior patterns (e.g., walking, swimming, food getting). Thus there is an unambiguous continuity in the development of overt behavior from the earliest embryonic stages to larval and adult stages. This developmental process is clearly of adaptive value for these forms, since they hatch at a rather immature stage of development yet must fend for themselves at that time.

With the possible exception of the ontogeny of sleep–wakefulness behavior in the chick (see Corner *et al.*, this volume), such a continuous development of overt motor behavior, with obvious relationships between embryonic and posthatching stages, does not appear to exist in birds. For example, Hamburger (1968) and his colleagues have characterized the behavior of chick embryos during much of the incubation period as being jerky, convulsivelike, and uncoordinated. Although no systematic studies have been done to statistically verify the uncoordinated nature of the chick embryo's motility, such a general characterization is, nevertheless, phenomenologically correct, especially when compared to fish and amphibians. This is not to say that there is a complete lack of coordination, or that behavior patterns resembling posthatching acts are entirely absent in the embryo (e.g., alternating leg movements, while rare, can be seen in the chick embryo). Nor does this mean that the simple muscular components of more complex posthatching motor action patterns are absent during the embryonic period. Certainly the muscles and the motoneurons, comprising the final common pathway, that are involved in a jerky or tonic leg movement of the embryo are the same as those involved in walking after hatching, for example. What such a statement does mean, however, is that on the basis of our present knowledge, it has not been possible to detect or describe any complex, overt posthatching behavior pattern in birds as having developed from early to later stages in a manner similar to that of swimming or walking in fish and amphibians, i.e., a gradual and continuous process with the behavior pattern becoming more and more adultlike. What seems to happen in the chick is that simple muscular components which have been occurring throughout most of the incubation period become organized rather suddenly into a unique and new behavior pattern at certain stages of development (e.g., hatching and posthatching behavior). This organization is probably due, in part, to exposure to, or an increased sensitivity to, new sensory stimuli (Kuo, 1932c)—such as when the bird stands for the first time—and in part to intrinsic neurogenetic events, such as maturation of specific morphological or functional patterns of neural connectivity at different brain levels. It also seems likely that in some cases, at least, the morphological and physiological substrate of complex posthatching motor behavior may have been developed to a high level during embryogeny, but remains actively inhibited by other systems until the time when actually needed, as perhaps determined by a change in sensory input upon hatching (e.g., posthatch arousal, see Corner *et al.*, this volume).

In spite of this basic difference in the overt development of behavior between fish and amphibians versus birds, one might still expect to detect some meaningful differences in the embryonic behavior of altricial versus precocial birds, especially since these two types of birds must do rather different things immediately after hatching in order to survive (e.g., see Anokhin, 1964). Some information concerning this problem is available from a recent comparative study of prehatching and hatching behavior discussed earlier (Oppenheim, 1972b). Special attention was directed in that study toward examining embryonic differences in oral activity that might be related to the striking differences in the method of obtaining food after hatching, e.g., the chick pecks, the duck dabbles or strains food through the bill, the cardinal gapes, and the pigeon places its beak deep inside the parents' mouth for regurgitation feeding.

The situation regarding oral activity in the duck embryo proved to be rather interesting. Not only was the *rate* of bill-clapping rather high in the duck (Table I), but the character of bill-clapping was remarkably similar to the straining or dabbling of newly hatched and adult ducks, namely, a rapid but low amplitude (i.e., the bill was not opened very wide) opening and closing of the mandibles. Whether this type of embryonic oral activity is a necessary precursor of dabbling or straining is not yet clear. Examination of embryonic oral activity in a species of diving duck might provide some insight into this question.

It was found that the spontaneous rate of gapinglike movements was remarkably low in all species during the last 2–6 days of incubation, and there were no differences observed between those altricial species that normally gape after hatching versus the other species (Oppenheim, 1972b). Furthermore, stimulation of the altricial embryos with stimuli known to reliably evoke gaping after hatching (Schaller & Emlen, 1961; Tinbergen & Kuenen, 1939) failed to elicit a true gaping response, even minutes before emergence, while the same stimulation, with the same birds, reliably evoked gaping shortly after emergence from the shell. One overt difference between these two periods is the posture of the embryo (i.e., a cramped hatching position *versus* being freely stretched out on its ventral surface after hatching). This observation suggests that sensory input concerning the position of the bird may play a role in inhibiting (or facilitating) the effectiveness of a specific stimulus in eliciting a complex behavior pattern such as gaping. Another, possibly related, difference between these two stages may be the level of arousal or wakefulness (Corner *et al.*, this volume).

It seems likely that similar kinds of environmentally or centrally (CNS) induced inhibition (as well as the lack of adequate stimulation or low levels of arousal, *in ovo*) might be responsible for the absence of clear embryonic precursors of complex posthatching behavior in birds, as compared to fish and amphibians (see Crain, Volume 2 of this serial publication; Foelix & Oppenheim, this volume, p. 124). Bird embryos may exhibit overt behavior merely to facilitate normal neuromuscular or skeletal development as first suggested by Coghill (1909) and later experimentally verified by Drachman (1968), or such behavior may be a specific adaptation to the intra-egg environment (e.g., prehatching and hatching behavior) which, for the most part, is the same in all birds. Thus, it may not be too surprising if, as with hatching behavior, early embryonic motor behavior (prior to prehatching stages) is also found to be basically similar in all birds, altricial and precocial, or if there is found to be a lack of species (or even higher order) specificity in the relationship between the embryonic and posthatching overt motor behavior of birds. This is not to say that the complex and adaptive behavior exhibited by a newly hatched bird is independent of prior motor behavior. It may well be that the normal organization and manifestation of postnatal behavior is somehow dependent upon prior prenatal activity. Even with this consideration in mind, however, overt embryonic behavior, in birds at least, may still be a rather specific and limited example of the Darwinian idea of *common embryonic resemblance*. If this suggestion is correct, then students of the development of behavior may have to modify their strategies if they hope to elucidate the way in which meaningful and complex patterns of posthatching motor behavior are developed during the embryonic period in different species of birds (see Hamburger, this volume).

V. Summary and Conclusions

The present consideration of prehatching and hatching behavior in egg-laying animals has shown that the common biological problem of escaping from the egg shell or capsule has been solved in a multitude of ways by the embryos of different groups of animals, ranging from invertebrates to monotremes, with the invertebrates showing the greatest diversity.

Although the evidence is rather limited for most forms, it would nevertheless appear that birds have evolved the most complex and specific *behavioral* mechanisms for use in escaping from the egg. With only one well-documented exception, the prehatching and hatching behavior of all birds that have so far been examined is remarkably similar, even between precocial and altricial forms. The one exception—and an interesting one indeed—is the family Megapodiidae, or the mound-building birds of Australia.

Whether the megapode habit of not physically incubating the eggs, as most birds do, is a primitive carry-over from a reptilian ancestor, or whether it represents a recent innovation that has evolved from the more typical avian incubation pattern is not yet clear. Frith (1962) appears to favor the latter, although his reasons for this choice are unclear. The recent study of Baltin (1969) suggests that the prehatching and hatching behavior of at least one species of Megapode, *Alectura lathami*, may more closely resemble reptiles than birds. A careful examination of the prehatching and hatching stages of both reptiles and megapodes might be quite fruitful in answering the nagging question of the taxonomic relationship of these forms.

It is also interesting that the embryos of many forms have developed a special appendage to aid in the hatching process (egg tooth, egg caruncle, egg burster). In many cases certain specialized movements have also evolved to aid in the use of this appendage (e.g., head- or snout-thrusting movements).

Except for a few invertebrates, the mechanisms underlying prehatching or hatching behavior have only been examined in birds, or more correctly, in the domestic fowl. The extensive use of the domestic fowl results primarily from its accessibility for laboratory studies at all times of the year. Although one should always be exceedingly cautious in using an animal that, due to its selective breeding history, is biologically maladaptive for studying general biological phenomena, the fact that the overt prehatching and hatching behavior of most other birds appears to be similar to that of the domestic fowl makes it likely that the underlying mechanisms, at least in broad outline, will also be similar.

As the present account has shown, we are still a long way from attaining a relatively complete understanding of any of the various prehatching and hatching mechanisms of even the domestic fowl. However, past and present studies (as reviewed and discussed here) have, if nothing else, certainly made it clear that the situation is exceedingly complex. Metabolic, hormonal, biochemical, neural, sensory, passive mechanical, and nonneural morphological factors are all involved in this behavior, and in at least certain instances (e.g., climax, synchronization, etc.), several of these factors may be subtly related to one another.

As stated in the Introduction, attempts to gain a better understanding of prehatching and hatching behavior involve a multitude of questions that are of interest for both general and developmental biology. It is hoped that, in the future, with the advantage of an increasingly sophisticated technology, and by combining a broadly comparative, physiological and behavioral approach, many of these questions can be answered. For, as recognized over 200 years ago by de Réaumur, it would indeed be intellectually satisfying—if not of immediate practical relevance—to know precisely how a bird embryo "comes out of his shell."

Acknowledgments

The research of the author presented in this article was supported by grants MH 15323 and MH 16598 from the National Institute of Mental Health, and by the continuing generous support of the North Carolina Department of Mental Health. I would like to thank Rubenia Daniels, Donna Goodwin, Marjory Jones, Hinda Levin, James Peterson, and John Reitzel for their help during various stages of the preparation of this article. Plate I, Fig. 2b was drawn by Mr. Frank Pearce, and many of the photos were taken by Mr. Gus Martin.

References

Abbot, U. K., & Craig, R. M. Observations on hatching time in three avian species. *Poultry Science*, 1960, **39**, 827–830.

Adams, E. A., & Bull, A. L. The effects of antithyroid drugs in chick embryos. *Anatomical Record*, 1949, **104**, 421–444.

Anokhin, P. K. Systemogenesis as a general regulator of brain development. *Progress in Brain Research*, 1964, **9**, 54–86.

Armstrong, P. B. Mechanism of hatching in *Fundulus heteroclitus. Biological Bulletin*, 1936, **71**, 407.

Asmundson, V. S. The position of turkey, chicken, pheasant, and partridge embryos that failed to hatch. *Poultry Science*, 1938, **17**, 478–489.

Balaban, M., & Hill, J. Perihatching behaviour patterns of chick embryos (*Gallus domesticus*). *Animal Behavior*, 1969, **17**, 430–439.

Balaban, M., & Hill, J. Effects of thyroxine level and temperature manipulations upon the hatching of chick embryos (*Gallus domesticus*). *Developmental Psychobiology*, 1971, **4**, 17–35.

Balfour-Brown, F. The life history of a water beetle. *Nature* (*London*), 1913, **92**, 20–24.

Baltin, V. Zur Biologie und Ethologie des Talegalla Huhns (*Alectura lathami, Gray*) unter besonderer Berücksichtigung des Verhaltens während der Brutperiode. *Zeitschrift für Tierpsychologie*, 1969, **26**, 524–572.

Beebe, C. W. *The bird, its form and function.* New York: Dover 1906. (Reprinted: 1965.)

Bellairs, A. d'A. *The life of reptiles.* New York: Universe Books, 1970.

Bent, A. C. *Life histories of North American wildfowl.* Part I. New York: Dover, 1923. (Reprinted: 1962.)

Bent, A. C. Life histories of North American Gallinaceous birds. *Bulletin of the U.S. National Museum*, 1932, **162**, xi–490.

Berrill, M. The embryonic behavior of *Caprella unica* (Crustacea: Amphipoda). *Canadian Journal of Zoology*, 1971, **49**, 499–504.

Bertling, A. E. L. On the hatching and rearing of the brush turkeys at the zoo. *Aviculture Magazine*, 1904, **2**, 294–297.

Betz, T. W. The pars distalis in avian development. In E. J. W. Barrington & M. Hamburg (Eds.), *Hormones in development.* New York: Appleton, 1971. Pp. 75–94.

Beyer, R. E. The effect of thyroxine upon the general metabolism of the intact chick embryo. *Endocrinology*, 1952, **50**, 497–503.

Bjärvall, A. The critical period and the interval between hatching and exodus in mallard ducklings. *Behaviour*, 1967, **28**, 141–148.

Bock, W. J., & Hikida, R. S. Turgidity and function of the hatching muscle. *American Midland Naturalist*, 1969, **81**, 99–106.

Bolotnikov, A., Kamenskii, Y., Afanaseva, L., & Yakonenko, L. Orientation and displacement of eggs in nests as factors of embryonic development of birds. *Ekologiya*, 1970, **5**, 85–87.

Bragg, A. N. Observations on the ecology and natural history of Anura. V. The process of hatching in several species. *Proceedings of the Oklahoma Academy of Science*, 1940, **20**, 71–74.

Brandstetter, W. E., Watterson, R. L., & Veneziano, P. Modified growth pattern of the musculus complexus ("hatching muscle") of chick embryos following thiourea treatment or hypophysectomy by partial decapitation. *Anatomical Record*, 1962, **142**, 299.

Breed, F. S. The development of certain instincts and habits in chicks. *Behavior Monographs*, 1911, **1**, 1–78.

Brooks, W. S., & Garrett, S. E. The mechanism of pipping in birds. *Auk*, 1970, **87**, 458–466.

Brooks, W. S. & Ungar, F. Effect of C_{21}-methyl steroids on the *musculus complexus* and hatching of the chick. *Proceedings of the Society for Experimental Biology and Medicine*, 1967, **125**, 488–492.

Buetter-Jansch, J. *Origins of man.* New York: Wiley, 1966.

Byerly, T. C., & Olsen, M. W. The influence of gravity and air-hunger on hatchability. *Poultry Science*, 1931, **10**, 281–287.

Byerly, T. C., & Olsen, M. W. Time and manner of determination of the malposition head-in-small-end-of-egg. *Poultry Science*, 1933, **12**, 261–265.

Byerly, T. C., & Olsen, M. W. Certain factors affecting the incidence of malpositions among embryos of the domestic fowl. *Poultry Science*, 1936, **15**, 163–168.

Byerly, T. C., & Olsen, M. W. Egg turning, pipping position and malpositions. *Poultry Science*, 1937, **16**, 371–373.

Choe, S. On the eggs, rearing, habits of the fry, and growth of same cephalopoda. *Bulletin of Marine Sciences of the Gulf and Caribbean*, 1966, **16**, 330–348.

Coghill, G. E. The reactions to tactile stimuli and the development of the swimming movements in embryos of *Diemyctylus torosus*, Eschscholtz. *Journal of Comparative Neurology*, 1909, **19**, 83–105.

Coghill, G. E. *Anatomy and the problem of behavior.* New York: Hafner, 1929. (Reprinted: 1964.)

Collias, N. E. The development of social behaviour in birds. *Auk*, 1952, **69**, 127–159.

Corner, M. A. Aspects of the functional organization of the central nervous system during embryonic development in the chick. In L. Jilek & S. Trojan (Eds.), *Ontogenesis of the brain.* Prague: Charles University Press, 1968. Pp. 85–96.

Corner, M. A., & Bakhuis, W. L. Developmental patterns in the central nervous system of birds. V. Cerebral electrical activity, forebrain function and behavior in the chick at the time of hatching. *Brain Research*, 1969, **13**, 541–555.

Corner, M. A., & Bot, A. P. C. Developmental patterns in the central nervous system of birds. III. Somatic motility during the embryonic period and its relation to behavior after hatching. *Progress in Brain Research*, 1967, **26**, 214–236.

Craig, W. Behavior of the young bird in breaking out of the egg. *Journal of Animal Behavior*, 1912, **2**, 296–298.

Cutting, W. C., & Tainter, M. L. Comparative effects of dinitrophenol and thyroxin on tadpole metamorphosis. *Proceedings of the Society for Experimental Biology and Medicine*, 1933, **31**, 97.

Cutting, W. C., Mehrtens, H. G., & Tainter, M. L. Actions and uses of dinitrophenol. *Journal of the American Medical Association*, 1933, **101**, 193–195.

Daniel, J. C. An embryological comparison of the domestic fowl and the redwinged blackbird. *Auk*, 1957, **74**, 340–358.

Davis, C. C. Mechanisms of hatching in aquatic invertebrate eggs. *Oceanography and Marine Biology Annual Review*, 1968, **6**, 325–376.

Dawes, C., & Simkiss, K. The acid-base status of the blood of the developing chick embryo. *Journal of Experimental Biology*, 1969, **50**, 79–86.

Decker, J. D. The influence of early extirpation of the otocysts on development of behavior of the chick. *Journal of Experimental Zoology*, 1970, **174**, 349–364.

Deraniyagala, P. E. P. *The tetrapod reptiles of Ceylon. Vol. I. Testudinates and crocodilians.* London: Dulau, 1939.

de Réaumur, M. *The art of hatching and bringing up domestick fowls of all kinds, at any time of year. Either by means of the heat of hot-beds, or that of common fire.* London: A. Millar & J. Nourse, 1750.

Drachman, D. B. The role of acetylcholine as a trophic neuromuscular transmitter. In G. E. W. Wolstenholme & M. O'Connor (Eds.), *Growth of the nervous system.* Boston: Little, Brown, 1968. Pp. 251–273.

Drent, R. H. Functional aspects of incubation in the herring gull (*Larus argentatus pont.*). *Behavior Monographs*, 1967 (Suppl. 17), 1–132.

Driver, P. M., Higgins, T., & Newman, D. "Pipping" mechanism and hatching in nidifuguous birds. *Nature (London)*, 1968, **219**, 394–395.

El-Ibiary, H. M., Shaffner, C. S., & Godfrey, E. F. Pulmonary ventilation in a population of hatching chick embryos. *British Poultry Science*, 1966, **7**, 165–176.

Fernandez-Marcinowski, K. Der Mechanismus des Schlüpfens bei den Amphibien-larven. *Biologisches Zentralblatt*, 1921, **41**, 423–432.

Fisher, H. I. The "hatching muscle" in the chick. *Auk*, 1958, **75**, 391–399.

Fisher, H. I. Hatching and the hatching muscle in some North American ducks. *Transactions of the Illinois Academy of Science*, 1966, **59**, 305–325.

Foote, F. M. Translocation of albumen in eggs and embryos of the quail, chick, and duck. *Poultry Science*, 1969, **48**, 304–306.

Freeman, B. M. Gaseous metabolism in the domestic chicken II. Oxygen consumption in the full-term and hatching embryo, with a note on a possible cause for "death in shell." *British Poultry Science*, 1962, **3**, 63–72.

Freeman, B. M. Studies on the oxygen requirements and hatching mechanisms of the domestic fowl. Unpublished doctoral dissertation, University of Leicester, 1964.

Freeman, B. M. The mobilization of hepatic glycogen in *Gallus domesticus* at the end of incubation. *Comparative Biochemistry and Physiology*, 1969, **28**, 1169–1176.

Freeman, B. M., & Misson, B. H. pH, pO_2, and pCO_2 of blood from the foetus and neonate of *Gallus domesticus. Comparative Biochemistry and Physiology*, 1970, **33**, 763–772.

Frith, H. J. *The Mallee-fowl, the bird that builds an incubator.* Sydney: Halstead Press, 1962.

George, J. C., & Iype, T. P. The mechanisms of hatching in the chick. *Pavo*, 1963, **1**, 52–56.

Gideiri, Y. B. A. The behaviour and neuroanatomy of some developing teleost fishes. *Journal of Zoology*, 1966, **149**, 215–241.

Gill. P. M. The effect of adrenaline on embryonic chick glycogen *in vitro* as compared with its effect *in vivo*. *Biochemical Journal*, 1938, **32**, 1792–1799.

Goethe, F. Experimentelle Brutbeendigung und andere Brutethologische Beobachtungen bei Silbermöwen (*Larus argentatus*). *Journal of Ornithology*, 1953, **94**, 160–174.

Goethe, F. Beobachtungen bei der Aufzucht junger Silbermöwen. *Zeitschrift für Tierpsychologie*, 1955, **12**, 402–433.

Gottlieb, G. *Development of species identification in birds: An inquiry into the prenatal determinants of perception.* Chicago: University of Chicago Press, 1971.

Gottlieb, G., & Kuo, Z.-Y. Development of behavior in the duck embryo. *Journal of Comparative and Physiological Psychology*, 1965, **59**, 183–188.

Gottlieb, G., & Vandenbergh, J. G. Ontogeny of vocalization in duck, and chick embryos. *Journal of Experimental Zoology*, 1968, **168**, 307–326.

Griffiths, M., McIntosh, D. L., & Coles, R. E. A. The mammary gland of the echidna, *Tachyglossus aculeatus*, with observations on the incubation of the egg and on newly hatched young. *Journal of Zoology*, 1969, **158**, 371–386.

Grossowicz, M. Influence of thiourea on development of chick embryo. *Proceedings of the Society for Experimental Biology and Medicine*, 1946, **63**, 151–152.

Hamburger, V. The beginnings of co-ordinated movements in the chick embryo. In *Growth of the nervous system.* G. E. W. Wolstenholme & M. O'Connor (Eds.), Boston: Little, Brown, 1968. Pp. 99–105.

Hamburger, V., & Narayanan, C. H. Effects of the deafferentation of the trigeminal area on the motility of the chick embryo. *Journal of Experimental Zoology*, 1969, **170**, 411–426.

Hamburger, V., & Oppenheim, R. W. Prehatching motility and hatching behavior in the chick. *Journal of Experimental Zoology*, 1967, **166**, 171–204.

Hartman, C. G. Studies in the development of the opossum *Didelphys virginiana*, *L*. V. The phenomena of parturition. *Anatomical Record*, 1920, **19**, 251–261.

Helfenstein, M., & Narayanan, C. H. Effects of bilateral limb-bud extirpation on motility and prehatching behavior in chicks, *Journal of Experimental Zoology*, 1969, **172**, 233–244.

Hill, J. P., & DeBeer, G. R. Development of the monotremata. VII. The development and structure of the egg-tooth and the caruncle in the monotremes and on the occurrence of vestiges of the egg-tooth and caruncle in marsupials. *Transactions of the Zoological Society of London*, 1950, **26**, 503–544.

Hsiao, C. Y., & Ungar, F. Lipid changes in the chick hatching muscle. *Proceedings of the Society for Experimental Biology and Medicine*, 1969, **132**, 1047–1051.

Hudson, W. H. *The naturalist in La Plata.* New York: Appleton, 1895.

Hughes, A. Spontaneous movements in the embryo of *Eleutherodactylus martinicensis. Nature (London)*, 1966, **211**, 51–53.

Hughes, A., Bryant, S. V., & Bellairs, A. d'A. Embryonic behaviour in the lizard, *Lacerta vivipara. Journal of Zoology*, 1967, **153**, 139–152.

Hughes, A., & Prestige, M. C. Development of behaviour in the hindlimb of *Xenopus laevis. Journal of Zoology*, 1967, **152**, 347–359.

Hutt, F. B., & Pilkey, A. M. Studies in embryonic mortality in the fowl. *Poultry Science*, 1934, **13**, 3–14.

Ishida, J. Hatching enzyme in the freshwater fish, *Oryzias latipes. Annotationes Zoologicae Japonenses*, 1944, **22**, 155–164.

Johnson, R. A. Hatching behavior of the bobwhite. *Wilson Bulletin*, 1969, **81**, 79–86.

Judson, C. L. The physiology of hatching of aedine mosquito eggs: carbon monoxide stimulation and ethyl chloride inhibition of hatching of *Aedes aegypti* (*L.*) eggs. *Journal of the Institute of Physiology*, 1963, **9**, 787–792.

Kaspar, J. L. The origin and development of motor patterns of the domestic chick. Unpublished doctoral dissertation, University of Wisconsin, 1964.

Kear, J. The calls of very young anatidae (*Cicero gew*). *Vogelwelt*, 1967, **1**, 93–113.

Key, B. J., & Marley, E. The effect of the sympathomimetic amines on behavior and electrocortical activity of the chicken. *Electroencephalography and Clinical Neurophysiology*, 1962, **14**, 90–105.

Kirkman, F. B. The birth of a black-headed gull. *British Birds*, 1931, **24**, 283–291.

Kleitman, N., & Koppányi, T. Body-righting in the fowl (*Gallus domesticus*). *American Journal of Physiology*, 1926, **78**, 110–126.

Klicka, J., Edstrom, R., & Ungar, F. Acid mucopolysaccharide changes in chick hatching muscle. *Journal of Experimental Zoology*, 1969, **171**, 249–252.

Klicka, J., & Kaspar, J. L. Changes in enzyme activities of the hatching muscle of the chick (*Gallus domesticus*) during development. *Comparative Biochemistry and Physiology*, 1970, **36**, 803–809.

Kobayashi, K., & Eiduson, S. Norepinephrine and dopamine in the developing chick brain. *Developmental Psychobiology*, 1970, **3**, 13–34.

Kollros, J. J. Endocrine influences in neural development. In G. E. W. Wolstenholme & M. O'Connor (Eds.), *Growth of the nervous system.* Boston: Little, Brown, 1968. Pp. 179–192.

Kovach, J. K. Spatial orientation of the chick embryo during the last five days of incubation. *Journal of Comparative and Physiological Psychology*, 1968, **66**, 283–288.

Kovach, J. K. Development and mechanisms of behavior in the chick embryo during the last five days of incubation. *Journal of Comparative and Physiological Psychology*, 1970, **73**, 392–406.

Kuo, Z.-Y. Ontogeny of embryonic behavior in Aves. II. The mechanical factors in the various stages leading to hatching. *Journal of Experimental Zoology*, 1932, **62**, 453–483.

Kuo, Z.-Y. Ontogeny of embryonic behavior in Aves. III. The structural and environmental factors in embryonic behavior. *Journal of Comparative Psychology*, 1932, **13**, 245–271.

Kuo, Z.-Y. Ontogeny of embryonic behavior in Aves. IV. The influence of embryonic movements upon the behavior after hatching. *Journal of Comparative Psychology*, 1932, **14**, 109–112. (c)

Kuo, Z.-Y., & Shen, T. C. Ontogeny of embryonic behavior in Aves. XI. Respiration in the chick embryo. *Journal of Comparative Psychology*, 1937, **24**, 49–58.

Kyûshin, K. The embryonic development and larval stages of *Hemitripterus villosus* (*pallas*). *Bulletin of the Faculty of Fisheries, Hokkaido University*, 1968, **18**, 277–289.

Landauer, W. The hatchability of chicken eggs as influenced by environment and heredity. *Storrs Agricultural Experiment Station, Bulletin,* 1967, (Revision of 1948 article, Bulletin No. 262.)

Laven, H. Beiträge zur Biologie des Sandregenpfeifers, *Charadrins hiaticula, L. Zeitschrift für Ornithologie*, 1940, **88**, 183–287.

Leibson, L. G., & Leibson, R. S. Neural and humoral regulation of the bood sugar content in ontogenesis. II. The effect of insulin, adrenalin and glucose on the blood sugar content in chick embryo. *Bulletin of the Academy of Sciences of the USSR, Biological Series*, 1943, **3**, 179–188.

Lind, H. Studies on the behavior of the blacktailed godwit, (*Limosa limosa* (*L*)). *Meddelelse fra Naturfredningsradets Reservatudualg* (*Copenhagen*), 1961, **66**, 1–157.

Lutz, B. Observations on frogs without aquatic larvae. *Boletim do Museu Nacional, Rio de Janeiro, Zoologia*, 1944, **15**, 4–7.

McCoshen, J. A., & Thompson, R. P. A study of clicking and its source in some avian species. *Canadian Journal of Zoology*, 1968, **46**, 169–172. (a)

McCoshen, J. A., & Thompson, R. P. A study on the effect of egg separation on hatching time and of the source of clicking sounds in the embryo of the domestic chicken. *Canadian Journal of Zoology*, 1968, **46**, 243–248. (b)

Moog, F., & Richardson, D. The influence of adrenocortical hormones on differentiation and phosphatase synthesis in the duodenum of the chick embryo. *Journal of Experimental Zoology*, 1955, **130**, 29–56.

Moore, R. T. The least sandpiper during the nesting season in the Magdalen Islands. *Auk*, 1912, **29**, 210–223.

Narayanan, C. H., & Oppenheim, R. W. Experimental studies on hatching behavior in the chick. II. Extirpation of the right wing. *Journal of Experimental Zoology,* 1968, **168**, 395–402.

Needham, J. *Chemical embryology.* Vol. 3. Cambridge, Eng.: Cambridge University Press, 1931.

Nice, M. M. Development of behavior in precocial birds. *Transactions of the Linnean Society of New York*, 1962, **8**, 1–211.

Noble, G. K. The hatching process in *Alytes*, *Eleutherodactylus*, and other amphibians. *American Museum Novitiates*, 1926, **229**, 1–7.

Olexa, A. Breeding of common musk turtles, *Sternotherus odoratus*, at Prague Zoo. *International Zoo Yearbook*, 1969, **9**, 28–29.

Oppenheim, R. W. Some aspects of embryonic behaviour in the duck (*Anas platyrhynchos*). *Animal Behaviour*, 1970. **18**, 335–352.

Oppenheim, R. W. Experimental studies on hatching behavior in the chick. III. The role of the midbrain and forebrain. *Journal of Comparative Neurology*, 1972, **146**, 479–505. (a)

Oppenheim, R. W. Prehatching and hatching behavior in birds: a comparative study of altricial and precocial species. *Animal Behaviour*, 1972, **20**, 644–655. (b)

Oppenheim, R. W., & Narayanan, C. H. Experimental studies on hatching behavior in the chick. I. Thoracic spinal gaps. *Journal of Experimental Zoology*, 1968, **168**, 387–394.

Patay, R. Sur un dispositif aerifere de l'embryon de *Pediculus vestimenti*, nitzsche. *Bulletin de la Société Zoologique, Française*, 1941, **64**, 182–189.

Pohlman, A. G. Concerning the causal factor in the hatching of the chick, with particular reference to the Musculus complexus. *Anatomical Record*, 1919, **17**, 89–104.

Pope, C. H. *Snakes alive and how they live*. New York: Viking 1946.

Portet, R. Effet de la thyroxine sur les échanges respiratoires de l'embryon de poulet. *Journal de Physiologie, France*, 1960, **52**, 200–201.

Portmann, A. Études sur la cérébralisation chez les Oiseaux. III. Cérébralisation et mode ontogénétique. *Alauda, Paris*, 1947, **14**, 161–171.

Portmann, A. Le développement postembryonnaire. In P. P. Grassé (Ed.), *Traité de zoologie*. Paris: Masson, 1950.

Poulsen, H. A study of incubation responses and some other behaviour patterns in birds. *Videnskabelige Meddlelser fra Dansk naturhistorisk Forening i Kjøbenhavn*, 1953, **115**, 1–131.

Preyer, W. *Specielle physiologie des embryo*. Leipzig: Grieben, 1885.

Provine, R. R. Hatching behavior of the chick (*Gallus domesticus*): Plasticity of the rotatory component. *Psychonomic Science*, 1972, **29**, 27–28.

Ramachandran, S., Klicka, J., & Ungar, F. Biochemical changes in the Musculus complexus of the chick (*Gallus domesticus*). *Comparative Biochemistry and Physiology*, 1969, **30**, 631–640.

Rennie, J. *The domestic habits of birds*. Vol. III. London: Charles Knight, 1833.

Rigdon, R. H., Ferguson, T. M., Trammel, J. L., Couch, J. R., & German, H. L. Necrosis in the "pipping" muscle of the chick. *Poultry Science*, 1968, **47**, 873–877.

Rogler, J. C., Parker, H. E., Andrews, F. N., & Carrick, C. W. The effects of iodine deficiency on embryo development and hatchability. *Poultry Science*, 1959, **38**, 398–405.

Romanoff, A. L., & Laufer, H. The effect of injected thiourea on the development of some organs of the chick embryo. *Endocrinology*, 1956, **59**, 611–619.

Romanoff, A. L., & Romanoff, A. J. *The avian egg*. New York: Wiley, 1949.

Romijn, C. Respiratory movements of the chicken during the parafoetal period. *Physiology and Comparative Oecology*, 1948, **1**, 24–28.

Romijn, C., Fung, K. E., & Lokhorst, W. Thyroxine, thiouracil, and embryonic respiration in white leghorns. *Poultry Science*, 1952, **31**, 684–691.

Romijn, C., & Roos, J. The air space of the hen's egg and its changes during the period of incubation. *Journal of Physiology* (*London*), 1938, **94**, 365–379.

Sacc, M. De l'oeuf pendant l'incubation. *Annales des Sciences Naturelles, Zoologie*, 1847, **8**, 150–192.

Sauer, E. G. F., & Sauer, E. M. The behavior and ecology of the south African ostrich. *Living Bird*, 1966, **5**, 45–75.

Schaller, G. B., & Emlen, J. T. The development of visual discrimination patterns in the crouching reactions of nestling grackles. *Auk*, 1961, **78**, 125–137.

Schifferli, A. Uber Markscheidenbildung im Gehirn von Huhn und Star. *Revue Zoologique*, 1948, **55**, 117–212.

Sikes, E. K., & Wigglesworth, V. B. The hatching of insects from the egg, and the appearance of air in the tracheal system. *Journal of Microscopical Science*, 1931, **74**, 165–192.

Sinha, M. D., Ringer, R. K., Coleman, T. H., & Zindel, H. C. The effect of injected thiouracil on body weight and hatchability of the chick embryo. *Poultry Science*, 1959, **38**, 1405–1409.

Smail, J. R. A possible role of the Musculus complexus in pipping the chicken egg. *American Midland Naturalist*, 1964, **72**, 499–506.

Smith, M. *The British amphibians and reptiles*. London: Collins, 1954.

Sollmann, T. *A manual of pharmacology and its applications to therapeutics and toxicology*. Philadelphia: Saunders, 1957.

Spalding, D. A. Instinct and acquisition. *Nature* (*London*), 1875, **12**, 507–508.

Sparber, S. B., & Shideman, F. E. Prenatal administration of reserpine: effect upon hatching behavior, and brainstem catecholamines of the young chick. *Developmental Psychobiology*, 1968, **1**.

Spooner, C. E., & Winters, W. D. Evidence for a direct action of monoamines on the chick central nervous system. *Experientia*, 1965, **21**, 256–258.

Steinmetz, H. Die Embryonalenentwicklung des Blässhuhns (Fulica atra) unter besonderer Berücksichtigung der Allantois. *Morphological Jahrbuch*, 1930, **64**, 275–338.

Steinmetz, H. Beobachtungen und Untersuchungen über den Schlüpfact. *Zeitschrift für Ornithologie (Leipzig)*, 1932, **80**, 123–128.

Stoddard, H. L. *The bobwhite quail, its habits, preservation and increase*. New York: Scribner's, 1931.

Sutter, E. Growth and differentiation of the brain in nidifugous and nidicolous birds. *Proceedings of the Xth International Ornithological Congress*, 1950, **10**, 636–644.

Taylor, L. W., Kreutziger, G. O., & Abercrombie, G. L. The gaseous environment of the chick embryo in relation to its development and hatchability. 5. Effect of carbon dioxide and oxygen levels during the terminal days of incubation. *Poultry Science*, 1971, **50**, 66–78.

Tinbergen, N. *The herring gull's world.* Garden City, N.Y., Doubleday, 1953. (Reprinted: 1967.)

Tinbergen, N., & Kuenen, D. J. Feeding behavior in young thrushes. Releasing and directing stimulus situations in *Turdus m. merula L.* and *T. e. ericetorum Turton.* In C. H. Schiller (Ed.), *Instinctive behavior, the development of a modern concept.* New York: International Universities Press, 1939. Pp. 209–238.

Tracy, H. C. The development of motility and behavior reactions in the toadfish (*Opsanus tau*). *Journal of Comparative Neurology*, 1926, **40**, 253–360.

Tremor, J. W., & Rogallo, V. L. A small animal acto-ballistocardiograph: description and illustrations of its use. *Physiology & Behavior*, 1970, **5**, 247–251.

Truman, J. W., & Sokolove, P. G. Silk moth eclosion: hormonal triggering of a centrally programmed pattern of behavior. *Science*, 1972, **175**, 1491–1493.

Tuge, H., Kanayama, Y., & Chang, H. Y. Comparative studies on the development of the EEG. *Japanese Journal of Physiology*, 1960, **10**, 211–220.

Vanderstoep, J., & Richards, J. F. The changes in egg shell strength during incubation. *Poultry Science*, 1970, **49**, 276–285.

Van Emden, F. I. Egg-bursters in some more families of polyphagous beetles and some general remarks on egg-bursters. *Proceedings of the Royal Entomological Society of London, Series A, 1946,* **21**, 10–12.

Vince, M. A. Artificial acceleration of hatching in quail embryos. *Animal Behaviour*, 1966, **14**, 389–394.

Vince, M. A. Embryonic communication, respiration and the synchronization of hatching. In R. A. Hinde (Ed.), *Bird vocalizations.* London & New York: Cambridge University Press, 1969. Pp. 233–260.

Vince, M. A. Some aspects of hatching behaviour. In B. M. Freeman & R. F. Gordon (Eds.), *Aspects of poultry behaviour*. Edinburgh: British Poultry Science, 1970. Pp. 33–62.

Vince, M. A. Foetal respiration rates and the development of a stable respiratory metabolism in quail. *Comparative Biochemistry and Physiology*, 1972, **40**, 555–569.

Vince, M. A. & Cheng, R. C. H. Effects of stimulation on the duration of lung ventilation in quail fetuses. *Journal of Experimental Zoology*, 1970, **175**, 477–486.

Vince, M. A., & Chinn, S. Effect of accelerated hatching on the initiation of standing and walking in the Japanese quail. *Animal Behaviour*, 1971, **19**, 62–66.

Vince, M. A., Green, J., & Chinn, S. Acceleration of hatching in the domestic fowl. *British Poultry Science*, 1970, **11**, 483–488.

Vince, M. A., Green, J., & Chinn, S. Changes in the timing of lung ventilation and late foetal development in quail. *Comparative Biochemistry and Physiology*, 1971, **39**, 769–783.

Vince, M. A., & Salter, S. H. Respiration and clicking in quail embryos. *Nature (London)*, 1967, **216**, 582–583.

Visintini, F., & Levi-Montalcini, R. Relaxione tra differenziazione strutturale e funcionále dei centri e delle vie nervose nell'embrione di pollo. *Schweizer Archiv Für Neurologie und Psychiatrie*, 1939, **43**, 1–45.

Visschedijk, A. H. J. The air space and embryonic respiration. 1. The pattern of gaseous exchange in the fertile egg during the closing stages of incubation. *British Poultry Science*, 1968, **9**, 173–184. (a)

Visschedijk, A. H. J. The air space and embryonic respiration. 2. The times of pipping and hatching as influenced by an artificially changed premeability of the shell over the air space. *British Poultry Science*, 1968, **9**, 185–196. (b)

Visschedijk, A. H. J. The air space and embryonic respiration. 3. The balance between oxygen and carbon dioxide in the air space of the incubating chicken egg and its role in stimulating pipping. *British Poultry Science*, 1968, **9**, 197–210. (c)

von Baer, K. E. *Ueber Entwicklungsgeschichte der Thiere. Beobachtung und Reflexion.* I. Königsberg: Theil-Bornträger, 1828.

von Orelli, M. Uber das Schlupfen von *Octopus vulgaris, Sepia officinalis* und *Loligo vulgaris. Revue Suisse de Zoologie*, 1959, **66**, 330–343.

Walker, A. D. New light on the origin of birds and crocodiles. *Nature* (*London*), 1972, **273**, 257–263.

Wesserman, G. F., & Bernard, E. A. Adrenalin content of the chick embryo adrenal gland during development. *Acta Physiologica Latinoamerica*, 1970, **20**, 171–173.

Waterman, A. J. Effects of 2,4-Dinitrophenol on the early development of the teleost, *Oryzias latipes. Biological Bulletin*, 1939, **76**, 162–170.

Waters, N. F. Certain so-called malpositions a normal occurrence in the normal development of the chick embryo. *Poultry Science*, 1935, **14**, 208–216.

Watterson, R. L., Zimmerman, T. J., & Johnson, A. T. Effects of hypophysectomy (by partial decapitation) and thiourea treatment on normal weight changes in the hatching muscle (Musculus complexus), gastrocnemius and heart of white Peking ducks during late and prolonged incubation stages. *Anatomical Record*, 1964, **148**, 409.

Wayre, P. Breeding rheas (*Rhea americana*) at Norfolk Wildlife Park. *International Zoo Yearbook*, 1966, **6**, 207–208.

Wetherbee, D. K., & Bartlett, L. M. Egg teeth and shell rupture of the American woodcock. *Auk*, 1962, **79**, 117.

Whitman, C. O. The behavior of pigeons. *Carnegie Institution of Washington, Publication*, 1919, **3**, 1–161.

Wigglesworth, V. B. *The principles of insect physiology.* London: Methuen, 1965.

Windle, W. F., & Barcroft, J. Some factors governing the initiation of respiration in the chick. *American Journal of Physiology*, 1938, **121**, 684–691.

Windle, W. F., & Nelson, D. Development of respiration in the duck. *American Journal of Physiology*, 1938, **121**, 700–707.

Windle, W. F., Scharpenberg, L. G., & Steele, A. G. Influence of carbon dioxide and anoxemia upon respiration in the chick at hatching. *American Journal of Physiology*, 1938, **121**, 692–699.

Wortis, R. P. The transition from dependent to independent feeding in the young ring dove. *Animal Behaviour Monographs*, 1969, **2**, 3–54.

Zamenhof, S., & van Marthens, E. Hormonal and nutritional aspects of prenatal brain development. In D. C. Pease (Ed.), *Cellular aspects of neural growth and differentiation.* Los Angeles: University of California Press, 1971.

SLEEP AND WAKEFULNESS DURING EARLY LIFE IN THE DOMESTIC CHICKEN, AND THEIR RELATIONSHIP TO HATCHING AND EMBRYONIC MOTILITY

MICHAEL A. CORNER, WALTER L. BAKHUIS, AND CORA van WINGERDEN

Netherlands Central Institute for Brain Research
Amsterdam, The Netherlands

I. Introduction

Due to the upsurge in recent years of interest in behavioral embryology, especially in birds (for reviews see Corner & Bot, 1967; Gottlieb, 1968; Hamburger, 1963, 1969, and this volume), a detailed quantitative picture has emerged for the chicken in particular. Hatching too, or "climax" behavior

(Hamburger & Oppenheim, 1967), has been extensively studied (see Oppenheim, this volume). Attention is now turning towards the question of neurophysiological and ultrastructural bases for motility in the chick embryo (see Foelix & Oppenheim; Provine; and Oppenheim, this volume). Even in this species, however, little is yet known about the relationship between the motor activity *in ovo* and the later behavior of the animal. With recent demonstrations that (especially in young animals) considerable movement takes place during sleep as well as during wakefulness (Scherrer, Verley, & Garma, 1970), the possibility must be seriously entertained that spontaneous embryonic motility—rather than being a specifically prenatal type of behavior or an early sign of "arousal"—has in fact something to do with the development of *sleep* mechanisms. It is this problem which forms the main subject of the present article. Some attention will also be given to the question of brain processes and structures underlying the various behavioral phenomena to be described, as these have been worked out so far in the chicken. Finally, a brief comparison will be made with information available from mammalian studies.

II. Descriptive Studies

A. *Differentiation between Sleep and Wakefulness*

A qualitative distinction between these two functional states, grossly different in behaviour, long antedates the rise of scientific methods of investigation. The usefulness of this distinction has survived into the present time, moreover, despite the extensive physiological and behavioral investigations made on this subject in modern times (for review see Kleitman, 1963). Although certain borderline conditions may become clarified, an increased knowledge of physiological correlates of sleep and wakefulness (arousal) also complicates the problem of satisfactorily defining these two states (Hebb, 1967). Dissociation can be produced experimentally, namely, among the variables which normally characterize each condition, and such dissociation is indeed the rule at very early stages of development (Scherrer *et al.*, 1970). Since the semantic difficulties in ontogenetic studies can therefore be well-nigh insurmountable, we might do well to consider the advisability of dispensing with attempts at overall state designations except when the stage-to-stage similarities are unequivocal. Beginning then with the essentially mature condition, where correlations among specific physiological variables are normally very high, any facet of each state could simply be projected back independently in order to reveal its specific pattern of manifestation at successive stages of development (e.g., Dreyfus-Brisac, 1968, 1970; Gramsbergen, Schwartze, & Prechtl, 1970).

The white leghorn chick shows a clear differentiation between sleep and wakefulness already on its first day of posthatch life (see Fig. 1). *Sleep* refers here to the state characterized behaviorally by a relaxed posture, especially of the head and neck, and by closed eyes. The cerebral electrogram (EEG) is dominated by slow waves of large amplitude, while the electromyogram (EMG) is relatively low at this time (e.g., Corner & Bakhuis, 1969; Klein, 1963). *Arousal* then is the opposite condition, involving sustained wide opening of the eyes together with elevation of the body and/or head. The EMG jumps to a higher amplitude, whereas the EEG slow waves become drastically reduced. These differences are the same as those characteristic of sleep and wakefulness in the full-grown fowl Key & Marley, 1962; Oökawa & Tuge, Kanayama, & Chang, 1960), as well as in reptiles and mammals Jouvet, 1965). It deserves mention in passing that all three chicks which were extensively monitored at $1\frac{1}{2}$ hours posthatch failed sometimes to show the expected cerebral flattening during episodes in which they were unmistake-

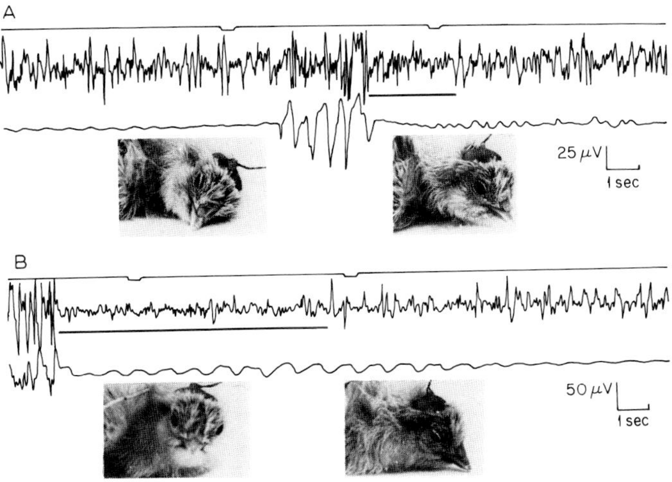

FIG. 1. Simultaneous recordings of spontaneous movements (bottom tracing of each group) together with the cerebral EEG after hatching in the domestic chicken. The photos showing the bird's behavioral state were made at the instants indicated by deflections in the marker channel (just above the brain recording channel). (The button on the polygraph was pressed manually at the same time the shutter of the camera was tripped.) (A) Recording made from a $1\frac{1}{2}$-hour-old chick, showing the EEG and behavioral arousal reaction which follows each burst of quasi-struggling movements as the bird tries to stand up. (B) The same chick $2\frac{1}{2}$ hours later, showing an example of the more pronounced and long-lasting arousal reaction at this age.

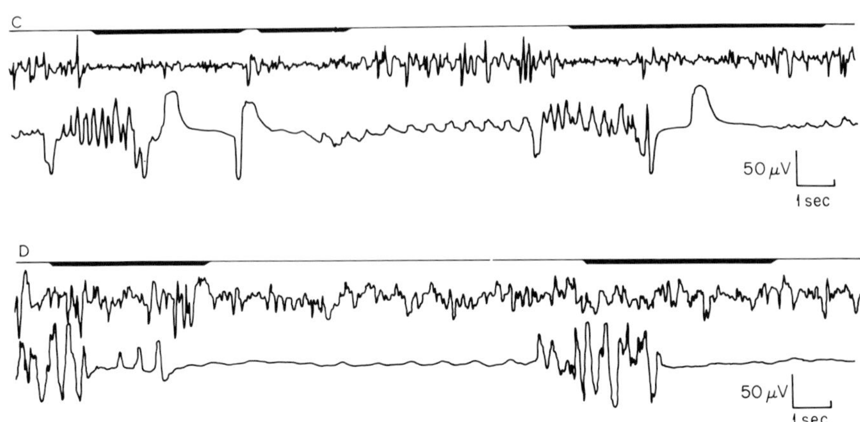

FIG. 1(C). A 13-hour-old chick, showing two relatively short arousal episodes. (The uppermost channel here indicates the precise time during which the head was lifted and the eyes were open, observed continuously.) Note that EEG flattening and the opening of the eyes clearly *follow* the onset of the body movements. (D) Two consecutive episodes where clear behavioral awakening (as in C, indicated on the uppermost channel) failed to be accompanied by the expected cerebral EEG changes in a chick at $1\frac{1}{2}$ hours posthatch.

ably awake according to the behavioral signs (Fig. 1D). The correspondence was 100% in each of them, however, when again registered 2–3 hours later. Such a lack of any cerebral reflection of behavioral arousal in the period shortly after hatching had earlier been reported (Corner & Bakhuis, 1969), but then in partially restrained birds.

B. Characterization of Sleep at Early Stages of Life

The "sleeping state" is by no means one of complete behavioral quietude (e.g., Corner & Bot, 1967). Jerky movements of various body parts occur frequently, as well as brief "startles," which may be followed by the eye opening and head lifting typical of behavioral arousal. If the young chick is restrained in the proper way, however, it soon falls asleep, and the startles become replaced by longer lasting stereotyped bursts of quasi-struggling movements (Fig. 2). These bursts are seldom accompanied by even transient opening of the eyes, and not by EEG flattening either, nor can the latter any longer be elicited by even intense sensory stimulation (Corner & Bakhuis, 1969; Corner, Peters, & Rutgers van der Loeff, 1966; Sheff and Tureen, 1962). Under such restraint, the chicks also could seldom be aroused by direct electrical brain stimulation which had been effective consistently while sleeping *un*restrained (van Wingerden, 1970). Approximately 15 locations throughout the hindbrain have been tested so far, each in a different prepara-

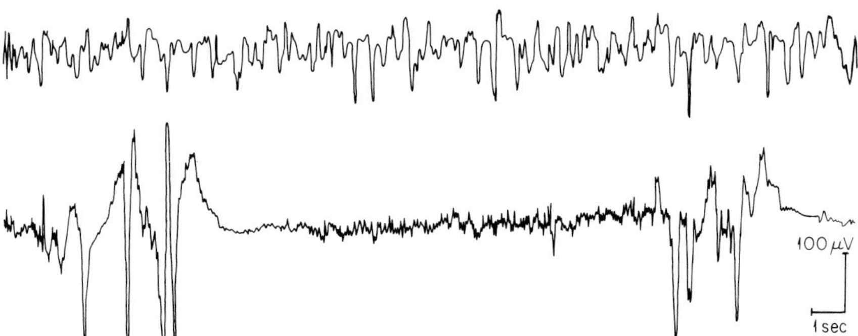

FIG. 2. (Above) methods used for immobilizing chicks sufficiently to induce a state of sleep. (left) Glass "eggshell", blown to the correct size; (middle) standard head-holder, according to Sedláček (1967); (right) "portable" holder, used for making the 1-day-old cerebral EEG shown below (upper trace); 1, silver ball electrode for recording from the brain surface or skull; 2, *idem*, for use as indifferent electrode (placed near comb); 3, ear pins; 4, beak holder; 5, leads to the amplifier. The simultaneous movement record (bottom trace) shows the stereotyped character of total body movement during sleep, as well as the brief loss of muscle tonus following many of such bursts (also see Fig. 6).

tion during the first week after hatching, but only three were found (all in the rostral pons, histologically checked) whose stimulation still led to EEG flattening plus eye opening once the bird was immobilized (Fig. 3). The sites which were able to elicit arousal only under free-moving conditions (also see Spooner, 1964), even with a 5-or 10-fold increase in the stimulus intensity, may therefore be presumed to have been only indirectly effective in this regard, e.g., secondary to the sensory input resulting from an evoked startle or other motor effect (also see Fig. 1C).

One of the most striking features about stereotyped motility in immobi-

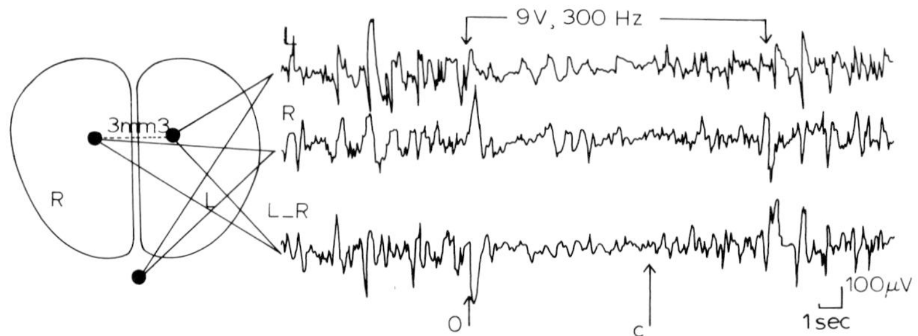

FIG. 3. Simultaneous monopolar and differential cerebral EEGs, recorded 3 mm from the midline (see insert at the left) in an immobilized 4-day-old chick, using chloridized silver ball electrodes resting directly upon the dorsal pial surface. An example of the short-lived arousal response (the eyes open and close respectively at o and c) produced consistently in this preparation by electrical stimulation 9 mm below the skull, at a point 3 mm from the midline. The electrode tips were found to be situated within the *anterior* basal pons.

lized sleeping chicks is its usually great regularity. The bursts have been recorded automatically (Fig. 4) in many cases, and interval histograms have been plotted for a number of these (Fig. 5). The distributions are all unimodal and, up to 3–4 days posthatch, the peak usually falls between 10 and 20 se-

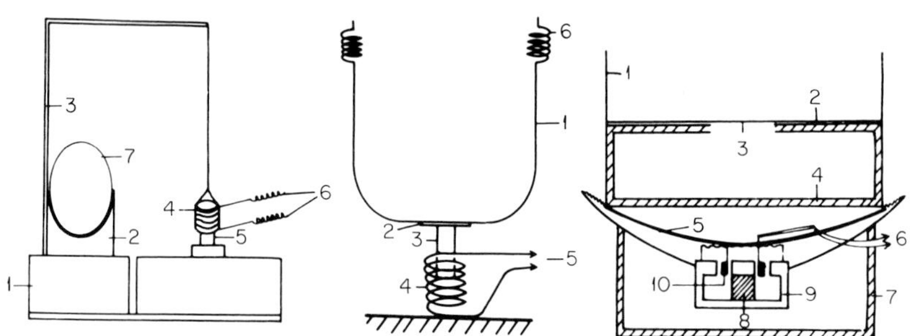

FIG. 4. Types of apparatus used for registration of pre- and poshatch movements in the chick. (Left) especially for movements inside the egg: (1) foam rubber cushion, (2) plastic egg cup, (3) stick cemented in upright position, (4) wire spools, (5) bar magnet, (6) leads to polygraph. (Middle) especially for posthatch sleep movements: (1) plastic cylinder of about 10 cm diameter, (2) metal plate, (3) bar magnet, (4) wire spool, (5) leads to polygraph, (6) steel springs. (Right) suitable for either type of preparation: (1) cardboard box, (2) removable floor for use with already hatched birds, (3) round opening for seating of egg, (4) bottom of the box, glued to (5) loudspeaker cone, (6) leads to polygraph, (7) supporting container for speaker, (8) magnet, (9) iron core, (10) wire spool. [Piezoelectric devices have been developed by others for recording motility from within the egg during incubation or hatching (Kovach, Callies, & Hartzell, 1970; Salter, 1966); these give an essentially identical picture to our own results, at least with respect to the stronger body movements.]

conds (cases B-J). From about 5 days of age, the distributions shifted toward higher values, with cases K and L (Fig. 5) representing only the two fastest preparations measured. In eight chicks examined at 4 days or more after hatching, there was a high correspondence in all cases between phasic episodes of EEG flattening and the stereotyped struggling movements, if present (Fig. 6B, C). The former are known to be accompanied by all the typical features of "paradoxical," or "activated," sleep (Hishikawa, Cramer, & Kuhlo, 1969; Klein, Michel, & Jouvet, 1964; Tradardi, 1966) and to occur in cycles of 5–10 minutes on the average (Corner *et al.*, 1966; Corner, Schadé, Sedláček, Stoeckart, & Bot, 1967; Jouvet, 1965; Oökawa & Takagi, 1968; Peters, Vonderahe, & Schmid, 1965). Even during portions of this cycle

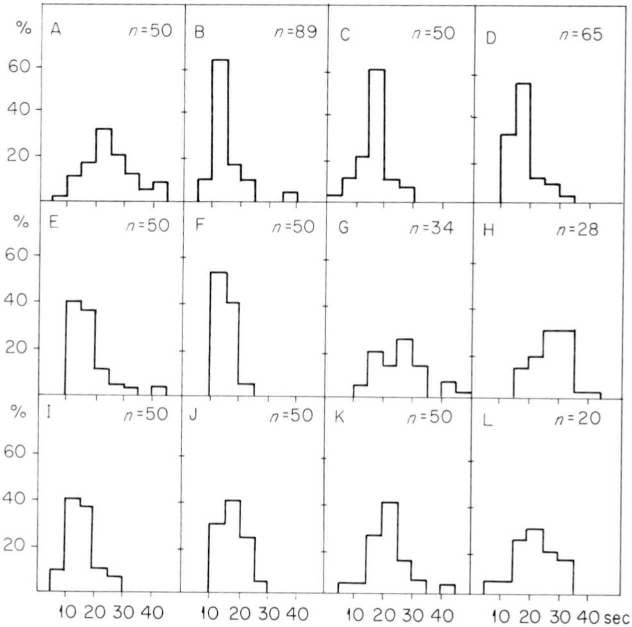

FIG. 5. Individual interval distributions between the onset of successive *stereotyped struggle movements* during sleep in the domestic chicken. In all cases except D the bird was immobilized in the standard holder (see Fig. 1), allowing the legs and wings to move freely. On the y-axis are shown the percentage of all measured intervals which fell in each 5-second bin (x-axis). The number of consecutive measurements (n) is given at the upper right of each box. (A, B) Chick immobilized 8 hours after hatching (room temperature), showing a trend from initially longer intervals to a highly stationary long-term pattern. (C) 1-day-old chick, recorded in the head-holder. (D) A different bird (1-day-old) recorded in a tightly fitting "glass egg." (E–G) Three 2-day-old chicks. (H–J) Three different chicks at 3 days of age. (K, L) The two older chicks, respectively 5 and 6 days after hatching, which showed the *highest* rate of spontaneous motility bursts.

where EEG activation was largely absent, the amplitude of the large slow potentials was usually clearly reduced during each motor burst (Fig. 6A). This was not so in younger birds (five cases, 0–2 days old), nor were their body movements coincident as a rule with the paradoxical EEG episodes when the latter were present (Fig. 6D–F).

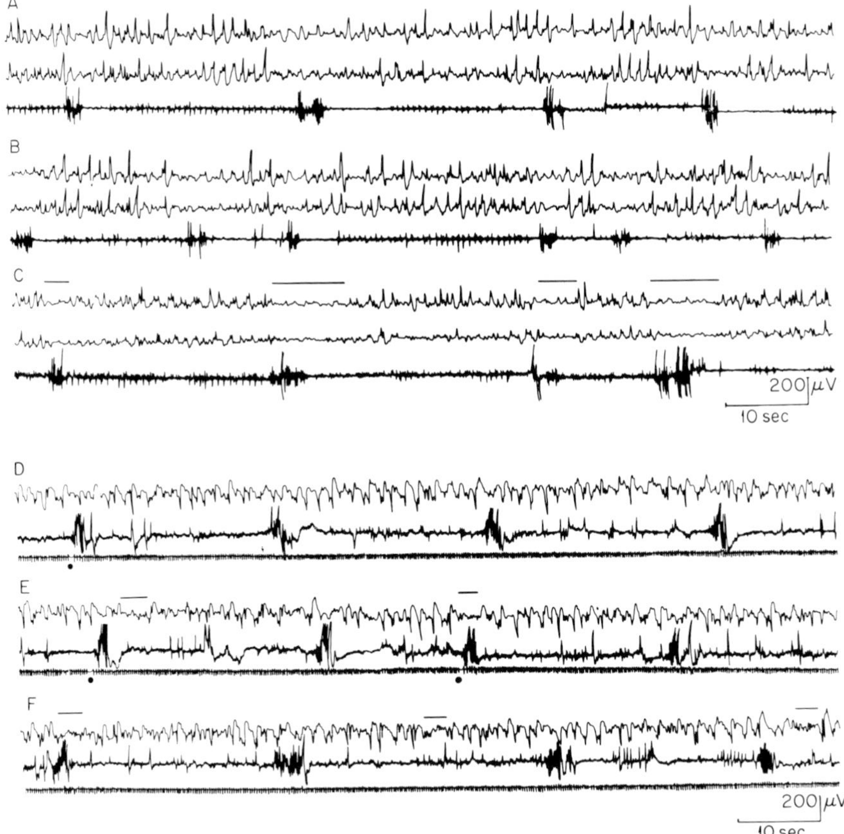

FIG. 6. The EEG and body movements during sleep under restraint in the young chicken (see text for details). A–C. 5-Day-old chick, illustrating the strong correspondence in older birds between motor bursts (lower trace) and paradoxical EEG episodes during the "activated" phase of the cerebral sleep cycle (C). The cerebral slow waves (left and right monopolar recordings) sometimes continue unchanged throughout a burst (A), but most often there is a clear amplitude reduction (B). D–F. 2-day-old chick (continuous registration), at which age no tendency exists for the cerebral slow waves (unilateral monopolar recording) to diminish during motor bursts (D) and where, during the "activated" phase (E, F), EEG flattening occurs largely independently of the movements (middle trace). In the electrocardiogram (EKG: bottom trace), examples can be seen of the *bradycardia* which intermittently preceded a burst.

C. Spontaneous Motility and Cerebral EEG during Hatching

The "climax" behavior pattern in birds (reviewed by Oppenheim, this volume) is strikingly reminiscent of the stereotyped motility pattern which we have just described for sleep (Fig. 7). Even the distribution of intervals between successive bursts under both natural and experimental conditions is essentially the same as during sleep, the peak values invariably falling between 10 and 20 seconds (Fig. 8). At incubation temperature (39–40° C) the bursts were found to slow to a stop within 5–6 hours after their onset (Corner & Bakhuis, 1969). More recent measurements on four chicks have shown, however, that the interval distribution is unchanged (based upon 42 values, having a range of 6–22 seconds) even after 24 hours at 35° C, which is the "metabolic equilibrium temperature" at this age (Freeman, 1963). Also, the cerebral bioelectric potentials during hatching are typical of sleep, but there is only an incidental coincidence between "paradoxical" EEG episodes and the struggling movements (Fig. 9). As observed under experimental conditions, the eyes remain closed during the motor bursts or else open only fleetingly, except if the head has been allowed too much freedom of movement. Thus, although a sleeplike cerebral EEG cannot be regarded as diagnostic at this age despite its mature appearance (since even clear behavioral arousal episodes are for some hours after hatching not necessarily accompanied by any change in the slow-wave pattern: see Fig. 1) the behavioral similarities alone seem sufficient to warrant the conclusion that the state of "sleep" is, in the chick, a direct posthatch continuation of the climax phase of embryonic development.

In striking contrast is the picture, to all appearances "wide awake," seen at the moment the shell is actually broken open and the head and neck become free to stretch (Fig. 10). The spontaneous motility pattern too changes sud-

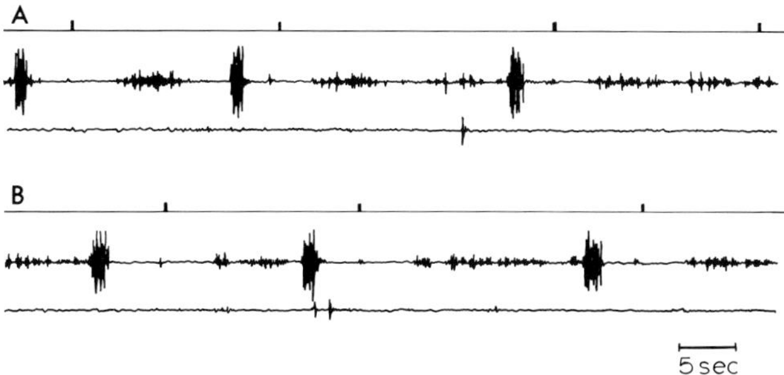

FIG. 7. Continuous recording (A, B: middle trace) from a 1-day-old chick, *immobilized, in a "glass egg"* (see Fig. 2), showing stereotyped struggling bursts against a background of weaker movements. The animal was observed continuously during the registration, and swallowing movements were indicated (top line) by pressing a button available for this purpose on the polygraph. The button trace is a control for floor vibrations in the experimental chamber.

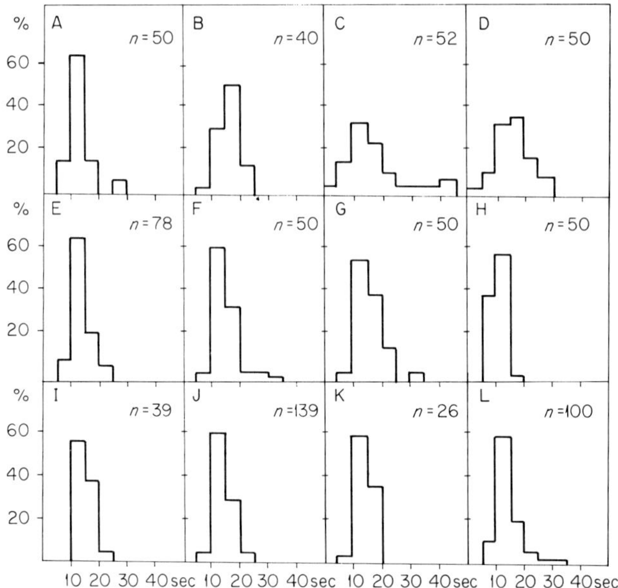

FIG. 8. The distribution of intervals measured between the onset of successive *movement bursts during hatching* in the domestic chicken. On the y-axis is indicated the percentage of all values falling in each of the 5-second bins (x-axis), with the number of measurements (n) being given in the upper right-hand corner of each box. (A–H) Individual histograms from naturally hatching chicks, recorded from intact eggs. (I) *Idem*, but with the shell broken just enough to allow the legs to move freely (and confirmed less extensively in three additional preparations; also see Kovach, 1970). (J–L Chicks transferred to glass "eggshells" just at the onset of the hatching movements.

denly in character, although a certain periodicity remains in evidence for some time (Fig. 11). The bursts of movement are now much more variable, however, and usually last longer than during climax. Furthermore, each is accompanied by extension and lifting of the head, with the eyes remaining wide open for several seconds. The periods of arousal grow progressively longer on the whole during the first hours following emergence (Bakhuis, 1971; Corner & Bot, 1967; Klein, 1963), and struggling-type movements become replaced by walking about, orientation reflexes, exploratory behavior, and the like. Once fully aroused, or if prevented from emerging for a day or so, the transition from the sleeping state to that of sustained wakefulness occurs rapidly (e.g., Corner *et al.*, 1966). Considering that even vigorous vestibular stimulation (shaking, rotating, etc.) has never succeeded in our experience in eliciting arousal either in a hatching or in a well-restrained sleeping chick (Corner, unpublished observations), and since free wing and/

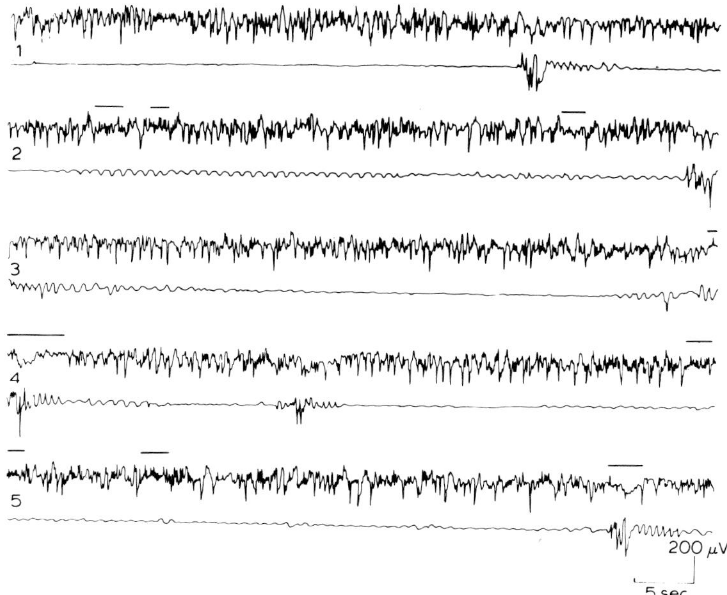

FIG. 9. Simultaneous recording of the cerebral EEG (top line of each pair) and body motility during hatching ("climax") in the domestic chicken (continuous recording 1–5). Large amplitude slow potentials often continue throughout the movement bursts, and are interrupted only by brief "paradoxical" episodes (indicated by horizontal stripes) which appear to be uncorrelated with the movements.

or leg movements also fail to do so (Bakhuis, 1971; Kovach, 1970; also see Fig. 8), it would appear to be proprioceptive neck afferent excitation which is needed for the induction and maintainance of the waking state in the young chick. In birds of 2 weeks of age or more, on the other hand, sustained wide

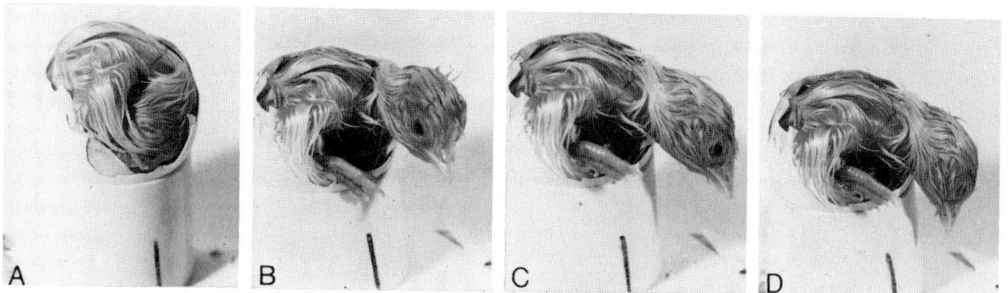

FIG. 10. The initial "arousal" behavior following upon emergence from the eggshell. Starting from the hatch position (A), the head becomes freed from constraint at a certain point and becomes lifted out and upward (and rightward), the eyes opening wide for the first time (B). After about 5 seconds, on the average, the head slowly sags (C) and the eyes close (D) until the next burst of movements.

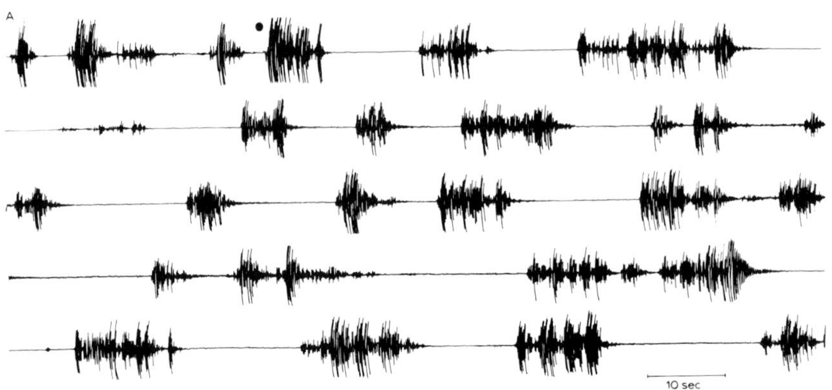

FIG. 11. Registrations of typical motility patterns immediately *following emergence* from the eggshell (behavior as in Fig. 10; also see Fig. 4 for the apparatus employed). Each case (A–C) is a continuous recording, and the first burst of movement which led to emergence is indicated by a blackened circle. (A) Example of pattern where the bursts continue without pause (but become much more variable, and on the whole longer, than during hatching).

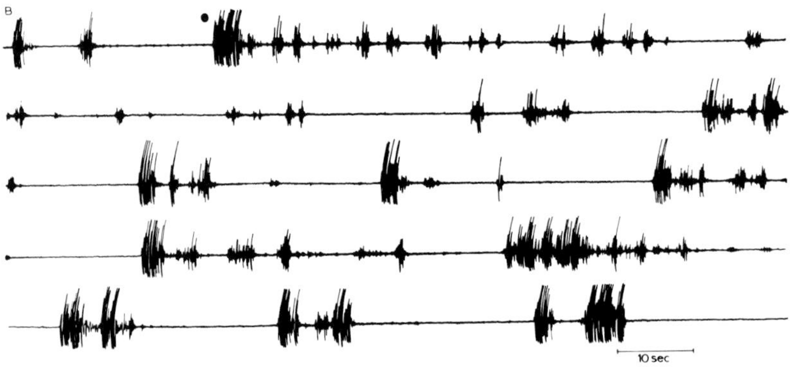

FIG. 11B. Example of pattern where a prolonged interruption of major movements occurs upon emergence, before the burst pattern as in case A commences.

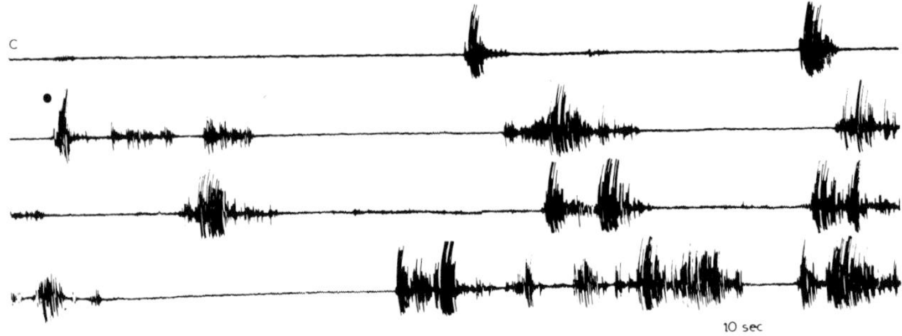

FIG. 11C. Typical posthatch motility pattern in a chick allowed to emerge right after the onset of climax behavior (top line).

opening of the eyes was commonly observed during motor bursts of sensory stimulation, even when equally well restrained (van Wingerden, 1970).

D. Motor and Cerebral Activity during Late Incubation

Embryos at early stage 45 (i.e., 1–2 days prior to climax, 20 embryos observed), while exhibiting wide eye opening in all except one case, both upon liberation of the head and during most of the successive motility bursts, fail to show the typical extension and elevation of the neck. The head is instead tucked down sharply with each vigorous body movement. During stage 44 (i.e., a day earlier) a partial opening of the eyes was observed in one out of six cases at the moment of head emergence, and in four cases during some of the subsequent movements. The head continued to hang down limply at this age during motor activity, its most characteristic pattern during each burst being a rapid side-to-side shaking. The behavioral arousal response thus develops gradually during the last 2 days or so of incubation, but it appears unlikely that this potentiality could normally express itself overtly *in ovo*.

The actual pattern of spontaneous motility in the chick embryo during the last week of incubation has been extensively studied in recent years (for reviews see Corner & Bot, 1967; Hamburger & Oppenheim, 1967; Oppenheim, this volume). One of the most striking features (also in the duck, see Gottlieb, 1965, 1968; Oppenheim, 1970) is the intermittent occurrence, against a background of weaker and irregular twitches, of stereotyped bursts of strong body movements and/or rhythmic beak-clapping (Fig. 12). These usually occur at intervals longer than a minute but sometimes, as is seen in Fig. 12, in a variable train of much shorter values. The interval histogram shows a unimodal distribution with a very long tail but a peak in the same range as during hatching, i.e., 10–15 seconds (Corner & Bot, 1967). As with hatching, the eyes remain closed or else open only fleetingly. Indeed, the strong qualitative resemblance between the late embryonic total body motility burst pattern—which has gone under different names: "stereotyped struggling" (Corner *et al.*, 1966), "co-ordinated rotatory," or "type III" movements (Hamburger & Oppenheim, 1967)—and the behavior characteristic of climax has not escaped the notice of other workers in this area (e.g., Kovach, 1968, 1970; also see Peters *et al.*, 1965). There is only some disagreement still as to whether these bursts appear *de novo*, at about 17 days of incubation (e.g., Hamburger, 1969; Hamburger & Oppenheim, 1967), or whether they show a progressive maturation from as early as 13–14 days (Corner, 1968; Corner & Bot, 1967).

The cerebral EEG of the chick embryo has been extensively studied (for reviews see Corner *et al.*, 1967; Ellingson & Rose, 1970), and is known to become essentially mature 2–3 days prior to hatching. By about 18 days *in ovo*, the large amplitude slow potentials become virtually continuous, and

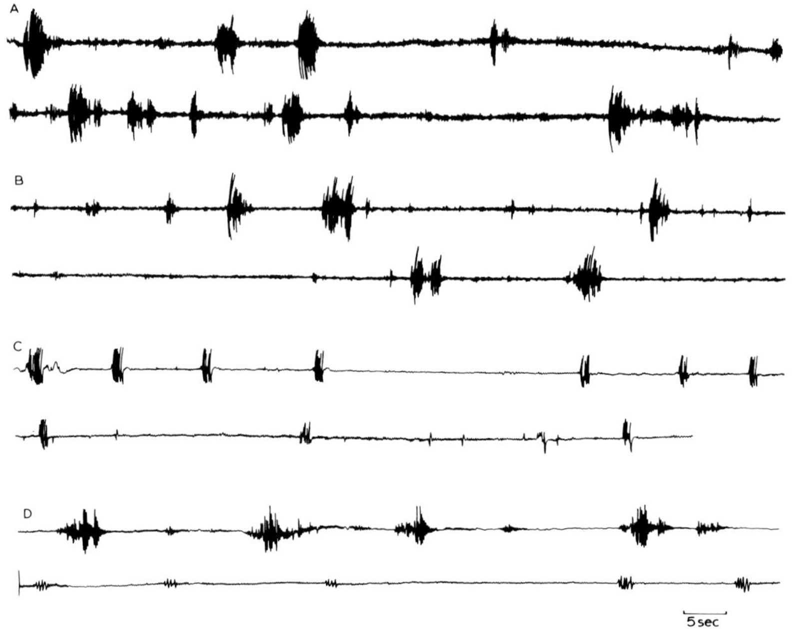

FIG. 12. Late *embryonic pattern* of spontaneous motility. (A) Recorded from an unopened egg (see Fig. 4) at 17 days of incubation (stage 43; see Hamburger & Hamilton, 1951), continuous tracing. (B) *Idem* at 19 days (late stage 44). (C) at 19 days, showing similar movements also when the electrodes are attached directly to the embryo (sweep speed here is $\frac{1}{2} \times$ the calibration indicated at the lower right, in order to show an entire, relatively long, train of stereotyped bursts. (D) (upper trace) Brief train of typical total body bursts in a 1-day-old "decerebrated" chick (see Fig. 14); (lower trace) segment taken from a train of bursts of beak-clapping only.

appear little different from those seen postnatally during sleep. The 5 to 10-minute cycle of "paradoxical" phasic EEG episodes appears only subsequent to this development, reaching a maximum prominence in the last stage prior to climax (Fig. 13). Movement bursts may occur during either phase of the cerebral cycle and have no noticeable effect upon the pattern of the bioelectric waves.

III. Experimental Analysis

A. Origins of Cerebral Bioelectric Slow Potentials

Periodic fluctuations in EEG activity are likely to be largely the reflection of an endogenous rhythm within the forebrain. Decerebration experiments

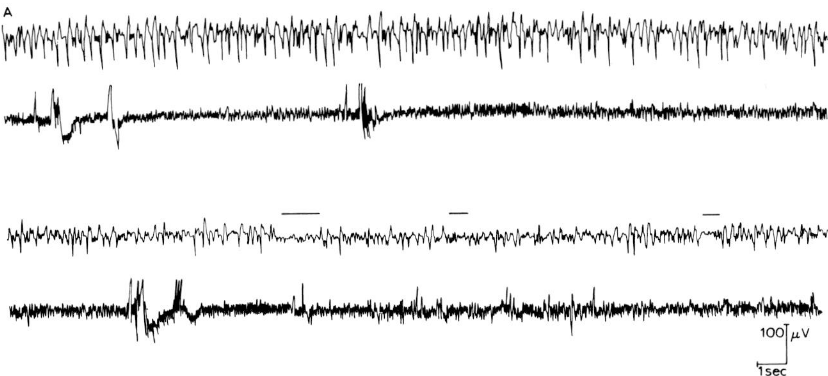

FIG. 13. The late *embryonic EEG* in the domestic chicken. (A) Continuous recording of the cerebral potentials at stage 45 (upper trace, unilateral monopolar registration) simultaneous with spontaneous body movements, showing the transition from high to low amplitude phases of the cycle. Especially flattened stretches of the EEG are indicated by horizontal bars.

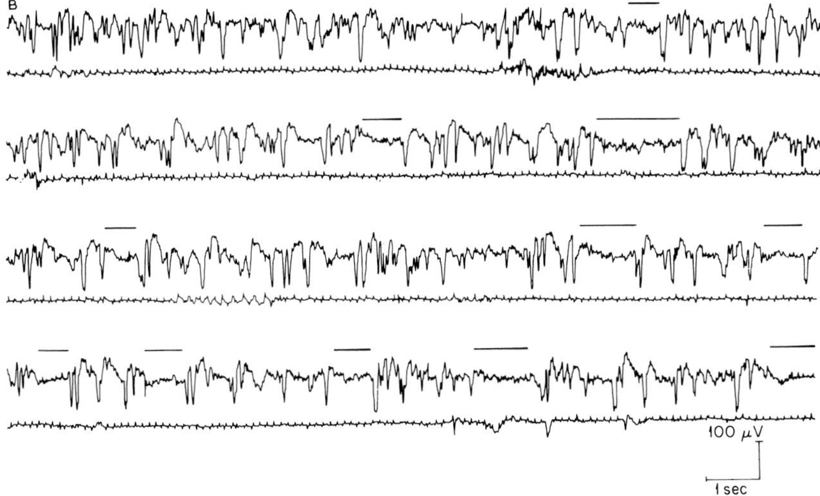

FIG. 13B. The EEG (upper trace) and the EKG recorded continuously in an early stage 44 embryo: the cyclic appearance of "paradoxical" episodes of EEG flattening (phase transition occurs on the second line) is just beginning to be evident.

performed after hatching (Marley & Stephenson, 1971; also see Spooner, 1964) have revealed that the "slow-wave sleep" pattern is generated within the forebrain and, as in mammals (Kellaway, Gol, & Proler, 1966; Villablanca, 1962; also see Jouvet, 1962), to a large extent within the cerebrum itself. A similar result was obtained by us for the embryonic period (Corner

& Bot, 1969) where, in addition, the persistence of cyclic fluctuations of the bioelectric activity could be demonstrated. Figure 14 shows a portion of a typical registration made from an operated stage 45 embryo, at the moment of transition from a high to a low amplitude portion of the cycle. Despite considerable variation in the durations of successive phases, it can be seen that a full cycle usually lasts about 10 minutes. This approximates closely the cycle length estimated for the intact preparation (see above), and suggests that the periodicity governing the appearance of "paradoxical" EEG episodes during sleep is determined at least partly within the cerebrum itself.

The cerebral oscillation described above falls at the upper limit of what has been called "infraslow" brain rhythms, and held to be a probably primitive and still widespread feature of nervous activity (Aladjalova, 1964). The existence of comparable rhythms in the brainstem and spinal cord, underlying

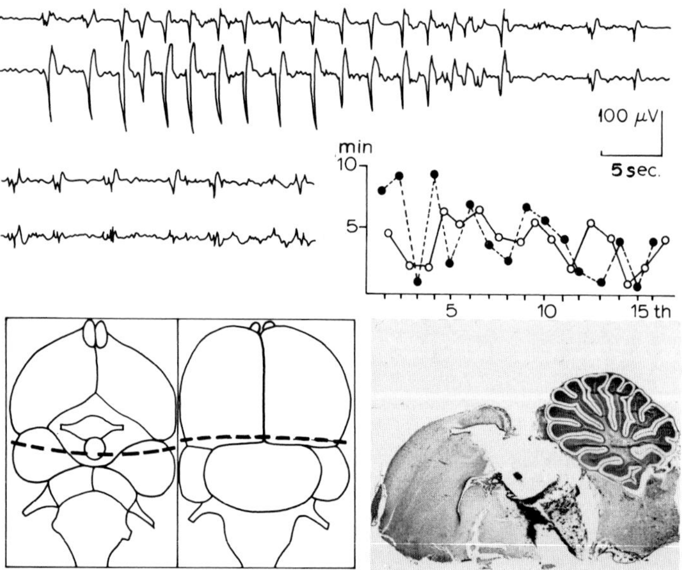

FIG. 14. Periodic changes in the cerebral EEG of the "isolated forebrain" *in situ*, 4 hours postop, in a chick embryo at stage 45 of Hamburger and Hamilton (1951). (A surface positive potential is reflected by an *upward* deflection in this figure.) (Above and middle-left) Continuous monopolar recording, simultaneously from two points separated by about 5 mm distance on the dorsal surface of one of the hemispheres, illustrating the synchronization typical for these fluctuations of activity; the transition is shown here between a short "activity phase" and a phase of relative "inactivity." (Middle-right) Durations, to the nearest half-minute, of *active* (closed circles) and of *inactive* (open circles) phases in a preparation for which 16 successive cycles were measured; there is apparent independence in both the successive and the alternate values, but a complete cycle usually lasts about 10 minutes. Below are shown: (left) respectively ventral and dorsal schematic views of the plane of transection, and (right) a parasagittal section through one of the operated brains.

motility cycles in the chick embryo (e.g., Decker & Hamburger, 1967; Hamburger & Balaban, 1963) and also in anuran larvae (Corner, 1964b), lends much credence to this hypothesis. Fluctuations of the same order of magnitude have, in fact, been discovered also in the spontaneous motility of late squid and cuttlefish embryos (Corner, unpublished observations). Nor need such cycles be imagined to necessarily depend upon large or heterogeneous networks nor to be driven by periodic hormonal or circulatory changes. Rhythmic changes in bioelectrical activity, falling in the 1 to 10-minute range, have been recorded from a high proportion of CNS fragments in tissue culture (Corner & Crain, 1972; also see Corner, 1964a; Corner & Crain, 1965; Cunningham, 1962), thus under steady environmental conditions and isolated from most of the normal extraneural influences which might have been a source of the periodicity. Finally, the intimate association noted between such potentials and the firing of single units—whether *in vitro* (Fig. 15; also see Crain, Volume 2 of this series) or *in situ* (e.g., Frost, & Gol, 1966; Purpura, 1969)—indicates that endogenous electrical rhythms may indeed reflect corresponding oscillations in the level, and/or the threshold, of functional activity within neuronal networks.

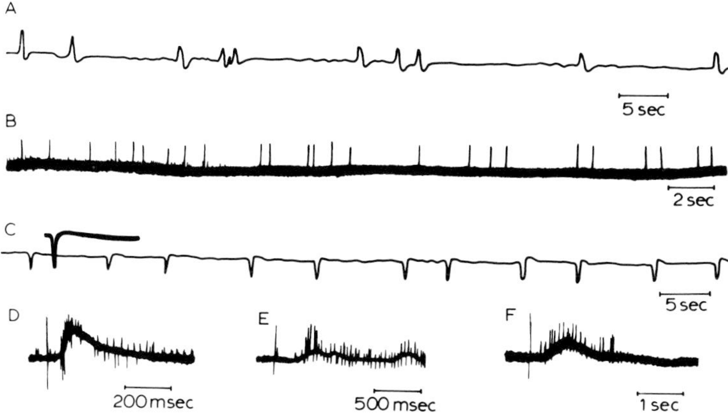

FIG. 15. Bioelectric activity in embryonic chick neural tissues cultured *in vitro* (for technique, see Crain, 1970). An upward deflection in this figure indicates *negativity* at the recording electrode, the amplitudes varying up to about 200μV. (A) Stereotyped spontaneous "complex" potentials from a *spinal cord* fragment, isolated at 4 days of incubation and cultured for 9 days *in vitro*. (B) Relatively regular spontaneous unit action potentials in a cord explant after 17 days i.v. (C) Complex potentials generated in a fragment of *cerebral hemisphere*, isolated at 11 days of incubation and cultured for 23 days i.v. (D) Long-lasting negative wave evoked in the same preparation as in (B), showing the associated polyneuronal spike discharge. (E) Long barrage of action potentials (triggered by a single shock) in the same cerebral explant as in (C) but at a different recording site. (F) Multiunit impulse barrage accompanying a long negative wave in a different cerebral explant, 4 weeks *in vitro*.

B. Participation of the Forebrain in Climax Behavior

Despite the highly mature pattern of their spontaneous electrical activity, the cerebral hemispheres play no noticeable role in determining the periodicity or character of the climax pattern of movements. Following bilateral removal by aspiration on the day before hatching was to begin, typical motility bursts occurred on schedule and with by and large the same interval distributions as in intact birds (Fig. 16). This was true whether the operated birds had been left in their shells and allowed to hatch normally (cases A and B), or whether they were immediately removed from the shell and immobilized the following day in the standard head holder (cases C-E). The same conclusion was reached when the hemispheres, instead of being removed, were surgically isolated (Fig. 17). In this series of experiments, the operations were carried out on the day of hatching and the measurements made a day later, with the birds in the standard holder (i.e., Fig. 2, above, middle). Perfectly normal interval distributions were found in all three cases in which the plane of transection passed just anterior to the optic chiasm (Fig. 17A).

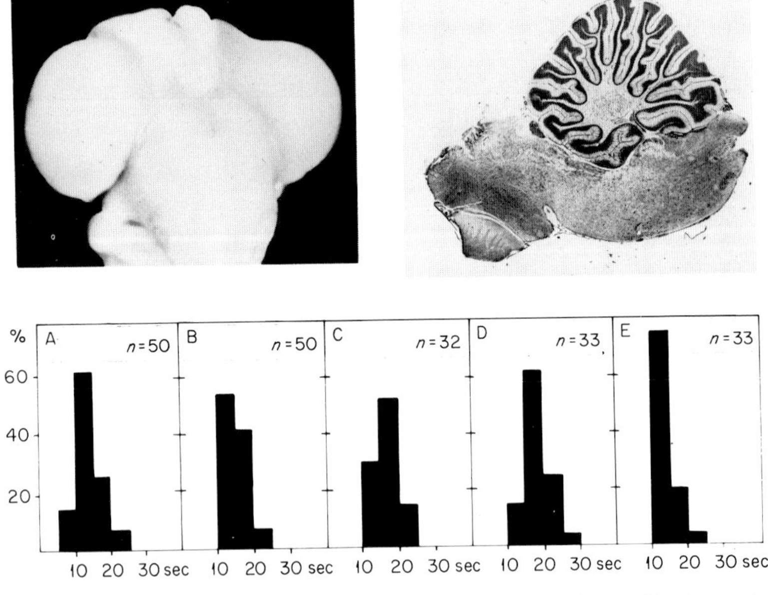

FIG. 16. Distributions of intervals between the onset of successive motility bursts, during hatching behavior in the five chicks (A–E, bottom) in which the cerebral hemisphere had been removed bilaterally by suction on the previous day. An example of the extent of the lesion, which was visually controllable during each operation, is also shown (top): (left) ventral view (anterior is upwards); (right) median sagittal section (hematoxylin-eosin staining).

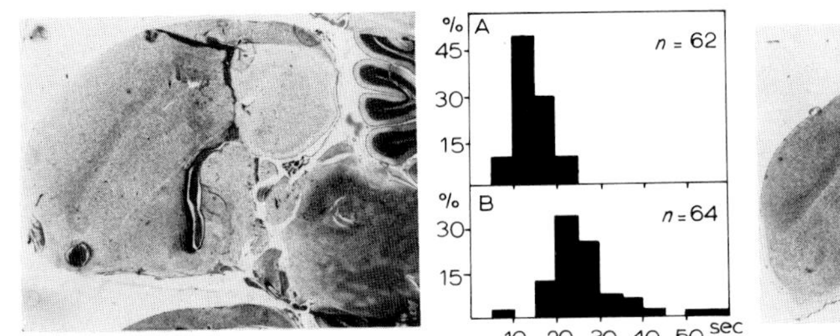

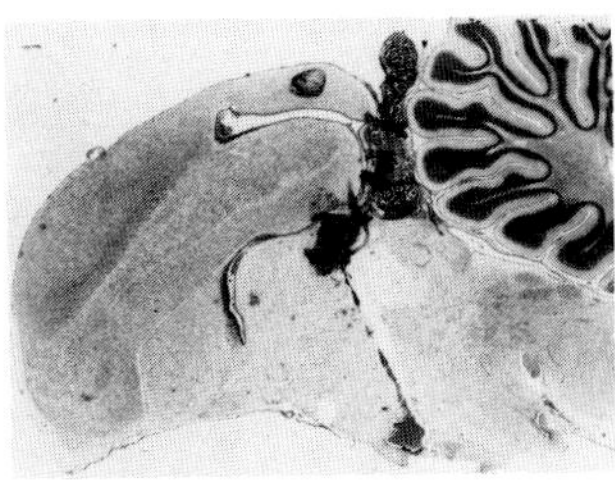

FIG. 17. Composite interval histograms from (A) preparations with the cerebral hemispheres eliminated by surgical transection (see insert at left), and (B) preparations where the lesion encroached upon the optic chiasm (insert at the right). In both groups three operated chicks were studied, with each of the individual distributions having its mode in the same range as the corresponding histogram, and the difference between them was statistically significant. (p = .05: Mann-Whitney test). The total number of intervals is given by n, and the percentage falling into each 5-second bin (x-axis) is shown on the y-axis.

and an almost normal sequence of movement bursts occurred even when the lesion was found to encroach slightly upon this structure (Fig. 17B). Since the difference between the two operated groups nevertheless turned out to be significant at the 5% level, a hint was furnished that the *diencephalon* might in some way be involved in hatching behavior.

This possibility is greatly strengthened by examination of the effects of still more caudal forebrain transections from the same experimental series. In all seven cases in which the plane of transection passed somewhere through the posterior half of the optic chiasm, there was a preponderance of intervals longer than 1 minute between bursts of body movements, with only intermittent trains of shorter values. This is the same result as was reported previously to follow a virtually total "decerebration" (Corner & Bakhuis, 1969; also see Figs. 12D & 14). Such a pattern mimics qualitatively the normal behavior of the late embryo, whether intact (Fig. 12) or decerebrated in the way illustrated in Fig. 14 (Corner, 1968). EEG recordings were used here to control against the possibility that loss of blood, rather than elimination of the forebrain, was responsible for the observed postoperative decline in spontaneous motility. Thus, only when the transection was practically completed (and most of the bleeding had already occurred) did the EEG show any obvious change, which then consisted of a selective disappearance of the low amplitude "background" waves (see Fig. 14), whereas hypoxia has just the opposite effect (Corner *et al.*, 1967). The forebrain, therefore, appears to play little or no part in determining the motility pattern of the chick embryo up until climax, at which point some new factor, requiring the participation of diencephalic structures, suddenly brings about a dramatic quantitative behavioral change (also see Oppenheim, 1972, and this volume).

C. Mechanisms Relating to Perihatch Motility Patterns

Whatever may be the trigger for the onset of climax movements, their occurrence almost certainly does not depend upon stimuli which reach the brain via ascending spinal pathways. This source of input was largely eliminated from consideration by means of surgical transection of the cord in the midcervical region—carried out in 1-day-old chicks, using overt motor responses to tactile stimuli as a test for persisting functional isolation across the lesion; six cases were kept for 4–5 days in vigorous condition—yet sparing the lateral tracts subserving respiration (Bakhuis, 1971). This operation had no noticeable effect upon the movement bursts themselves during sleep under restraint (also see Oppenheim & Narayanan, 1968), although the intervals between them were significantly shorter than normal (Fig. 18). In addition, the vestibular and auditory modalities can most likely be discounted, on the basis of their inhibitory rather than excitatory effect upon spontaneous bursts (Fig. 19). The movements also continued indefinitely (in the same set of preparations) when the last conceivable sound source,

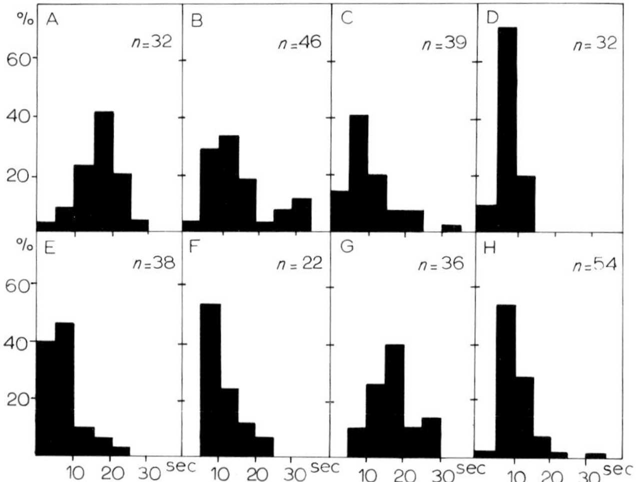

FIG. 18. Interval histograms for consecutive motility bursts during sleep in chicks with *high spinal lesions*, operated on the first day after hatching (see text for details). (A) 1-day-old chick, recorded on the day of operation. (B–C) Two other birds, at 2 days of age (i.e., first day postop). (D–E) another 2-day-old chick, the second set of (consecutive) measurements being made 20 minutes after the first set. (F) 3-day-old chick (i.e., recorded on the second day postop). (G) The same preparation as in (B), now 3 days old. (H) The same bird as in (F), at 5 days of age. Comparison of the above eight distributions, from five preparations over 3 days, with the equivalent control cases (i.e., Fig. 5, cases C–J) by means of the Mann-Whitney test, *two-tailed*, revealed a difference between the two groups which was statistically significant at the 1% level.

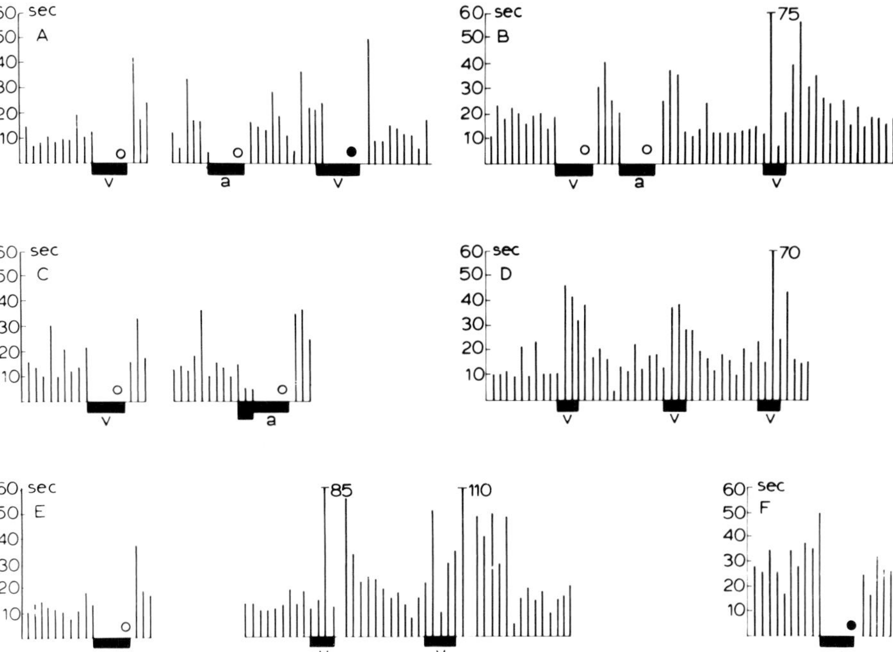

FIG. 19. *Sensory-evoked inhibition* of climax (hatch) motility in the domestic chicken. Time-amplitude plots, or "tachograms," display intervals measured (in seconds, as shown on the y-axis) between the terminations of successive bursts of body movement (x-axis), in five different preparations, A–E. (F) is one of the same five birds, but immediately after having been moved gently within the incubator. Periods of stimulation (lasting 2 minutes whenever the bursts were completely suppressed, and otherwise as indicated) are shown by thick bars under the abscissa. An open circle indicates No-response upon terminating the stimulation, and a filled circle indicates an Off-response (in the former case, the next "interval" is thus actually the *latency* to the first poststimulus burst). A number to the right of a given vertical line signifies an interval longer than 60 seconds. "v" Indicates continuous *vestibular* stimulation (shaking), whereas "a" means continuous *acoustic* stimulation. In cases A and B the latter was furnished by the loud peeping of a nearby chick, and in case C by a fluttering whistle tone, which was lowered in intensity, as indicated by the reduced thickness of the bar, after it had initially *evoked* two unusually short intervals. (This type of stimulation proved to be ineffective in cases D and E, and is not included in the figure.)

that of the incubator motor, had been turned off. Finally, the fact that chicks are able to hatch in total darkness (our observations) or in the absence of both optic lobes (Oppenheim 1972, and this volume) seems to eliminate visual input as a requirement for this type of behavior. The fact too that the motility pattern in question has been shown earlier in this article to be a manifestation of *sleep* [a state which occurs quite normally in all other respects, even in blinded chicks (Oökawa, 1971)], the bursts thus occuring together with a minimal muscle tonus, is further indirect grounds for suspecting an endogenous rather than a reflex origin. This picture is fully analogous to the one presented for the cat (Jouvet, 1965), in which phasic episodes of activated sleep normally occur independent of sensory input, although they can also be evoked as a reflex. If confirmed for the chick, such a conclusion would furnish still another example of continuity between embryonic motility, which is known now to be essentially nonreflexogenous up until a few days prior to climax (see Hamburger, this volume), and the mechanisms underlying behaviors seen still later in maturation. (Also see Oppenheim, this volume, for further experiments on the role of sensory input during hatching stages.)

Regardless of the source of excitation for spontaneous stereotyped body movements, their patterning into bursts which occur at regular intervals is very likely accomplished by an intrinsic process. This can be concluded from a series of experiments in which the motor responses to *tonic* stimulation were investigated in six hatchlings which, still at incubator temperature, had been prevented for several hours from emerging from the shell. Five of these had already ceased making climax movements (see Corner & Bakhuis, 1969), but typical hatch behavior could be induced in four of them by means of a steady train of electric shocks applied to one of the wings (Fig. 20). When stimulation was started only slightly supraliminal for evoking a burst, the intervals between successive bursts fell initially in the same range as normally during hatching (cases A-C). Relatively more intense stimulation resulted at first in still shorter intervals (see case D, and the later trials in A, B, and C). These effects, while consistent, were partly obscured by a secondary lengthening of the intervals as stimulation continued, although this last phenomenon could usually be temporarily reversed by turning up the stimulus parameters still more (see Fig. 20). The other two birds in this group provided evidence that such "refractoriness" had a peripheral cause, rather than being due to a central build-up of inhibition, etc. In case E, for example, the climax motor bursts (which in this preparation were still occurring spontaneously) could be caused to accelerate by applying tonic electrical stimulation, but the intervals soon restabilized at their original values instead of increasing steadily. The last preparation failed to show more than a single burst or two so long as the shocks continued at a fixed rate, but typical bursts could be evoked at any desired frequency simply by regularly interrupting the pulse train for a fraction of a second.

The refractory effect was still more pronounced in three embryos tested at 17 days *in ovo*, so that, although here too the typical response to tonic input

excitation, motor barrages are cut off sharply after $\pm$ 2 seconds. The patterning of such discharges into a train of bursts at more or less regular intervals is apparently accomplished by means of a postdischarge elevation, with subsequent decay, of the threshold of the network. Examination of the interburst interval distribution suggests that the excitability returns to normal in approximately 20 seconds, which is also the time found by Kuo (1939) during which there is a depression of excitability following total body movements in the chick embryo (also see Kovach, 1970). Since neither the absence of both optic lobes (Oppenheim, 1972, and this volume) nor the elimination of most of the spinal cord (Bakhuis, 1971; Oppenheim & Narayanan, 1968) has been found to interfere with climax behavior, the ability of decerebrated chicks, i.e., operated as shown in Fig. 14, to generate highly similar bursts of movement (see Fig. 12; also see Corner, 1968) implicates the *brainstem* as the location of the above "relaxation oscillator" mechanism. Its presence even in very small neural fragments, taken from several parts of the central nervous system and cultured *in vitro* (Fig. 15), further suggests that a highly specific or elaborate functional organization is not required. The role of the diencephalon with respect to the hatching movements would then need to consist solely of providing a *tonic* source of excitation, thus mimicking the wing stimulation experiments described above, with the patterning of the output being accomplished by lower motor circuits (Corner, 1968).

The bursts of *arousal* behavior seen just after hatching (see Fig. 11) may well be of brainstem origin too, since they can be evoked, and also may occur spontaneously, in decerebrated chicks, and since the spinal cord is not required for wakefulness (Bakhuis, 1971). The periodicity (Fig. 11) of the startles which initiate each of these bursts in intact "neonate" chicks suggests, in addition, that the forebrain's hatch influence may continue for a time thereafter to trigger brainstem motor discharges but that, given the new behavioral possibilities following emergence from the shell, sleep-type bursts become transformed into the more complicated movement pattern called "waking." Only with an at least partially intact cerebrum do these brief arousals gradually become sustained waking behavior, making possible the variety of actions and reactions which, understandably but mistakenly, have dominated attention in the past when parallels have been sought between pre- and posthatch behavior (see Kuo, 1967). Without the forebrain, the bird continues in a motility pattern which so closely resembles that seen during normal sleep that the possibility must be seriously entertained that this state is, in fact, largely brought about by a process of "*functional* decerebration" (Corner, 1968; Corner & Bakhuis, 1969). The earlier described projection of the state of sleep back into prehatch stages, where forebrain removal was found to have no obvious effect upon the movements, increases the plausibility of this interpretation. (A highly schematic analysis of the various systems and processes involved in the control of perihatch behavior patterns in the chick is offered in Fig. 21.)

IV. Discussion

A. Related Phenomena in Forms Other than Birds

Considerable information is available about the early ontogeny of sleep in mammals, and there are several indications that the same continuity with prenatal life exists as has been portrayed for the chick in the present paper. The cerebral EEG, for instance, matures in precocial forms such as the sheep (Bernhard, Kaiser, & Kolmodin, 1959; Dawes, Fox, Leduc, Liggens, & Richards, 1972; Ruckebusch, 1972) and the guinea pig (Bergström, 1962; Jasper, Bridgman, & Carmichael, 1937; Rosen & McLaughlin, 1966) into a high amplitude, "slow wave sleep"-like pattern already prior to birth. Judging from the published records and descriptions, this is a continuous, or at least a predominating, condition throughout late gestation. In addition, bursts of rapid eye and other movements typical of the activated phases during sleep—and associated with flattening of the EEG tracings—are reported to occur *in utero* throughout the last third of gestation (Dawes *et al.*, 1972). These events are accompanied by "periods of rhythmic respiratory movements," thus resembling the earliest breathing pattern in the chick embryo (Corner & Bot, 1967). Another recently published paper (Ruckebusch, 1972) has confirmed the existence of seemingly quite typical episodes of paradoxical sleep in the sheep fetus. On the other hand, the possibility suggested by this author that brief periods of *waking* might also appear *in utero* warrants further investigation in this and other species.

As we have seen for the chick, relevant, *behavioral* data would be the most convincing for these late fetal stages, but none are available to our knowledge. Studies on very early premature babies, in which a wide range of variables was monitored (Dreyfus-Brisac, 1968, 1970), have suggested that no state can be distinguished at first which corresponds to "wakefulness," and that there instead exists a condition best described as "primitive, or undifferentiated, sleep" (also see Parmelee, Wenner, Akiyama, Schultz, & Stern, 1967a; Parmelee, Wenner, Akiyama, Stern, & Flescher, 1967b; Petre-Quadens, Hardy, & DeLee, 1969). This state consists of an almost continuous high level of irregular and variable jerky movements involving most parts of the body, in this respect resembling a chick embryo of about 2 weeks *in ovo* (Corner & Bot, 1967; Provine, 1972, and this volume). A similar behavior pattern has also been described for the sheep fetus (Barcroft & Barron, 1937), as well as for several *altricial* mammalian species during sleep (e.g., Garma & Verley, 1969; Gramsbergen *et al.*, 1970; Valatx, Jouvet, &

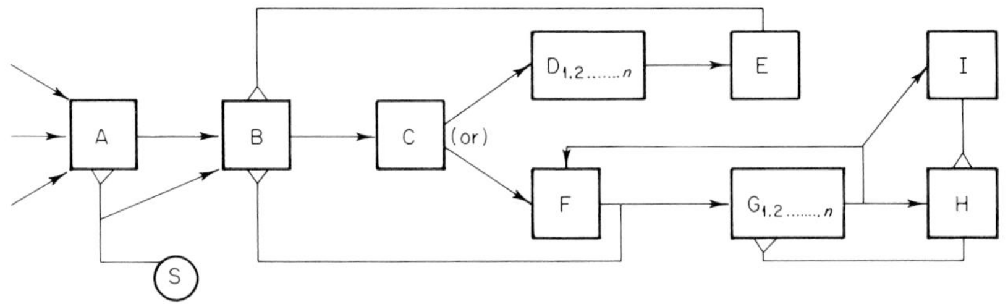

FIG. 21. A diagrammatic representation of the repetitive sequences of events seeming to underly rhythmical behavior patterns in *perihatching* and young chicks (see text for the empirical basis of the illustrated interactions; *excitation* is shown by a pointed arrow, *inhibition* by a reversed arrowpoint). (A) *Forebrain trigger* circuit, probably anterior diencephalic, tonically activated by "internal milieu" factors such as hormones, and presumably also by "external" factors such as spontaneous neural excitation originating elsewhere; the total effect is a (tonic) output to lower CNS regions. (B) *Lower generator* circuit, probably located in the brainstem: intense reverberatory neural firing is set off by the excitation coming from (A) or by sensory stimuli. (Sensory inputs, S, are inhibitory on A at relatively low intensities but excitatory on B if more intense). (C) *Initial behavioral effect* ("startle"), mediated by the various motor elements driven by B. (D) (If the head is immobilized) a *continued stereotyped sequence* of movements (struggling) for as long as the generator network B continues to reverberate. (E) *Refractoriness*, built up locally by continued neuromotor activity, terminating the discharges in B after 2–3 seconds and raising its threshold for reexcitation. [Because of the *tonic* activity from A, this negative feedback must result in repetition of such bursts (at a rate set by the decay time course ("relaxation") of the postdischarge "relative refractory period") until the neck finally becomes free to stretch itself.] (F) (If *un*restrained) *proprioceptive input*, possibly from neck stretch receptors, inhibiting the circuit B immediately after its initial motor effect (i.e., C). (G) Lower, possibly reticular, *arousal* network producing a different sequence of motor effects ("waking") than in D (including maintainance of the head and body erect: thus a positive feedback on F). (H) "*Refractoriness*," possibly built up locally, terminating arousal activity (G) within 5–10 seconds on the average. This leads to the cessation of input, F, and thus termination of the inhibition on B. (Again, because of the tonic activity from A, the cycle would restart as soon as the excitability of B then approaches close enough to its initial level.) (I) The posthatch onset of a *forebrain*, possibly hypothalamic, arousal mechanism (I) which is gradually incorporated during the first few hours following normal hatching. For convenience, it is shown as being excited directly from the lower arousal circuit, G, and its reciprocal excitation (positive feedback thus) mediated via a *disinhibition* of G from the effects of H. This last component accounts for longer maintained arousal, and would be terminated either by certain *environmental* changes (e.g., warmth) or by *endogenous* excitability fluctuations ("sleep/waking cycle"). It has a variety of complex effects, both on sensory-motor functions and on brain biolectric activity, which are not schematized here (see Corner, 1968; Corner & Bakhuis, 1969).

Jouvet, 1964). As in the chick embryo (see above; also see Corner & Schadé, 1967), in the cat the subsequent drastic decline of such spontaneous movements does not seem to be caused by inhibitory influences descending from the forebrain (Valatx, 1964), as has often been postulated (e.g., Bergström,

1969; Volokhov, 1961), but it may be true for the fetal sheep (Barcroft & Barron, 1937).

Since the above type of motility, as in the chick embryo (see Hamburger, 1964, 1969, and this volume), has a diffuse origin including the spinal cord (Barcroft & Barron, 1937; Corey, 1937), it is hardly suitable for use as a criterion for the existence in young animals of "paradoxical, or activated, sleep" (PS) in the strict physiological sense (Jouvet, 1965, 1967), as has been done at times in the past (Jouvet, 1965, 1967; Jouvet-Mounier, Astic, & Lacote, 1969; Sterman, 1967). Indeed, spontaneous startles and other jerky movements occur frequently during "quiet" sleep, as defined for example by the type of breathing, cerebral EEG, etc., as well as during "active" sleep in neonatal mammals (Garma & Verley, 1969; Gramsbergen *et al.*, 1970; Prechtl, 1969; Wolff, 1966). Characteristic bursts of rapid eye movements (REMs) are present, as expected, in the latter phase of sleep but are virtually absent at earlier stages of development, where distinctive defining criteria for PS are thus often lacking, despite the very high level of spontaneous muscular twitching (Dreyfus-Brisac, 1968, 1970; Garma & Verley, 1969). If amount of REMs is used as the criterion for PS in the human (Petre-Quadens *et al.*, 1969) one finds that, instead of a simple decline of PS with development (e.g., Jouvet-Mounier *et al.*, 1969; Roffwarg, Muzio, & Dement, 1966; Shimizu & Himwich, 1968), there is first a rise to a peak level prior to the decline. This approach would give a broader significance to our own findings on the chick (i.e., with respect to [a] the ontogeny of "slow wave" relative to paradoxical sleep, and [b] the presence of spontaneous local or generalized jerky movements even during the slow-wave phase), and in fact is supported by quantitative data for the amount of PS as function of age in two nonhuman mammalian species (Gramsbergen *et al.*, 1970; Meier & Berger, 1965).

The existence of stereotyped bursts of motor activity during sleep has also been reported for early human development, usually taking the form of cyclic trains of sucking, rather than of total body movements (e.g., Dreyfus-Brisac, 1968, 1970; Wolff, 1966). These bursts of mouth movements strongly resemble those typical of bird embryos (e.g., Gottlieb, 1965, 1968). "Spontaneous startles," on the other hand, have a generalized motor involvement (Prechtl, 1969) but are relatively short-lasting, like those during unrestrained sleep in young chicks. Recent unpublished observations (M. A. Corner) indicate, however, that spontaneous total-body motility takes on a stereotyped burstlike character in infant rats too if they are immobilized in the way done for the chicks (see above). Bursts of total movements occur spontaneously also in the early behavior of many lower vertebrates (Armstrong & Higgens, 1971: Coghill, 1931; Corner, 1964b; Herrick, 1937; Tracy, 1926), taking the form of repetitive brief bouts of swimming movements or of gill-depression movements only. Such bursts have more recently been observed to characterize the embryonic motility of decapod molluscs as well (squid and cuttlefish (M. A. Corner, unpublished observations). This type of discharge is thus very widespread phylogenetically, at least in *developing* nervous systems, and leads to some striking similarities in early behavior.

In fish and anuran larvae (Armstrong & Higgins, 1971; Corner, 1964b), the spontaneous motor bursts were found to be generated by the medulla and, in one species of frog at least, also by the spinal cord. This pattern of function was largely preserved in isolated fragments of these same tissues cultured *in vitro* (Corner, 1964a; Corner & Crain, 1965), often with regular interburst intervals quantitatively comparable to those seen in *intact* chicks or larval amphibians. In addition, similar bursts of activity have been described for fetal CNS tissue cultures taken from *warm*-blooded species such as the chick (see Fig. 15, and Cunningham, 1961a, 1961b) and the rat or mouse (e.g., Crain, 1970, and Volume 2 of this serial). Endogenous generation of action potentials and a postbarrage relative refractory period, the latter having a "relaxation time" of between 15 and 30 seconds, were found in such preparations (Corner & Crain, 1969, 1972). The same temporal patterning mechanism thus appears to be operative *in vitro* as was deduced earlier for the intact chick embryo.

In addition to brainstem and cord material, the spontaneous bursting pattern has been observed in cultures of cerebrum and of cerebellum from the chick embryo (Cunningham & Rylander, 1961; Cunningham & Stephens, 1961; also see Fig. 15) and from rodent fetuses (see Crain, Volume 2 of this series; Crain & Bornstein, 1964; Schlapfer, 1969). The cerebral cortex *in situ* can be induced to mimic the functioning of cortical tissues in culture, either by means of surgically isolated "slabs" or by X-irradiation, both of which techniques result in intensive collateral sprouting and, presumably therefore, more intensive excitatory interactions within the neuronal network (Purpura, 1969; Purpura & Shofer, 1967). Widespread barrages of action potentials also typify the spontaneous discharges in very immature normal cerebral cortex (Huttenlocher, 1967). Stereotyped bursts of maximally intense firing may therefore well be the characteristic mode of function which tends to develop in neural circuits having diffuse (i.e., "reticular") and dense interconnections among its elements. It would not be surprising if these conditions were commonly met both in "primitive" embryonic networks and in small explants, even when the latter are made using tissues normally characterized by a more highly ordered structure. Since strychnine applied to the explants increases their overall excitability but fails to noticeably alter the time course of the postbarrage elevation of threshold (Corner & Crain, 1969), it is not certain if this bursting pattern *per se* requires the presence of inhibitory units in the circuit.

B. Conclusions and Summary

Although the possible significance of the various findings reviewed in this paper have been mentioned throughout, in the appropriate places, it will perhaps do well to bring together here the major conclusions, tentative to some degree as they are. The interpretations suggested by the available data will be listed as a series of postulates, together with some seemingly plausible approaches for testing them. Much of the following set of working ideas is also presented pictorially in Fig. 21.

Postulate 1. The general state which is denoted in higher vertebrates by the word "sleep" is the direct postnatal continuation of the condition existing permanently prior to birth (or hatching). It is characterized by cyclic fluctuations of activity, and, especially in very immature animals, by both quasi-random and highly stereotyped movements which are generated at least partly by nonreflexogenic neural discharges. Motor responses to sensory input are similar in kind to those occuring spontaneously, and discrimination among different stimuli or orientation toward them is at a minimum.

Additional functional criteria—such as biochemical or electrophysiological—are desirable for establishing the actual degree of *continuity* of prenatal life with, respectively, sleep and wakefulness. Furthermore, additional details about the patterning of movements during sleep and their possible relationship to specific stimulus configurations, as well as to past "experiences," would help to ascertain the degree of *automatism* truly existing. Modern electronic techniques would make possible multiple simultaneous registrations of even weak muscular activity, and also precisely controlled programs of stimulation.

Postulate 2. "Sleep behavior" is typically produced by relatively primitive (i.e., diffusely interconnected) nerve nets, extending throughout the brainstem and spinal cord. These tend to be readily triggered into maximally intense excitation, similar to what in more highly structured networks would require a pathological degree of hyperexcitability, but also possess mechanisms for quickly quenching such discharges. Slower cycles of excitability exist in addition, making possible fairly complicated output rhythms in the face of constant tonic inputs. Higher brain systems would influence this pattern only quantitatively, e.g., the onset of hatch movements in birds, or the reduction of late fetal or early neonatal movements in mammals.

The question of (a) the *locations* of "pacemaker" neurons, motor patterning mechanisms, and sensory distribution pathways, and (b) the cellular *processes* underlying autorhythmicity and functional interactions within the network, is also an important area which ought to be amenable to modern methods, e.g., by intracellular and multiple extracellular microelectrode registrations.

Postulate 3. "Sleep-type" motility, as portrayed above, is an evolutionarily ancient mode of function, reflected already in the embryonic or early larval behavior of aquatic vertebrates, and also in certain invertebrates. The activated phases during sleep in amniotes (i.e., "rhombencephalic, or paradoxical sleep") would represent a vestige of this originally diffuse early system, which became specialized with further evolution into a progressively important, but as yet poorly understood, life-long function.

More extensive *comparative* studies, both into the detailed patterning and the underlying neurophysiological processes of very early behavior, would enable us to properly assess the real generality of the motility which we have been emphasizing so far. Many of the basic observations could undoubtedly be carried out by relatively simple means, so that an intelligent choice of the species to be studied would seem to be the chief problem in this particular program of research.

Postulate 4. Wakefulness (arousal) is a state which, in precocial species, develops with "forward reference" to its later function, since the conditions for its manifestation are not normally present until some time after the potentiality has appeared. It requires sensory input, including specific modalities, in order to achieve or sustain dominance over the sleep-motor system, with which it is linked by reciprocal inhibition. Discriminative awareness of the environment and goal-directed behavior distinguish this (set of) state(s), made possible by functional integration of forebrain networks with relatively primitive brainstem ones.

The localization of lower arousal circuits might be approached by means of direct brain stimulation, combined with defect experiments employing the same stereotactically positioned electrodes for both purposes. Specific patterns of electrical activity would then be predicted to be recorded at such sites, according to the predominating source of the input: from *sensory* pathways, from *forebrain* excitation (or inhibition), or from *reciprocal* inhibition from the sleep systems.

The question of the influence which functional activities such as we have described might have upon the further maturation of the nervous system should not be overlooked as a field of future inquiry (Roffwarg *et al.*, 1966). Precisely controlled reductions in the level of neuronal function during ontogeny have been applied in tissue culture studies (see Crain, Bornstein, & Peterson, 1968) and this approach warrants extension to the intact animal. Similarly, the converse experiment of producing controlled *augmented* bioelectric activity in the course of development would also be likely to help answer this question, and is certainly feasible using currently available electrophysiological tools.

References

Aladjalova, N. A. *Slow electrical processes in the brain.* Amsterdam: Elsevier, 1964.

Armstrong, P. B., & Higgins, D. C., Behavioral encephalization in the bullhead embryo and its neuroanatomical correlates. *Journal of Comparative Neurology*, 1971, **143**, 371–384.

Bakhuis, W. L. The emergence from the egg of the chicken (*Gallus domesticus*) and the role therein of the central nervous system (in Dutch). *Unpublished master's thesis*, Amsterdam Municipal University, 1971.

Barcroft, J., & Barron, D. H. Movements in midfoetal life in the sheep embryo. *Journal of Physiology (London)*, 1937, **91**, 329–351.

Bergström, R. M. Brain and muscle potentials from the intra-uterine fetus in un-narcotized conscious animals. *Nature (London)*, 1962, **195**, 1004–1005.

Bergström, R. M. Electrical parameters of the brain during ontogeny. In R. J. Robinson (Ed.), *Brain and early behaviour: Development in the fetus and infant.* London: Academic Press, 1969. Pp. 15–36.

Bernhard, C. G., Kaiser, I. H., & Kolmodin, G. M. On the development of cortical activity in fetal sheep. *Acta Physiologica Scandinavica* , 1959, **47**, 333–349.

Coghill, G. W. Corrolaries of the anatomical and physiological study of *Amblystoma* from the age of earliest movements to swimming. *Journal of Comparative Neurology*, 1931, **53**, 147–168.

Corey, A. Development of the fetal rat following electro-cautery of the brain. *American Journal of Physiology*, 1937, **115**, 599–603.

Corner, M. A. Localization of capacities for functional development in the neural plate of *Xenopus laevis. Journal of Comparative Neurology*, 1964, **123**, 243–256. (a)

Corner, M. A. Rhythmicity in the early swimming of anuran larvae. *Journal of Embryology and Experimental Morphology*, 1964, **12**, 665–671. (b)

Corner, M. A. Aspects of the functional organization of the central nervous system during embryonic development in the chick. In L. Jílek & S. Trojan (Eds.), *Ontogenesis of the brain.* Prague: Charles University Press, 1968. Pp. 85–96.

Corner, M. A., & Bakhuis, W. L. Cerebral electrical activity, forebrain function and behavior in the chick at the time of hatching. *Brain Research*, 1969, **13**, 541–555.

Corner, M. A., & Bot, A. P. C. Somatic motility during the embryonic period of birds and its relations to behavior after hatching. *Progress in Brain Research*, 1967, **26**, 214–236.

Corner, M. A., & Bot, A. P. C. Electrical activity in the isolated forebrain of the chick embryo. *Brain Research*, 1969, **12**, 473–476.

Corner, M. A., & Crain, S. M. Spontaneous contractions and bioelectric activity after differentiation in culture of presumptive neuromuscular tissues of the early frog embryo. *Experientia*, 1965, **21**, 422–424.

Corner, M. A., & Crain, S. M. The development of spontaneous bioelectric activities and strychnine sensitivity during maturation in culture of embryonic chick and rodent central nervous tissues. *Archives Internationales de Pharmacodynamie et de Therapie*, 1969, **182**, 404–406.

Corner, M. A., & Crain, S. M. Patterns of spontaneous bioelectric activity during maturation in culture of fetal rodent medulla and spinal cord tissues. *Journal of Neurology*, 1972, **3**, 25–45.

Corner, M. A., Peters, J. J., & Rutgers van der Loeff, P. Electrical activity patterns in the cerebral hemisphere of the chick during maturation, correlated with behavior in a test situation. *Brain Research*, 1966, **2**, 274–292.

Corner, M. A., & Schadé, J. P. Developmental patterns in the central nervous system of birds. IV. Cellular and molecular bases of functional activity. *Progress in Brain Research*, 1967, **26**, 237–250.

Corner, M. A., Schadé, J. P., Sedláček, J., Stoeckart, R., & Bot, A. P. C. Development of electrical activity patterns in the cerebral hemisphere, optic lobe and cerebellum of birds. *Progress in Brain Research*, 1967, **26**, 145–192.

Crain, S. M. Bioelectric interactions between cultured fetal rodent spinal cord and skeletal muscle after innervation *in vitro*. *Journal of Experimental Zoology*, 1970, **173**, 353–370.

Crain, S. M., & Bornstein, M. B. Bioelectric activity of neonatal mouse cerebral cortex during growth and differentiation in tissue culture. *Experimental Neurology*, 1964, **10**, 425–450.

Crain, S. M., Bornstein, M. B., & Peterson, E. R. Maturation of cultured embryonic CNS tissues during chronic exposure to agents which prevent bioelectric activity. *Brain Research*, 1968, **8**, 363–372.

Cunningham, A. W. B. Spontaneous potentials from explants of chick embryo pons in culture. *Experientia*, 1961, **17**, 233–234. (a)

Cunningham, A. W. B. Spontaneous potentials from explants of chick embryo spinal cord in tissue culture. *Naturwissenschaften*, 1961, **48**, 719–720. (b)

Cunningham, A. W. B. Qualitative behavior of spontaneous potentials from explants of 15 day chick embryo telencephalon *in vitro*. *Journal of General Physiology*, 1962, **45**, 1065–1076.

Cunningham, A. W. B., & Rylander, B. J. Behavior of spontaneous potentials from chick cerebellar explants during 120 hours in culture. *Journal of Neurophysiology*, 1961, **24**, 141–149.

Cunningham, A. W. B., & Stephens, S. G. Qualitative effect of strychnine and brucine on spontaneous potentials from explants of telencephalon. *Experientia*, 1961, **17**, 569–571.

Dawes, G. S., Fox, H., Leduc, B., Liggens, G., & Richards, R. T. Respiratory movements and paradoxical sleep in the foetal sheep. *Journal of Physiology (London)*, 1970, **210**, 47P–48P.

Dawes, G. S. Fox, H. E., Leduc, B. M. Liggens, G. C., & Richards, R. T. Respiratory movements and rapid eye movement sleep in the foetal lamb. *Journal of Physiology*, 1972, **220**, 119–143.

Decker, J. D., & Hamburger, V. The influence of different brain regions on periodic motility of the chick embryo. *Journal of Experimental Zoology*, 1967, **165**, 371–384.

Dreyfus-Brisac. C. Sleep ontogenesis in early human prematurity from 24 to 27 weeks of conceptional age. *Developmental Psychobiology*, 1968, **1**, 162–169.

Dreyfus-Brisac, C. Ontogenesis of sleep in human prematures after 32 weeks of conceptional age. *Developmental Psychobiology*, 1970, **3**, 91–122.

Ellingson, R. J., & Rose, G. H. Ontogenesis of the electroencephalogram. In W. A. Himwich (Ed.), *Developmental neurobiology*, Springfield, Ill. Thomas, 1970. Pp. 441–474.

Freeman, B. M. The effect of temperature on the resting metabolism of the fowl during the first month of life. *British Poultry Science*, 1963, **14**, 275–278.

Frost, J. D., & Gol, A. Computer determination of relationships between EEG activity and single unit discharges in isolated cerebral cortex. *Experimental Neurology*, 1966, **14**, 506–519.

Garma, L., & Verley, R. Ontogenèse des états de veille et de sommeil chez les mammifères. *Revue de Neuropsychiatrie Infantile*, 1969, **17**, 479–504.

Gottlieb, G. Prenatal auditory sensitivity in chickens and ducks. *Science*, 1965, **147**, 1596–1598.

Gottlieb, G. Prenatal behavior of birds. *Quarterly Review of Biology*, 1968, **43**, 148–174.

Gramsbergen, A., Schwartze, P. & Prechtl, H. F. R. The postnatal development of behavioral states in the rat. *Developmental Psychobiology*, 1970, **3**, 267–280.

Hamburger, V. Some aspects of the embryology of behavior. *Quarterly Review of Biology*, 1963, **38**, 342–365.

Hamburger, V. Ontogeny of behavior and its structural basis. In D. Richter (Ed.), *Comparative neurochemistry*. Oxford: Pergamon, 1964. Pp. 21–34.

Hamburger, V. Emergence of nervous coordination. Origins of integrated behavior. In M. Locke (Ed.), *The emergence of order in developing systems*. New York: Academic Press, 1969. Pp. 251–271.

Hamburger, V., & Balaban, M. Observations and experiments on spontaneous rhythmical behavior in the chick embryo. *Developmental Biology*, 1963, **7**, 342–365.

Hamburger, V., & Hamilton, H. L. A series of normal stages in the development of the chick embryo. *Journal of Morphology*, 1951, **88**, 49–92.

Hamburger, V., & Oppenheim, R. W. Prehatching motility and hatching behavior in the chick. *Journal of Experimental Zoology*, 1967, **166**, 171–204.

Hebb, D. O. Cerebral organization and consciousness. In S. S. Kety, E. Evarts, & H. L. Williams (Eds.), *Sleep and altered states of consciousness*. Baltimore: Williams & Wilkins, 1967. Pp. 1–7.

Herrick, C. J. Development of the brain of *Amblystoma* in early functional stages. *Journal of Comparative Neurology*, 1937, **67**, 381–422.

Hishikawa, Y., Cramer, H., & Kuhlo, W. Natural and melatonin-induced sleep in young chickens—a behavioral and electrographic study. *Experimental Brain Research*, 1969, **7**, 84–94.

Huttenlocher, P. R. Development of cortical neuronal activity in the neonatal cat. *Experimental Neurology*, 1967, **17**, 247–262.

Jasper, H. H., Bridgman, C. S., & Carmichael, L. An ontogenetic study of cerebral electrical potentials in the guinea pig. *Journal of Experimental Psychology*, 1937, **21**, 63–67.

Jouvet, M. Recherches sur les structures nerveuses et les mechanismes responsables des differentes phases du sommeil physiologique. *Archives Italiennes de Biologie*, 1962, **100**, 125–206.

Jouvet, M. Paradoxical sleep—a study of its nature and mechanisms. In K. Akert, C. Bally, & J. P. Schadé (Eds.), *Sleep mechanisms*. Amsterdam: Elsevier, 1965. Pp. 20–62.

Jouvet, M. Neurophysiology of the states of sleep. *Physiological Reviews*, 1967, **47**, 117–177.

Jouvet-Mounier, D., Astic, L., & Lacote, D. Ontogenesis of the states of sleep in rat, cat and guinea pig during the first post-natal month. *Developmental Psychobiology*, 1969, **2**, 216–239.

Kellaway, P., Gol, A., & Proler, M. Electrical activity of the isolated cerebral hemisphere and the isolated thalamus. *Experimental Neurology*, 1966, **14**, 281–304.

Key, B. J. & Marley, E. The effect of the sympathomimetic amines on behavior and electrocortical activity of the chicken. *Electroencephalography and Clinical Neurophysiology*, 1962, **14**, 90–105.

Klein, M. *Étude polygraphique et phylogènique des états de sommeil*. Thèse de Medecine. Lyon: Bosc Frères, 1963.

Klein, M., Michel, F., & Jouvet, M. Étude polygraphique du sommeil chez les oiseaux. *Comptes Rendues de la Société de Biologie*, 1964, **158**, 99–103.

Kleitman, N. *Sleep and wakefulness*. Chicago: University of Chicago Press, 1963.

Kovach, J. K. Spatial orientation of the chick embryo during the last five days of incubation. *Journal of Comparative and Physiological Psychology*, 1968, **66**, 283–288.

Kovach, J. K. Development and mechanisms of behavior in the chick embryo during the last five days of incubation. *Journal of Comparative and Physiological Psychology*, 1970, **73**, 392–406.

Kovach, J. K., Callies, D., & Hartzell, R. Procedures for the study of behavior in avian embryos. *Developmental Psychobiology*, 1970, **3**, 169–178.

Kuo, Z-Y. Studies on the physiology of the embryonic nervous system. II. Experimental evidence on the controversy over the reflex theory in development. *Journal of Comparative Neurology*, 1939, **70**, 437–459.

Kuo, Z-Y. *The dynamics of behavior development.* New York: Random House, 1967.

Marley, E., & Stephenson, J. D. Actions of dexamphetamine and amphetamine-like amines in chickens with brain transections. *British Journal of Pharmacology*, 1971, **42**, 522–542.

Meier, G. W., & Berger, R. J. Development of sleep and wakefulness in the infant rhesus monkey. *Experimental Neurology*, 1965, **12**, 257–277.

Oökawa, T. Electroencephalograms recorded from the telencephalon of the blinded chicken during behavioral sleep and wakefulness. *Poultry Science*, 1971, **50**, 731–736.

Oökawa, T., & Gotoh, J. Electroencephalogram of the chicken recorded from the skull under various conditions. *Journal of Comparative Neurology*, 1965, **124**, 1–14.

Oökawa, T., & Takagi, K, Electroencephalograms of free behavioral chicks at various developmental ages. *Japanese Journal of Physiology*, 1968, **18**, 87–99.

Oppenheim, R. W. Some aspects of embryonic behavior in the duck (*Anas platyrhynchos*). *Animal Behaviour*, 1970, **18**, 335–352.

Oppenheim, R. W. Experimental studies on hatching behavior in the chick. III. The role of the midbrain and forebrain. *Journal of Comparative Neurology*, 1972, **146**, 479–506.

Oppenheim, R. W., & Narayanan, C. H. Experimental studies on hatching behavior in the chick. I. Thoracic spinal gaps. *Journal of Experimental Zoology*, 1968, **168**, 387–394.

Parmelee, A., Wenner, W., Akiyama, Y., Schultz, M., & Stern, E. Sleep states in premature infants. *Developmental Medicine and Child Neurology*, 1967, **9**, 70–77. (a)

Parmelee, A., Wenner, W., Akiyama, Y., Stern, E., & Flescher, J. Electroencephalography and brain maturation. In A. Minkowski (Ed.), *Regional development of the brain at early life.* Springfield, Ill. Thomas, 1967. Pp. 459–480. (b)

Peters, J. J., Vonderahe, A. R., & Schmid, D. Onset of cerebral electrical activity associated with behavioral sleep and attention in the developing chick. *Journal of Experimental Zoology*, 1965, **160**, 255–262.

Petre-Quadens, O., Hardy, J. L., & DeLee, C. Comparative study of sleep in pregnancy and in the newborn. In R. J. Robinson (Ed.), *Brain and early behaviour.* London: Academic Press, 1969. Pp. 177–189.

Prechtl, H. F. R. Brain and behavioral mechanisms in the human newborn infant. In R. J. Robinson (Ed.), *Brain and early behaviour*. London: Academic Press, 1969. Pp. 115–130.

Provine, R. R. Ontogeny of bioelectric activity in the spinal cord of the chick embryo and its behavioral implications. *Brain Research*, 1972, **41**, 365–378.

Purpura, D. P. Stability and seizure susceptibility of immature brain. In H. Jasper, A. Ward, & A. Pope (Eds.), *Basic mechanisms of the epilepsies.* London: Churchill, 1969. Pp. 481–505.

Purpura, D. P., & Shofer, R. J. Modification of ontogenetic patterns in mammalian brain. *Progress in Brain Research*, 1967, **22**, 458–479.

Roffwarg, H. P., Muzio, J. N., & Dement, W. C. Ontogenetic development of the human sleep-dream cycle. *Science*, 1966, **152**, 604–619.

Rosen, M. G., & McLaughlin, A. Fetal and maternal electroencephalography in the guinea pig. *Experimental Neurology*, 1966, **16**, 181–190.

Ruckebusch, Y. Development of sleep and wakefulness in the foetal lamb. *Electroencephalography and Clinical Neurophysiology*, 1972, **32**, 119–128.

Salter, S. H. A note on the recording of egg activity. *Animal Behaviour*, 1966, **14**, 41–43.

Scherrer, J., Verley, R., & Garma, L. A review of French studies in the ontogenetical field. In W. A. Himwich (Ed.), *Developmental neurobiology*. Springfield, Ill.: Thomas, 1970. Pp. 528–549.

Schlapfer, W. T. Bioelectric activity of neurons in tissue cultures: synaptic interactions and effects of environmental changes. Unpublished doctoral dissertation, Lawrence Radiation Laboratory, Berkeley, California, 1969.

Sedláček, J. *Prenatal development of electrical properties of the cerebral tissue.* Doctoral dissertation. Prague: Academia, 1967.

Sheff, A. G., & Tureen, L. L. EEG studies of normal and encephalomalacic chicks. *Proceedings of the Society for Experimental Biology and Medicine*, 1962, **111**, 407–409.

Shimizu, A., & Himwich, H. E. The ontogeny of sleep in kittens and young rabbits. *Electroencephalography and Clinical Neurophysiology*, 1968, **24**, 307–318.

Spooner, C. E. *Observations on the use of the chick in the pharmacological investigation of the central nervous system.* (Doctoral dissertation, University of Michigan) Ann Arbor Mich.: University Microfilms, 1964. No. 64–6387.

Sterman, M. B. Relationship of intra-uterine fetal activity to maternal sleep stage. *Experimental Neurology*, 1967, **19** (Suppl. 4), 98–106.

Tracy, H. The development of motility and behavior reactions in the toadfish (*Opsanus tau*). *Journal of Comparative Neurology*, 1926, **40**, 254–360.

Tradardi, V. Sleep in the pigeon. *Archives Italiennes de Biologie*, 1966, **104**, 516–521.

Tuge, H., Kanayama, Y., & Chang, H. Y. Comparative studies on the development of the EEG. *Japanese Journal of Physiology*, 1960, **10**, 211–220.

Valatx, J-L. *Ontogénèse des différents états de sommeil: Étude polygraphiques chez le chaton.* Thèse de Médicine. Lyon: Imprimerie des Beaux-Arts, 1964.

Valatx, J-L., Jouvet, D., & Jouvet, M. Evolution électroencephalographique des différents états de sommeil chez le chaton. *Electroencephalography and Clinical Neurophysiology*, 1964, **17**, 218–233.

van Wingerden, C. Investigations into the function of the reticular activating system in connection with sleep and waking behavior in the chick (in Dutch). Unpublished master's thesis, Amsterdam Municipal University, 1970.

Villablanca, J. Electroencephalogram in the permanently isolated forebrain of the cat. *Science*, 1962, **138**, 44–46.

Volokhov, A. A. On the significance of various levels of the central nervous system in the formation and development of motor reaction in embryogenesis. In P. Sobotka (Ed.), *Functional and metabolic development of the central nervous system.* Pilzen: University Medical Bulletin, 1961. Pp. 141–147.

Wolff, P. H. The causes, controls and organization of behavior in the neonate. *Psychological Issues*, 1966, No. 27.

Section 4

SENSORY PROCESSES: EMBRYONIC BEHAVIOR IN BIRDS

INTRODUCTION

Sensory stimulation can have a number of different influences during embryonic development, among which are the following. (1) It can affect the ongoing overt motility of the embryo, either exciting movement or inhibiting it. (2) Particular patterns of sensory stimulation encountered at an early age—independent of their immediate effect on motor movement—may influence the perception and behavior of the embryo or neonate at a later age. Of course, these two effects of sensory stimulation are not mutually exclusive, but it is important to realize that (1) can occur without (2) and vice versa. So these are rather different propositions which require different experimental techniques for their verification or analysis—each effect must be documented in its own right, and that is what we shall see in the ensuing articles by Margaret Vince and by Monica Impekoven and Peter Gold.

The most important defining feature of autogenous or spontaneous embryonic motility, as described for the chick and other species in Section One, is its independence of sensory stimulation, in the sense that such motility is primarily self-generated. Although we now realize, as Dr. Hamburger notes in his article in Section Two, that even the chick embryo, the "classic" subject for such studies, may respond now and then to sensory stimulation without undermining the notion of the primacy of spontaneous movement, there has been some possibility that the role of sensory stimulation might be regarded as an insignificant factor in the embryonic behavior of birds because of the tremendous emphasis placed on the spontaneity of chick embryonic motility in recent years (and in this volume!). If that is the case, the succeeding articles by Miss Vince and by Drs. Impekoven and Gold should go far in redressing the balance. Applying the comprehensive investigative framework of Zing-Yang Kuo to the embryonic behavior of quail, chicks, ducks, and other avian species, Miss Vince's discussion goes well beyond a concern merely for the role of sensory factors in the early development of behavior. On the other hand, Drs. Impekoven and Gold examine the immediate response of gull embryos to different patterns of sensory stimulation relatively early in development, and they also demonstrate the influence of late embryonic sensory stimulation on postnatal behavior in gulls. One has the impression from these studies, as well as from the work of others which is reviewed by the authors, that the various roles of sensory stimulation in embryonic behavior will come in for ever-increasing analysis in a widening range of species in the near future.

SOME ENVIRONMENTAL EFFECTS ON THE ACTIVITY AND DEVELOPMENT OF THE AVIAN EMBRYO

MARGARET A. VINCE

Cambridge Psychological Laboratory
University of Cambridge
Cambridge, England.

I. Introduction

We know well that the development of the avian embryo is dependent on external conditions—the physics and chemistry of incubation (Landauer, 1961; Lundy, 1969). Incubating conditions must be kept within certain levels. Heat, humidity, and the gases surrounding the shell, as well as the shell itself, all contribute toward development. If these conditions vary for too long beyond certain limits, development becomes abnormal and finally comes to an end.

Anatomical and physiological development, therefore, depends on external conditions. In its activity, however, the embryo was for a long time believed to develop in isolation from its environment. For example, Lehrman (1953), in dealing with behavior, refers to the avian embryo as "the perfect isolation experiment." More recently, considering the initiation of lung ventilation, Dawes and Simkiss (1969) have stated that "the avian embryo develops within a cleidoic egg which largely insulates it from the environment [p. 79]."

In their work on embryonic motility, Hamburger and his colleagues have tended to present a similar viewpoint. Considering motility in the chick embryo, they have shown that the major part of this is "spontaneous," that is, nonreflexogenic and originating from random self-generated discharges of neurones in the spinal cord, at least up to day 17 of incubation. At later incubation stages, these random activities continue, although the movements leading up to hatching from about day 17 are observed to be coordinated activities believed to originate *de novo* from that time (Hamburger & Oppenheim, 1967). These are assumed to depend on centers of integration within the nervous system which "probably differentiate and mature in earlier stages and are activated by endogenous and exogenous stimuli [Hamburger, 1968, p. 104]." Thus, although a part played by sensory information is not ruled out at any stage, the main interest and emphasis in this work has been in establishing the primary role in behavioral development of the maturation of the nervous system.

In the present paper I propose to consider another aspect of avian embryonic development and activity: the extent to which this is dependent on, or controlled by, the environment, either via sensory processes or by other means.

During the last 3 or 4 days before hatching there is now ample evidence of an interaction between the embryo and its environment outside the shell. On the more behavioral level, Tschanz (1968) has shown in the guillemot (*Uria a. aalge*) that the embryo and the incubating parent respond to each other's calls and other activities in ways which are important for survival (described) further in Section III). In addition, but at a level where a distinction between physiology and behavior becomes meaningless, it has been shown that embryos of the quail and other precocial species respond differentially to sounds and/or vibrations produced by neighboring eggs. Also, this response affects individual developmental rates and times of hatching (Vince, 1964, 1970). and Section IV.

There is, therefore, evidence that stimulation from outside the egg can affect development at a fairly late stage of incubation, that is, from about the equivalent of day 17 in the domestic fowl. Sensory mechanisms involved in an effect of this kind are known to be functional at this stage (see review by Gottlieb, 1968), and it is believed that the coordinated movements from 17 days onward may be activated by stimulation (Hamburger & Oppenheim, 1967). Thus, the general picture at this stage appears relatively consistent.

Nevertheless, problems remain to be solved or to be explored fully. We do not yet know what is the earliest stage at which postural and developmental changes can be affected, when embryos are given the type of accelerating stimulation which results in the synchronization of hatching. There is also the possibility of environmental effects on embryonic ontogeny which are only indirectly dependent on sensory input. This latter view may conveniently be presented by considering the work of Kuo (1932a, 1932b, 1932c, 1967). Kuo, who initiated the method of observing the avian embryo's activities through windows in the shell and who analyzed its behavioral development, considered this in the light of an interaction between embryonic activity, structure, and the immediate environment, in this way rejecting the views of Coghill (1929): "We wish to express our doubt concerning the view that morphological development of the nervous system can be used as the absolute index of behavioral development [Kuo, 1932c, p. 267]." Unfortunately, Kuo did not always present quantitative data, and some of his ideas have been found wanting. Nevertheless, his observations include a detailed account of possible effects on embryonic activity of changes in its own structure and the structure and movement of the remaining contents of the shell, which tend to be ignored elsewhere and now need reexamination. The view that the direction, extent, and patterning of embryonic movement can be to some extent forced on the organism by the movement or changing size and relationship of structures within the shell could be a commonsense one with much positive potentiality, and it will be considered later (see Section V).

Before discussing these three topics, and as a background to them, I will give a brief summary of work on the developmental stages undergone by the avian embryo during the incubation period.

II. The Development of Activity and Stages of Development in the Embryo of the Domestic Fowl

The work of Hamburger and his colleagues has shown that from days $3\frac{1}{2}$–4 of incubation, the chick embryo shows a type of periodic motility (Hamburger and Balaban, 1963). These movements begin 3 or 4 days before the reflex circuits are completed, and according to numerous studies (Decker, 1970; Hamburger, Balaban, Oppenheim & Wenger, 1965; Hamburger & Narayanan, 1969; Hamburger & Oppenheim, 1967), they remain independent of sensory input at least up to day 17. According to Oppenheim (1970), who counted the number of movements per minute in the chick and duck embryo, this spontaneous activity increases gradually with age until about two-thirds of the way through the incubation period, when it begins to fall off; it persists, however, at least until the time of hatching and consists of independent movements of different parts of the body, such as head, wings, legs, eyelid, beak, etc., occurring independently of each other in constantly changing and unpredictable combinations.

Kuo (1932a, 1932b, 1932c) observed both active and passive embryonic movements and considered them in their relationship with changes in embryo size and structure and with the extraembryonic contents of the egg. For example, the first perceptible movements he describes as very minute passive vibrations associated with heart beats and circulating movements of the large blood vessels. Later (days 4–6), while the embryo is small and light, passive movements continue and change into large "swinging" movements when amnion contractions and the consequent passive movements of the yolk sac begin. Amniotic activity and the swinging movements produced by it become more rapid, regular, and extensive between days 6 and 9. Then they begin to fall off as movements of the amnion become more sluggish, and they gradually cease by about day 13. During this time the embryo has become larger and heavier, and by about day 10 it comes to lie in a more fixed position at right angles to the long axis of the egg near the air chamber membrane with its back on the yolk sac (Fig. 1). According to Kuo (1932b), the orientation and direction of growth of the allantoic stalk and the yolk stalk may in part be responsible for the embryo's taking up this position as the swinging movements come to an end. At this stage the albumen, now heavier than the yolk, sinks into the narrow end of the egg, while the partly depleted and hence lighter yolk sac is displaced upward and comes to cover the ventral surface of the embryo, holding its legs, now grown much longer, folded over the breast. From this position and between about 10–16 days the embryo slowly turns to take up its prehatching position (said to be lengthwise of the egg, but see Section V), with the beak buried under the yolk sac [see Figs. 2A (2) and (3), 2B (2) and (3) for bobwhite embryos, and Fig. 2 on Pp. 292 and 293 for a comparison between late developmental stages in

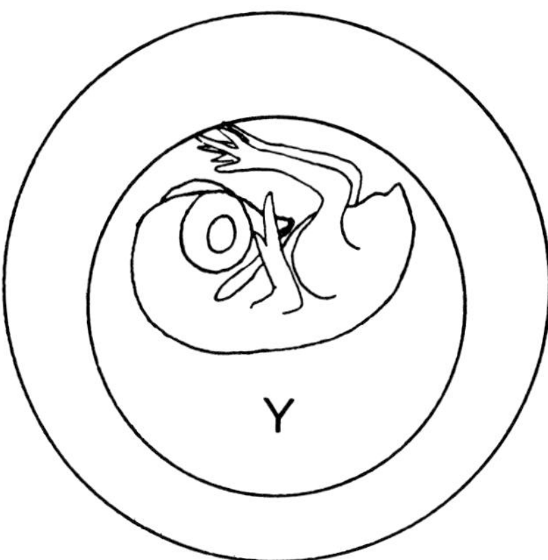

FIG. 1. Embryo of the domestic fowl. Normal position after fixation. Inner circle represents window cut in shell over the airspace. The inner shell membrane was made transparent by the melted vaseline technique. Y = yolk. From Kuo (1932b) by courtesy of the Press of the Wistar Institute, Philadelphia.

quail and the domestic fowl; also Oppenheim, this volume, for an extensive review of hatching in a number of species]. According to Kuo, this turning is accomplished as a result of wriggling movements of the body and thrusting movements of the legs. The leg movements gradually push the yolk upward toward the membrane, while the relatively heavy embryo tends at the same time to slip beneath it while the position of the neck is fixed by being squeezed between the yolk, the membranes, and the shell. Turning is said to be completed by day 17, and at this stage (see Fig. 2B for the quail) the movements leading to hatching begin.

These are described by Hamburger and Oppenheim (1967), who regard them as qualitatively different from the random motility referred to earlier. They are now fully coordinated movements, which alternate with the random motility and result in the series of postural changes and activities which end in the emergence of the chick.

First the head is lifted out from under the yolk sac, and after many movements of different intensity, mostly toward the right (but occasionally also to the left), the beak is turned upward toward the membranes and comes to rest under the now outstretched right wing ("tucking"), with the beak pointing toward the membranes separating the embryo from the air space (Fig. 2C). At about this stage the yolk sac begins to be withdrawn into the body cavity. The beak pushes upward against the air space membranes which become draped over its tip. By a series of head and beak movements, the membrane is worn thin and the smaller blood vessels are drained. Finally

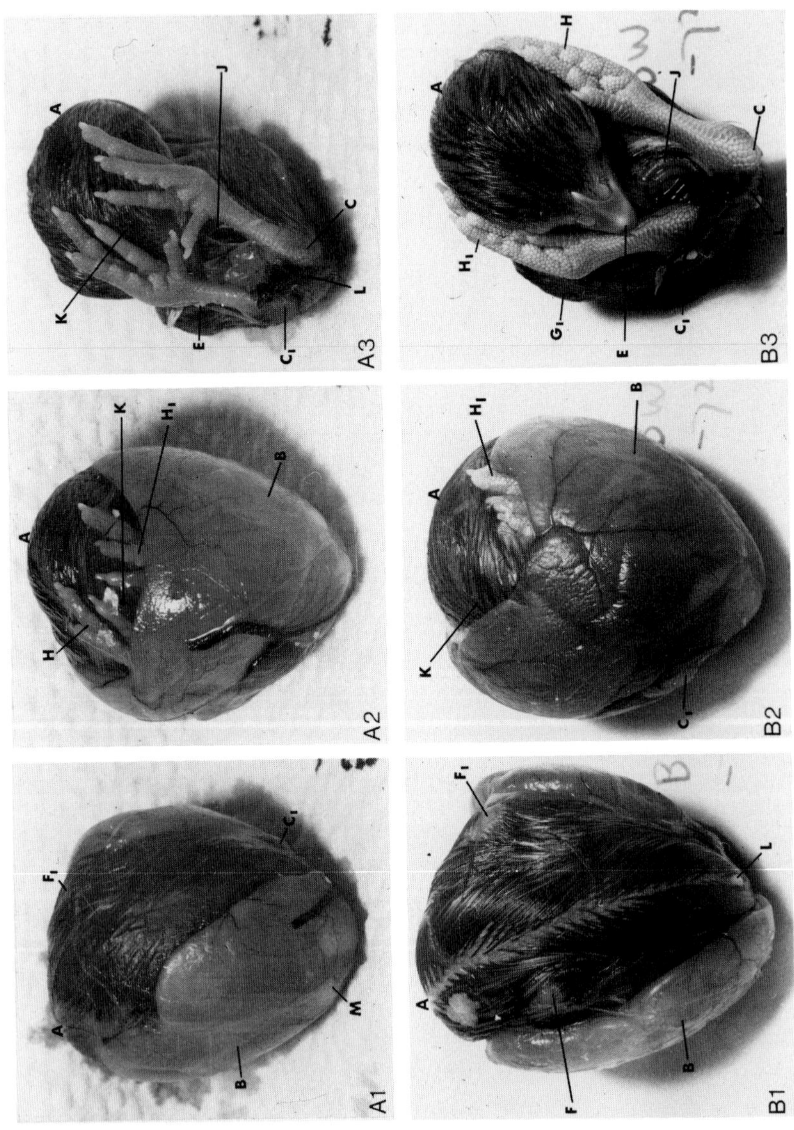
A1
A2
A3
B1
B2
B3

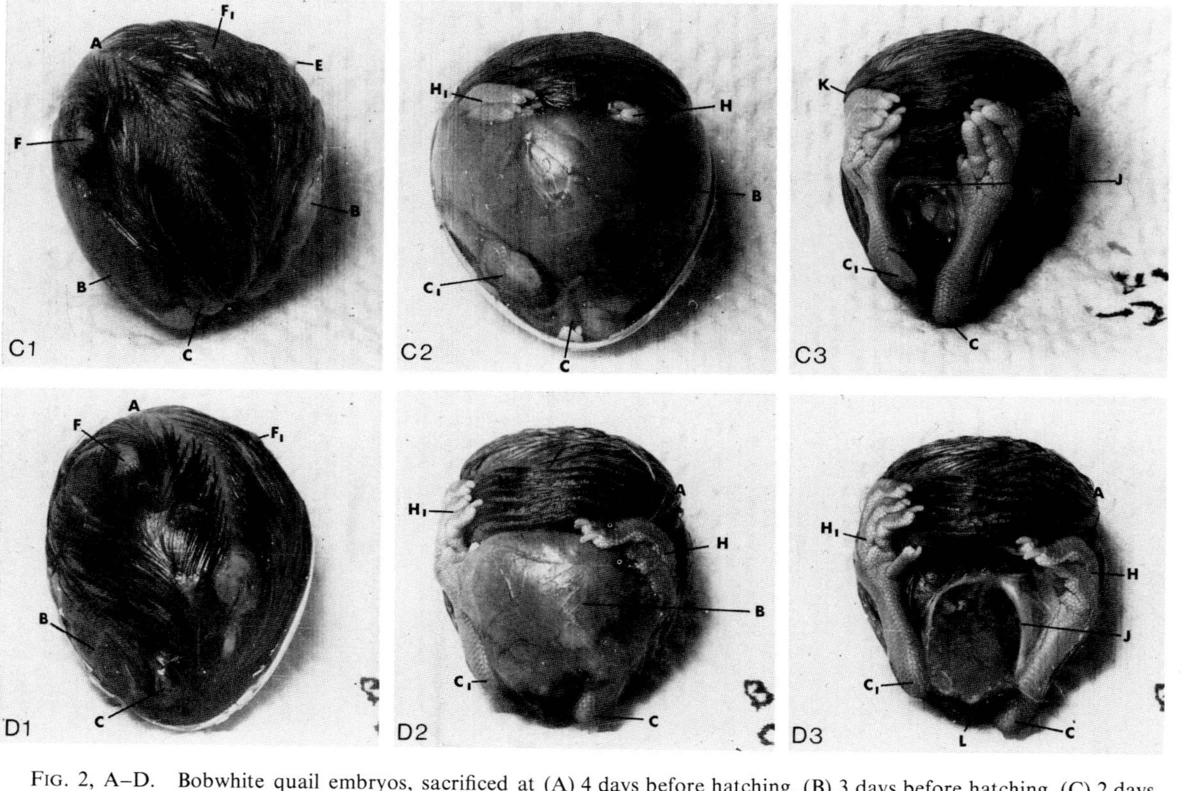

FIG. 2, A–D. Bobwhite quail embryos, sacrificed at (A) 4 days before hatching, (B) 3 days before hatching. (C) 2 days before hatching, and (D) during the period of clicking (between 12 and 0 hours before hatching). Eggs were fixed in formalin. For each stage (1) shows the dorsal aspect, (2) the ventral aspect, and (3) the ventral aspect with the yolk sac still outside the body removed. For eggs, A, B, and C, the yolk sac was still entirely outside the body, while for egg (D) it was partially withdrawn. A, line of spine; B, yolk sac; C, left tarsal joint; E, tip of beak; F, left wing joint; F_1, right wing joint; G, left wing; G_1, right wing, H, left foot; H_1, right foot; J, umbilicus; K, top of head; L, tail; M, albumen.

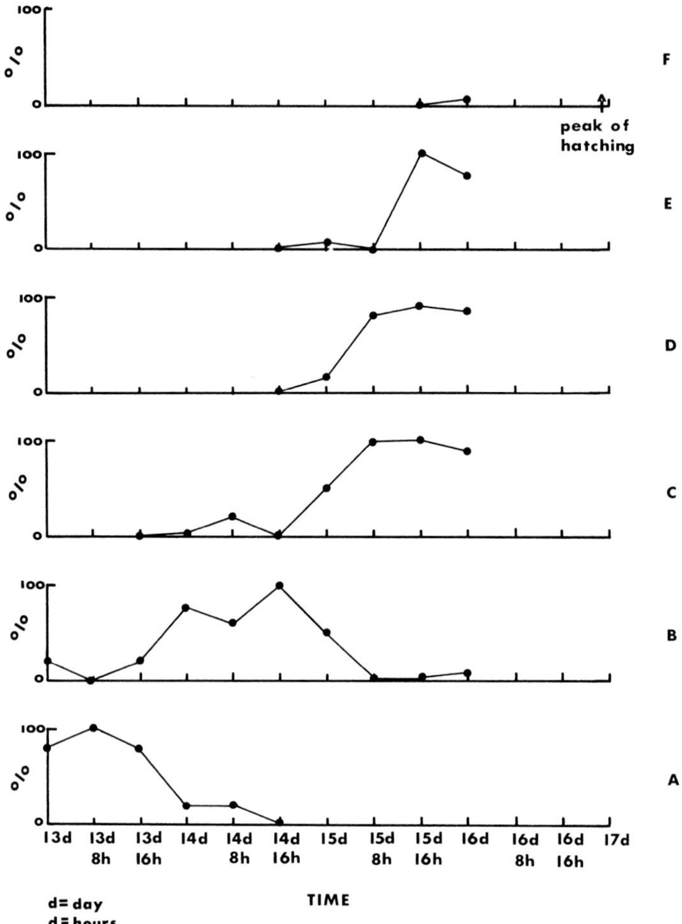

FIG. 2. *Continued.* Japanese quail. Timing of developmental stages leading to hatching. A, percentage of embryos with the beak buried in the yolk sac. B, the percentage making tucking movements. C, the percentage already tucked. D, the percentage which has pierced the membrane. E, the percentage pipped & F, the percentage hatched. Reproduced from Green & Vince (1973) by courtesy of *British Poultry Science*, Edinburgh, Scotland.

the membranes are pierced, and the beak enters the air space. Either before or shortly after membrane penetration, the movements of lung ventilation begin, resulting in slow inflation of the lungs (M. Vince & B. Tolhurst, unpublished observations). At this stage the air within the air space is depleted of oxygen and heavily loaded with carbon dioxide (Visschedijk, 1968), as indeed, is embryonic blood (Dawes & Simkiss, 1969; Freeman & Missen, 1970). The hatching muscle has by now enlarged almost to its full extent (e.g., see Fisher, 1958; Ramachandran, Klicka, & Ungar, 1969), and a few

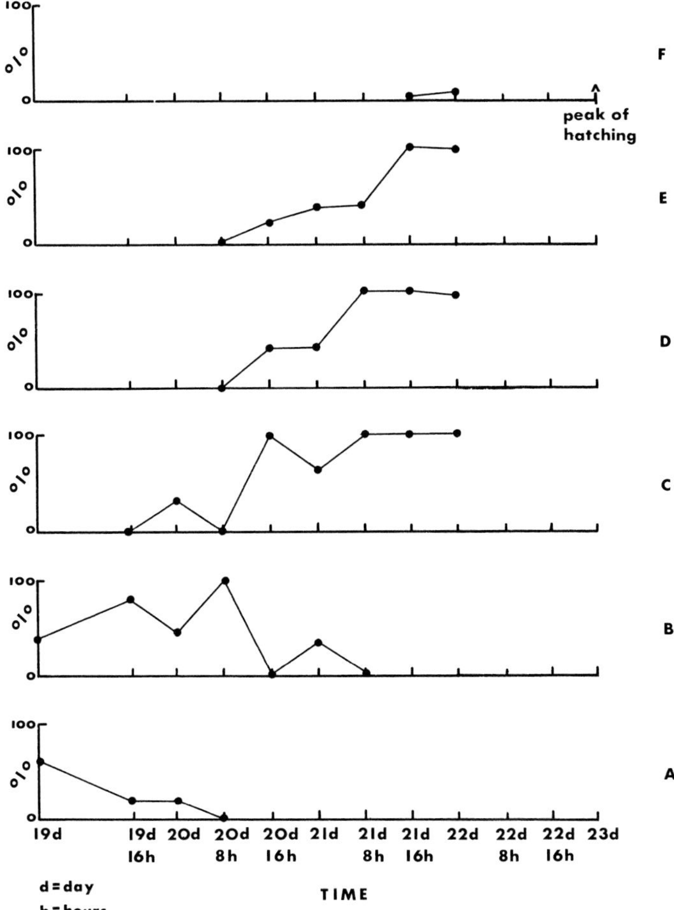

FIG. 2. *Continued.* Bobwhite quail. (For explanation see p. 292).

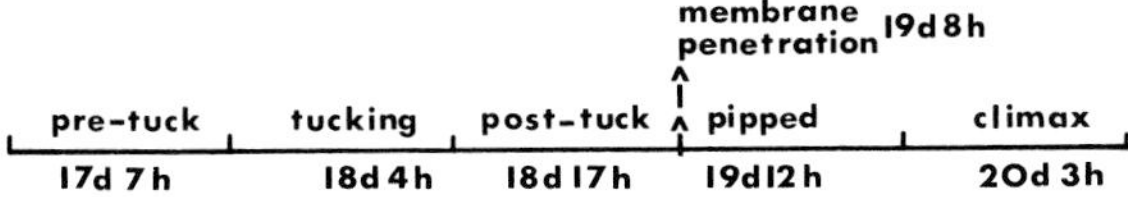

FIG. 2. *Continued.* Comparable stages in the domestic fowl. From Hamburger and Oppenheim (1967).

hours after membrane penetration "pipping" movements begin. As a result of these the egg tooth is brought sharply against the shell at intervals until the shell is cracked and the membranes pierced. The embryo can then begin to breathe atmospheric air. During this time respiratory movements become more regular, and the rate increases (Romijn, 1948). This increase occurs quite suddenly, when there is a change in the amplitude and pattern of breathing as each breath begins to produce a loud click (Horner, 1853; Vince & Salter, 1967). Click-breathing continues while the withdrawal of the yolk sac is completed and until an hour or two before hatching, when the rate and amplitude of breathing tend to drop. Finally, by a series of coordinated movements the shell begins to be cut round while the embryo turns within it; the shell "cap" is then pushed off, and the chick emerges.

III. Interaction between the Embryo and its Environment outside the Shell by Means of Vocalizations

An outstanding study of the embryo/environment interaction is that of Tschanz (1968) on the guillemot (*Uria a. aalge*).

The guillemot is a colonial cliff-nesting species breeding on narrow rocky ledges overhanging the sea. Each pair incubates a single egg, and on hatching the chick either hides within the brooding parent's plumage (or a cleft in the rock) or responds to the foraging parent's feeding call by emerging from its hiding place and approaching the parent to be fed. To approach the parent the chick must often thread its way along the ledge between neighboring guillemot parent/chick groups and ignore the feeding calls of these other adults.

Tschanz has shown that essential parent/chick ties are established during the embryonic period. From field observations he divided the process of hatching into four phases: (1) a preliminary phase (as part of the embryonic activity consists of vocalizing, it seems certain that this phase begins with the initiation of lung ventilation); (2) formation of the pipping hole; (3) formation of the "breathing" hole [like the gull embryo (Drent, 1967), the guillemot pips the shell many times, making an actual hole in it before cutting round]; and (4) cutting off the shell cap before emergence. He reports that periods of inactivity alternate with periods of activity (breaking the shell and calling). Either activity stimulates the incubating parent to rise, to roll the egg, and to utter the precursor of the feeding call. In this way the embryo stimulates parental activity and becomes familiar with the characteristics of its own parent's vocalizations. At the same time, parental activities such as egg-rolling, nest relief, and vocalization, intensify the hatching activities and calling of the young. Tschanz found also, in the preliminary phase of hatching activity, that all parental calls affected the activity of the young in the same way. At later stages the embryo responded only to the call which it heard most frequently when it was itself most active, that is, the feeding call as given by one or the other of its parents.

In experimental work (also see Impekoven, this volume), Tschanz demonstrates that the guillemot learns the feeding calls of its own parents in this brief period before hatching.

IV. Response to Siblings

In the work of Tschanz the stimuli and responses have been identified, and the phenomena described are consistent with the picture of embryonic development briefly outlined above. The auditory system is already operative by the time lung ventilation begins (see review by Gottlieb, 1968), and at this late embryonic stage reflex activities are recognized as operating (see Hamburger, 1968). In the work now to be described on the synchronization of hatching in quail, the mode of interaction is less clear-cut but can nevertheless be understood within the same framework.

A. The Synchronization of Hatching in Quail

The bobwhite and Japanese quail (*Colinus virginianus* and *Coturnix coturnix japonica*) are both species where the eggs in a clutch all hatch at the same time. [According to Stoddard (1931), a clutch of the bobwhite quail will hatch within an hour or less.] It has been demonstrated that this synchronized hatching results from embryos stimulating each other. As a result of this stimulation, the more advanced ones are held back and hatch later than they would if kept in isolation, while retarded embryos are brought on and their hatching time advanced. It is clear that hatching cannot be advanced unless the developmental stages leading to it have also been speeded up; normally the yolk sac is not fully withdrawn and the chorioallantoic circulation sealed off until shortly before emergence. Thus, the synchronization of hatching must bring with it changes in the prehatching developmental rate.

The embryos synchronize their hatching when eggs are in contact, and some idea of the difference between the times of hatching of small groups of eggs in contact and isolated eggs in the same incubator at the same time is given for the bobwhite quail in Fig. 3 and for the Japanese quail in Fig. 4. (Since in the quail incubation begins after the laying of the last egg, it seems likely that Figs. 3 and 4 give a good idea of the amount of adjustment required for synchronization under natural conditions also.) Although the precise time of hatching of each group will depend on the amount of advancement or retardation of the individuals comprising it, the groups (in Figs. 3 and 4) on the whole tend to hatch toward the middle of the hatching time of the isolates, supporting the view that stimulation acting between the eggs reduces the spread of hatching. Thus, the rate of development and the time of hatching of individuals in these species are to some extent controlled by signals acting between the eggs. I will deal separately with the two effects which have been observed, advancement and retardation.

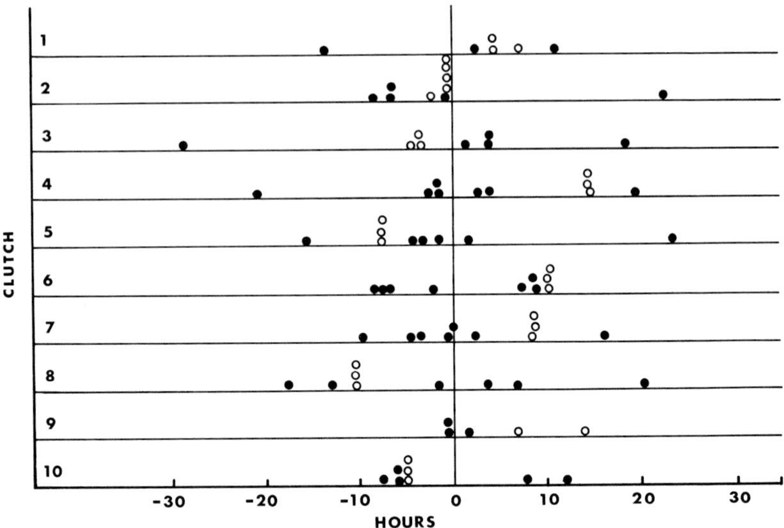

FIG. 3. Bobwhite quail. Hatching time of eggs in contact compared with isolated siblings. 0 = mean hatching time of the isolates. Closed circles indicate isolates and open circles, eggs in contact. From Vince (1968b), by courtesy of Baillière, London.

B. The Acceleration of Hatching

The advancement of development and hatching time is more easily understood than retardation, and the general picture is a simpler one. It seems almost certain that an important (but probably not the only) stimulus which can act to speed up development is some aspect of the regular loud clicking which precedes hatching by about 12–15 hours in the quail. This clicking is produced by a specialized and rather complex type of breathing which, in these species, normally begins many hours after the first onset of lung ventilation. The association between clicking and breathing (there are slightly different types of click, the loudest coinciding with expiration) was first described by Horner (1853), and it has been confirmed by recordings made by Vince and Salter (1967, see also Vince, 1970). Although the function of clicking (if any) is unknown, it has been heard in all species tested (Driver, 1965; McCoshen & Thompson, 1968a; Vince, 1966) and would appear to be linked in some way with the respiratory physiology of the developing embryos and to have acquired a communicating function in certain species. It has been reported from species which lay a single egg, such as the guillemot (Tschanz, 1968) and the gannet (J. B. Nelson, personal communication), where, it is true, it could play a part in the parent/embryo interaction, if not in interaction between siblings. Recordings showing the earlier, regular,

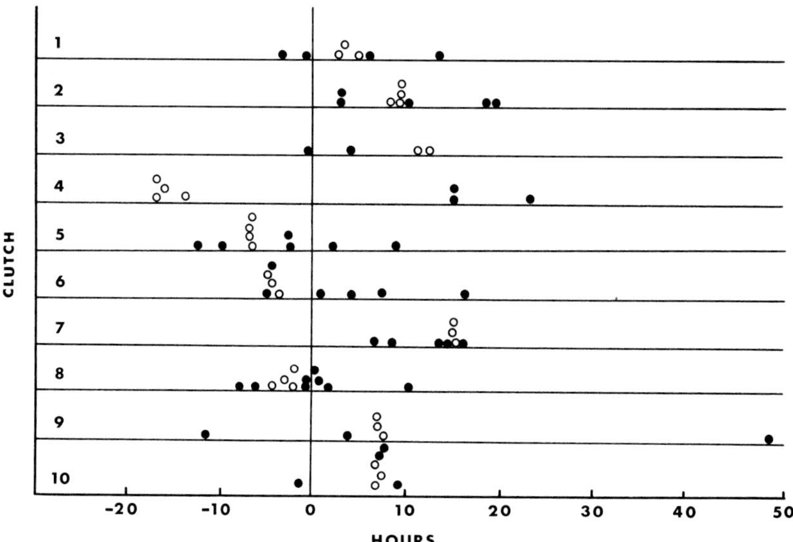

FIG. 4. Japanese quail. Hatching time of eggs in contact compared with isolated siblings. 0 = mean hatching time of the isolates. Closed circles indicate isolates and open circles indicate eggs in contact. From Vince (1968b), by courtesy of Baillière, London.

silent breathing, and the later more complex type of breathing which includes a click are reproduced in Figs. 5 and 6. It may be noted, however, that whereas all avian species appear to breathe through their lungs for a time before hatching, not all follow this pattern of a long period of slow, silent breathing preceding a period where almost every breath produces a loud click. Ducks, for example, breathe silently for only a very short period before beginning to click with every breath, and the same appears to be true for at least certain small passerines (M. Vince & R. Adkins, unpublished observations).

There are two types of evidence which suggest strongly that the beginning of regular loud clicking can have an accelerating effect on neighboring eggs. Bobwhite quail clutches where one egg had been put into the incubator 24 hours late were observed at 4-hour intervals and the time of pipping, the onset of regular clicking, and the time of hatching noted for each egg. In this experiment (results from three clutches are shown in Fig. 7), the time of pipping varied greatly between the individuals within a clutch (as the incubation period depends on the length of storage, development is inevitably variable), but the onset of clicking was fairly well synchronized and the time of hatching closely synchronized (Vince, 1964). In this experiment also those eggs which were the most advanced and which pipped the shell

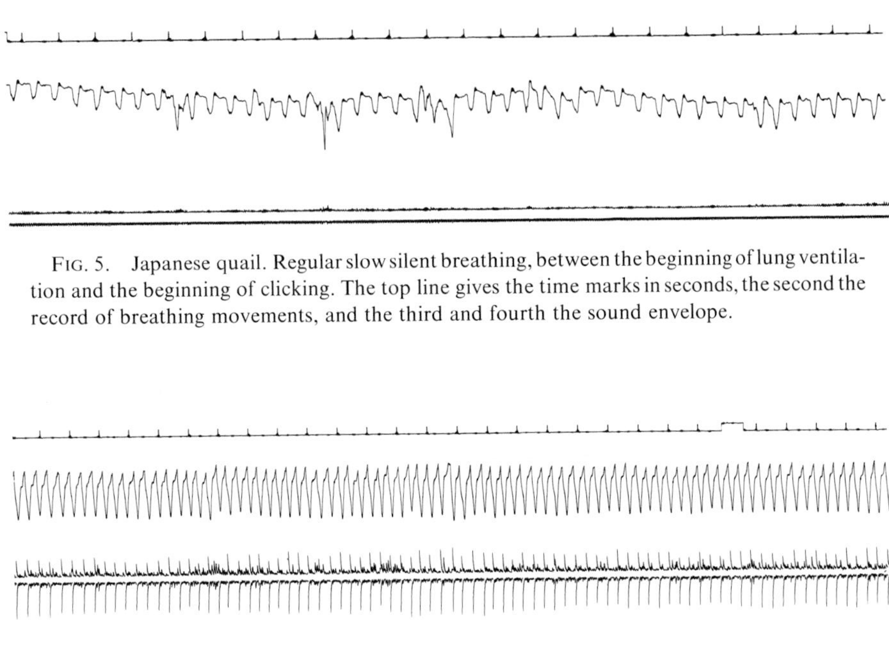

FIG. 5. Japanese quail. Regular slow silent breathing, between the beginning of lung ventilation and the beginning of clicking. The top line gives the time marks in seconds, the second the record of breathing movements, and the third and fourth the sound envelope.

FIG. 6. Japanese quail. Rapid breathing during the period of regular loud clicking. The top line gives the time marks in seconds, the second line gives the record of breathing movements, the third and fourth the sound envelope showing the clicks, and the fifth line shows the waveform of the sounds.

early breathed silently for the longest period of time before beginning to click, while the more retarded embryos, which pipped the shell late, breathed silently for very short periods before the beginning of clicking. Thus, the beginning of clicking was roughly synchronized in all the eggs, being held back in some and hastened on in others. We do not yet know how this synchronization of the beginning of clicking is achieved. It may be that short bursts of clicks which precede regular clicking have an effect on neighboring eggs. The beginning of the regular loud clicking is often quite sudden.

There is more evidence that clicking can act as an accelerating stimulus. Artificial clicks (in the experiments reported here square waves from an oscillator) can be fed into an isolated egg by means of a vibration generator (when the signals are presumably felt as vibration) or a loudspeaker (clearly, providing auditory stimulation). The hatching time of the stimulated egg

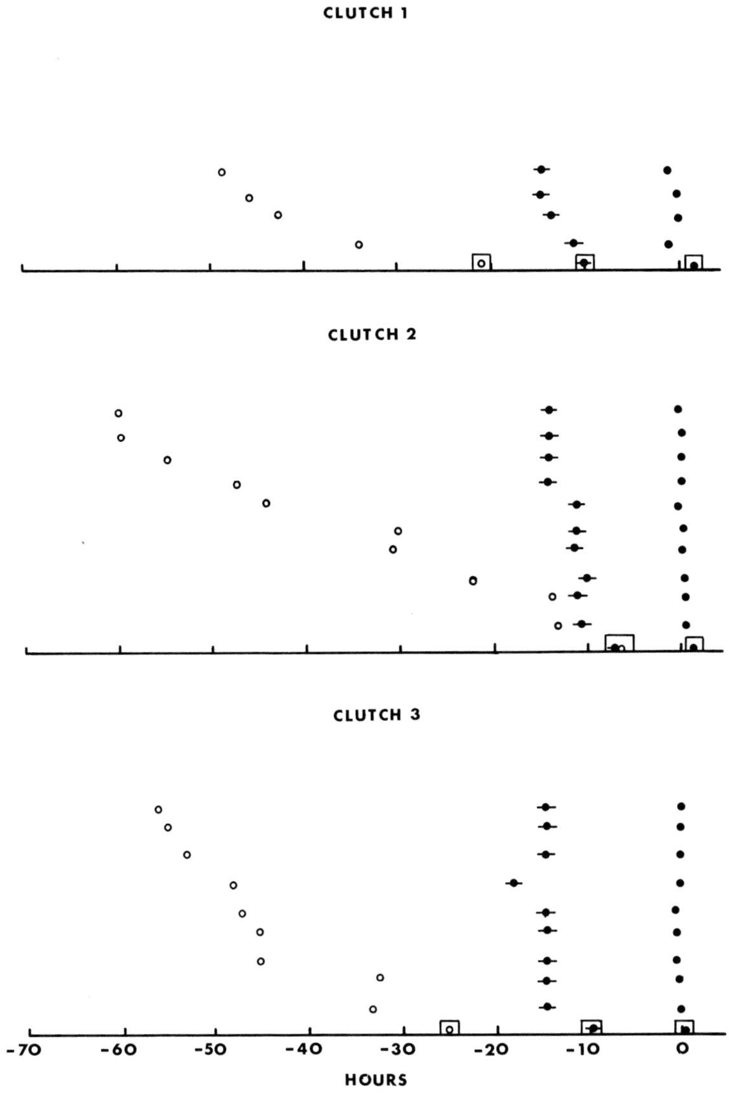

FIG. 7. Timing of final developmental stages in individual eggs of the bobwhite quail. Three clutches are shown in each of which the eggs were kept in contact, and in each clutch one egg (the bottom one in each clutch as shown here) was put into the incubator 24 hours late. Each line represents one egg, and ○ indicates the time of pipping, -●- indicates the time when regular loud clicking began, and ● indicates the time of hatching. From Vince (1969), by courtesy of the Cambridge University Press, London and New York.

is then compared with the means of unstimulated controls. Such artificial stimulation provided at the appropriate rate and amplitude resulted in early hatching (Vince, 1966, 1968a), the vibrator producing a slightly greater effect than the loudspeaker. Amplitudes used were consistent with those produced by an embryo while clicking, and the rates which resulted in the greatest amounts of acceleration were between $1\frac{1}{4}$ and about 8 clicks per second in the Japanese quail, which is comparable with a normal respiration and click rate in that species of $2–3\frac{1}{2}$ per second (Fig. 8). It is true, however, that we cannot be certain that clicking is the normal accelerating stimulus until more work has been carried out.

C. *The Retardation of Hatching*

We know less about retardation than about acceleration. We know, however, that it is more difficult to hold back development that to speed it up, and that it is even more difficult in the Japanese quail than in the bobwhite quail. In the latter species an egg incubated in contact with a single sibling given 24 hours less incubation will hatch after the normally expected time of hatching, while contact with a single egg of this kind will have no effect on a Japanese quail egg (Vince, 1968b). In this species, retardation only occurs when an egg is incubated in contact with 3, 4, 5, or more retarded siblings (Vince & Cheng, 1970b) (Fig. 9). The nature of retarding stimuli is now beginning to emerge. In an experiment where the rate at which artificial clicks were provided was varied for different eggs, a high click rate (between about 90–300 clicks per second) resulted in Japanese quail eggs hatching late (Vince, 1968a; see also Fig. 8), while in the bobwhite quail there were indications of a retarding effect when the click rate was slowed down to about .9 clicks per second (Vince, unpublished observations) (Fig. 10). As a slow respiration rate of about 1 breath per second precedes the faster breathing associated with clicking, it seems possible that the embryonic respiration rate may be paced by the breathing of neighboring eggs. We have found, however, that embryos in contact do not actually breathe in time with each other, although this would not rule out the pacing hypothesis entirely. Alternatively or, indeed, in addition to this, the faster click rates which are associated with retardation in the Japanese quail produce a low frequency sound, and there is a possibility that a retarding stimulus could be sounds of low frequency (between about 20 and 80 Hz in the quail, fowl, and duck) which can be picked up by specially constructed transducers (R. Adkins & S. Salter, unpublished observations). Using this equipment the clicks exhibit a waveform typically of about 500 dynes peak to peak, and the low frequency sounds have an amplitude of about one-third of this. They occur in short bursts and at irregular intervals and can be picked up from about 4 days

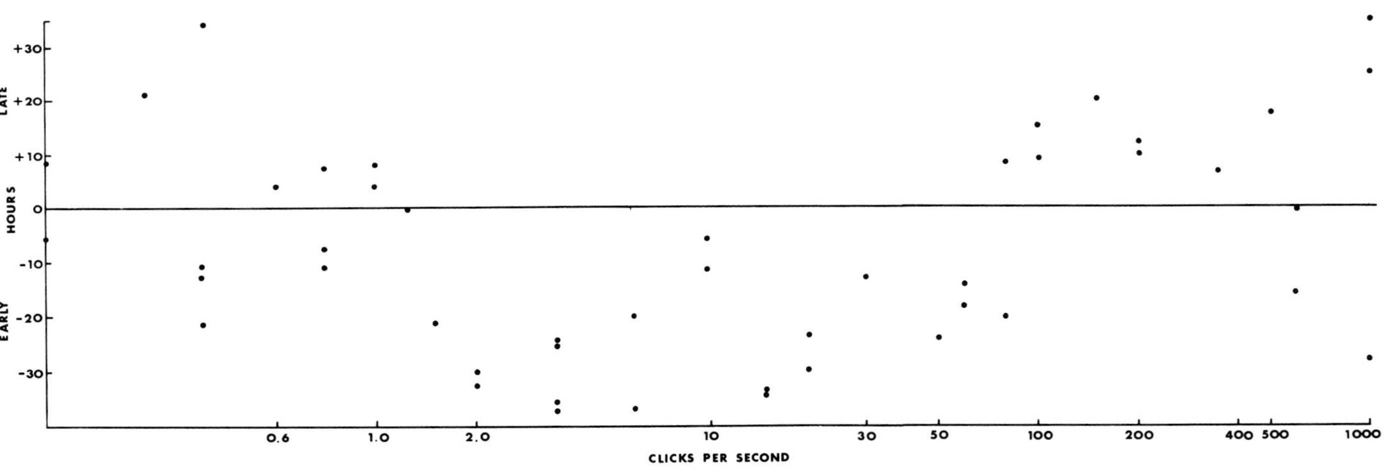

FIG. 8. Japanese quail. Stimulation with artificial clicks. The click rate has been varied between 1 click every 10 seconds and 1000 clicks per second. Each point represents the hatching time of one stimulated egg compared with the mean hatching time of its unstimulated siblings. + indicates that the stimulated egg hatched late, and – indicates that it hatched early. The horizontal line at 0 hours represents the mean hatching time of the isolates. From Vince (1968a) by courtesy of *British Poultry Science*, Edinburgh, Scotland.

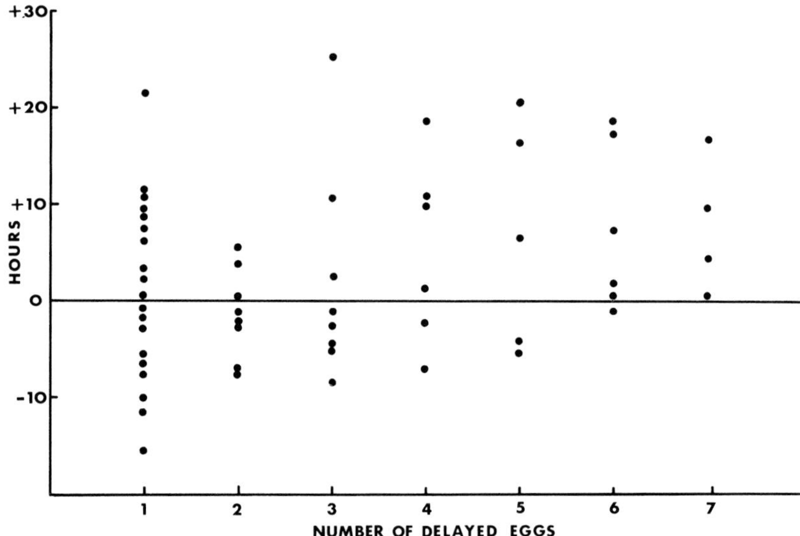

FIG. 9. Japanese quail. The hatching time of individual eggs stimulated by contact with one, two, three, four, five, six, or seven siblings, given 24 hours less incubation. The hatching time of each egg is compared with the mean of isolated siblings. + indicates that the stimulated egg hatched late, and – that it hatched early. Zero hours represents the mean hatching time of the isolates. From Vince & Cheng (1970b), by courtesy of Baillière, London.

before hatching in the Japanese quail but increase in loudness shortly before the onset of lung ventilation. These sounds (illustrated in the quail in Fig. 11) can be observed to accompany embryonic movements. It seems arguable that they could become audible to the human ear, when amplified, and also begin to have an effect as sound or more likely as vibration, on neighboring embryos at the time when the yolk sac surrounding the embryo is becoming smaller and the growing embryo is forced, at least in its dorsal aspect, more closely against the shell [see Fig. 2, compare A(1) and B(1)]. Work on the effect of recorded sounds of this type on the hatching time of an isolated egg shows a small but consistent (and statistically significant) effect of retardation (Fig. 12). Again, it is clear from the work reviewed by Gottlieb (1968) that the sensory systems involved in receiving signals of this type are operating at the requisite time.

D. *Stages of Development when Accelerating and Retarding Stimuli Become Effective*

That both accelerating and retarding signals must become effective before the beginning of regular loud clicking may be seen from Fig. 7 above, which shows that, within each clutch, the more and the less advanced eggs

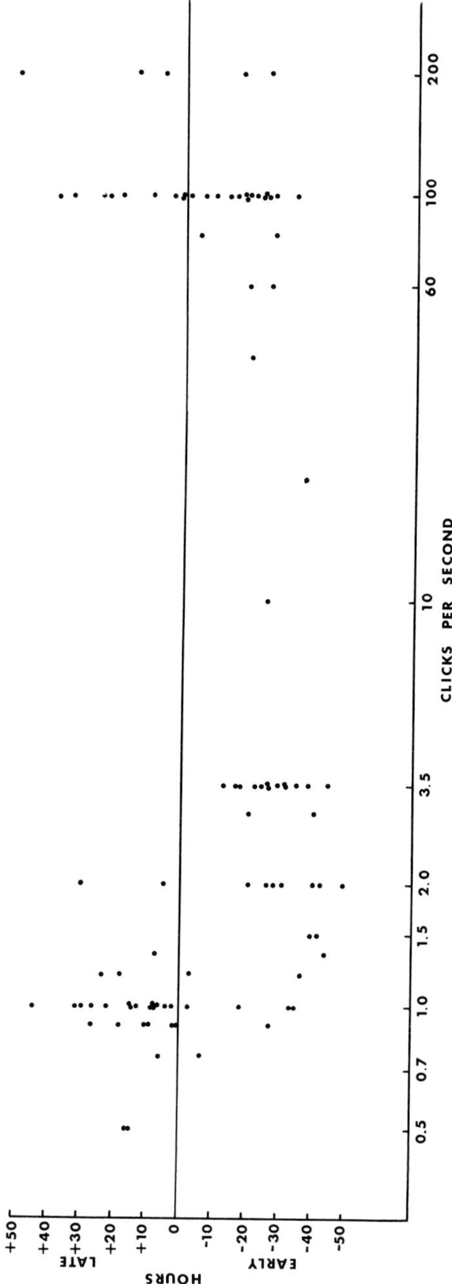

FIG. 10. Bobwhite quail. Stimulation with artificial clicks. The click rate has been varied between 1 click every 2 seconds and 200 clicks per second. Each ● represents the hatching time of one stimulated egg compared with the mean hatching time of its unstimulated siblings. − indicates that it hatched early, and + that it hatched late. The horizontal line at 0 hours represents the mean hatching time of the isolates.

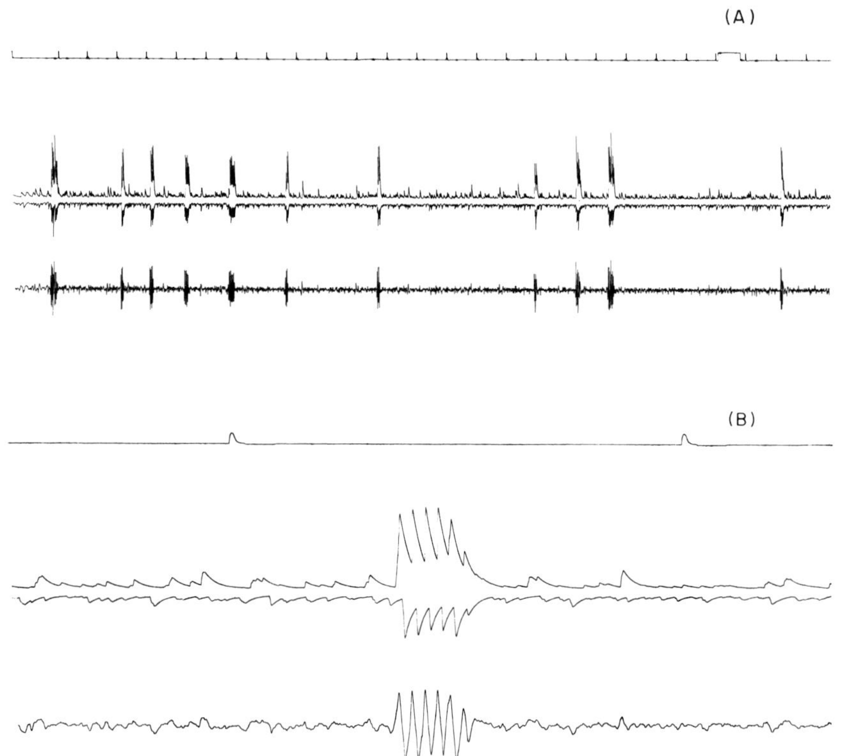

FIG. 11. Japanese quail. Low frequency sounds which become audible about 24 hours before the beginning of lung ventilation. The top line gives the time marks in seconds, and the bottom one the waveform of the sounds. The middle recording shows the sound envelope. In this recording part of the shell over the air space was removed to increase the amplitude of signals. (A) recording made at a paper speed of 10 mm per second and; (B) recording made at paper speed of 100 mm per second.

began to click at very nearly the same time. Apart from this we have very little evidence concerning the stage or stages of development when retarding stimuli are effective. In one experiment where recording was carried out from a series of eggs each incubated in contact with two siblings given 24 hours less incubation, it was evident that the more advanced egg tended on occasions to begin clicking and, in fact, to click for periods of 10 seconds or longer and then to revert to slow silent breating again. Finally, all three eggs began clicking more or less simultaneously (Vince & Cheng, 1970a). Thus, part, at least, of the retarding effect could occur during this period of slow, silent lung ventilation. (It is not possible to explain this effect as a response to a *lack* of signals from the neighboring eggs, as isolates recorded did not behave in this way.)

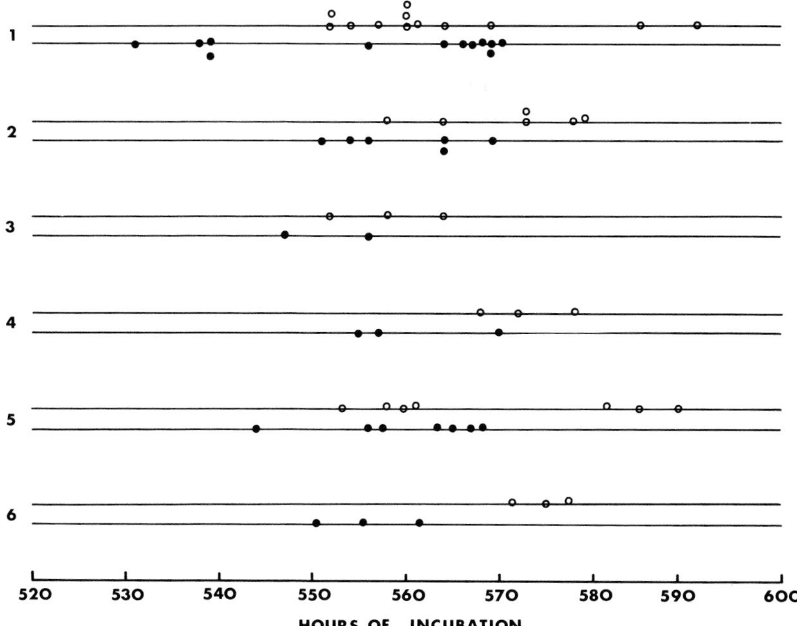

FIG. 12. Bobwhite quail. Effect of low frequency sounds on the time of hatching; six clutches. For each clutch one group of eggs was stimulated by means of a tape loop of recorded low-frequency sounds. Hatching times of stimulated (○) eggs compared with unstimulated (●) eggs incubated at the same time, but in a different incubator. Stimulation was begun between 2–3 days before the expected time of hatching.

We know a little more about the time when accelerating stimuli are effective, as it has been possible to show a number of developmental stages which are advanced by such stimuli. In addition to the beginning of regular clicking, we have evidence that accelerated Japanese quail penetrated the membranes and also pipped the shell before unstimulated controls. However, in the same experiment there was no evidence of the advancement of these two stages in bobwhite quail tested under comparable conditions. In this experiment (Vince, Green, & Chinn, 1971), stimulated eggs which were allowed to hatch did so about 24–30 hours before the unstimulated controls. In the same experiment it was shown that lung ventilation began earlier in the stimulated Japanese quail eggs than in unstimulated controls, although again, not in the bobwhite quail (where hatching, however, was advanced). In both species the duration of lung ventilation was curtailed (although there were certain exceptions, as mentioned below) in those stimulated embryos which were allowed to hatch. This finding confirms that of the earlier experiment shown in Fig. 7.

Similarly, Pani, Coleman, Georgis, Kulenkamp, and Kulenkamp (1969) have shown that yolk sac withdrawal is completed early in stimulated eggs. Under normal conditions we find that eggs frequently begin to click before the yolk sac is fully withdrawn although, of course, it is fully withdrawn in a normal chick at the time of hatching. Very rarely an accelerated chick hatches with a small piece of yolk sac not yet withdrawn into the body cavity, but accelerated chicks are almost always anatomically indistinguishable from controls, even when hatching 24 hours or more before their unstimulated siblings.

In an experiment carried out, not on the quail, but on the domestic fowl, it was found that the increase in weight of the hatching muscle (*Musculus complexus*), which rises to a peak at the time of pipping (Fisher, 1958), takes place much more rapidly in accelerated eggs than in unstimulated controls (Fig. 13) (J. Green, unpublished observation). In this experiment, measures for both groups were made before the hatching muscle reached its full size.

Thus, it would begin to appear that development from about the time when lung ventilation begins can be accelerated as the result of stimulation. In addition, as accelerated embryos go through the normal series of postural changes and begin to breathe in the normal "tucked" position, it would appear that development earlier than this also takes place more rapidly, for example, in the Japanese quail, where the beginning of lung ventilation can be advanced. These earlier postural changes must include lifting the beak out from under the yolk sac, bringing the beak up under the right wing, and tucking.

This does not mean, however, that the rate of all development is speeded up quite to the same extent. For example, it may be seen from Fig. 7 that the duration of clicking remains roughly constant in clutches where some eggs are being speeded up and others slowed down, a finding which has since been confirmed in groups of eggs incubated under different stimulating conditions (Vince & Cheng, 1970b. also see Fig. 14). Similarly, it has recently been found that the interval between the beginning of lung ventilation and full inflation of the lungs as indicated by a lung flotation technique is roughly constant in stimulated and unstimulated Japanese quail and unstimulated Khaki Campbell duck embryos (M. Vince & B. Tolhurst, unpublished observations).

Although we cannot fully understand the mechanisms underlying the synchronization of hatching until we know more about retardation, it would appear that, from at least the equivalent of 17 days in the domestic fowl (see Fig. 2, Pp. 292, 293), developmental stages of neighboring quail embryos are brought slowly more and more closely together until the eggs begin to click roughly synchronously and hatch at the same time. This view, that the embryos respond to external stimulation from incubation ages equivalent to

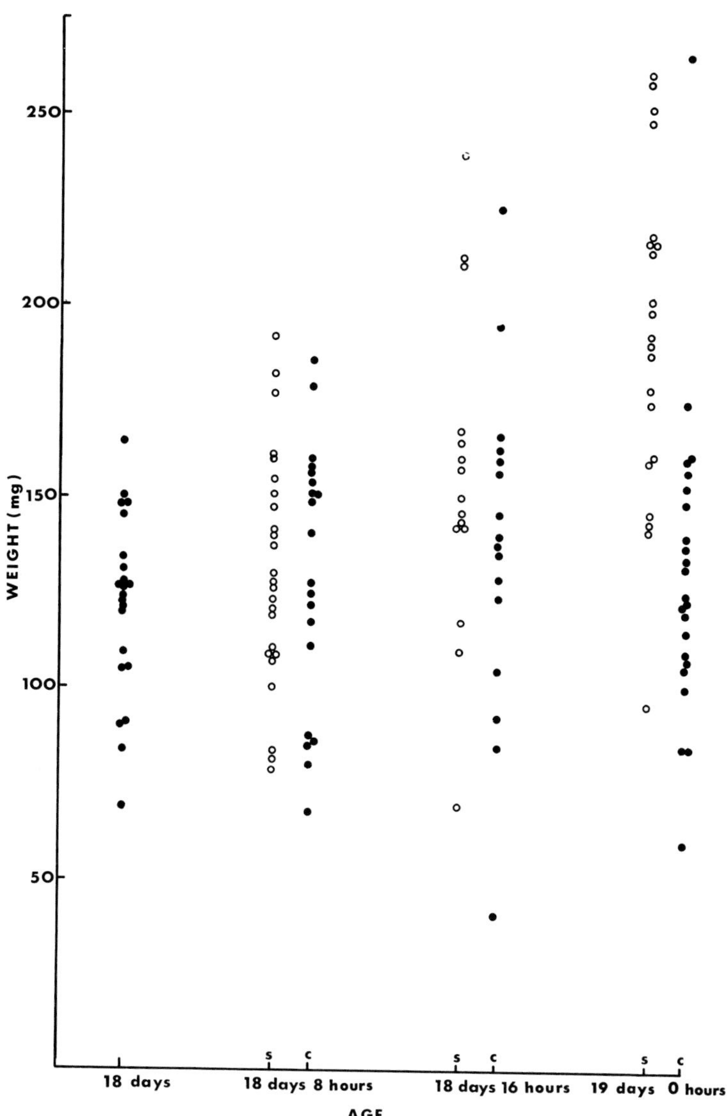

FIG. 13. The domestic fowl. Hatching muscle weights in stimulated and unstimulated embryos. Stimulation began at 18 days for all groups. ○ Indicates muscles from embryos given accelerating stimulation; ● indicates muscles from unstimulated embryos.

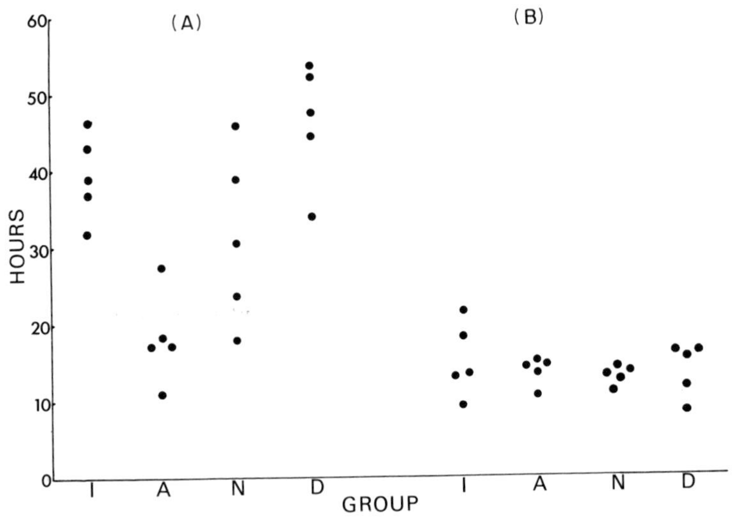

FIG. 14. Bobwhite quail. Duration of lung ventilation under four different conditions of incubation: I, in isolation, A, in contact with two eggs given 24 hours more incubation (potentially accelerating stimulation), N, in contact with two eggs given the same amount of incubation (normal type of stimulation), and D, in contact with two eggs given 24 hours less incubation (potentially delaying stimulation). Each dot represents the duration of breathing of one embryo during (A) the silent period before regular loud clicking begins and (B) the period of regular loud clicking. From Vince & Cheng (1970a), by courtesy of the Press of the Wistar Institute, Philadelphia.

day 17 in the domestic fowl, would be consistent with the prevailing views on hatching behavior, as described in Section II.

E. *Effects of Stimulation in Other Precocial Species*

The discussion above shows that the two quail species do not respond to stimulation in precisely the same way under natural or under artificial conditions. Experiments carried out on other precocial species gave results which show a similar variation between species and an effect of stimulation, although in these species hatching is not actually synchronized.

Eggs of the domestic fowl are known not to hatch at the same time in incubators, but McCoshen and Thompson (1968b) have pointed out that eggs kept in contact have a smaller spread of hatching than do eggs incubated in isolation. More recently it has been shown both that eggs of the White Leghorn respond to natural accelerating signals[contact with another egg given 24 hours more incubation; Vince (1973), and that artificial clicks provided at the rate which accelerates quail eggs have a similar effect on eggs of the domestic fowl, in this case Light Sussex (Fig. 15; also see Vince, Green,

& Chinn, 1970)]. Therefore, in these domesticated breeds a measure of synchronization, or at least some measure of advancement, occurs with the appropriate stimulation. That this includes not only the time of hatching but also the advancement of the developmental processes on which hatching depends is shown in Fig. 16. Here groups of four eggs were incubated with each one in isolation, while the beginning of breathing and also hatching was recorded in each one. One of the four was given accelerating stimulation (artificial clicks at a rate of 3 per second). Almost without exception the stimulated embryos began breathing earlier and hatched sooner than their nonstimulated controls. In all stimulated eggs the duration of breathing was less than that of the controls.

In the mallard duck it has been reported by Bjärvall (1967) that, although hatching is not precisely synchronized, wild broods hatch within a relatively short time (3–8 hours). We have found a measure of synchronization in eggs of the Khaki Campbell duck kept in contact (the spread of hatching within clutches had a range between 2 hours 41 minutes and 6 hours 54 minutes) in comparison with isolated controls in the same incubator (where the spread of hatching ranged between 14 hours 10 minutes and 27 hours 39 minutes) (Vince 1973). Mallard eggs tested in the same way as in the previously described experiment gave results similar to those of the domestic fowl (Fig. 17).

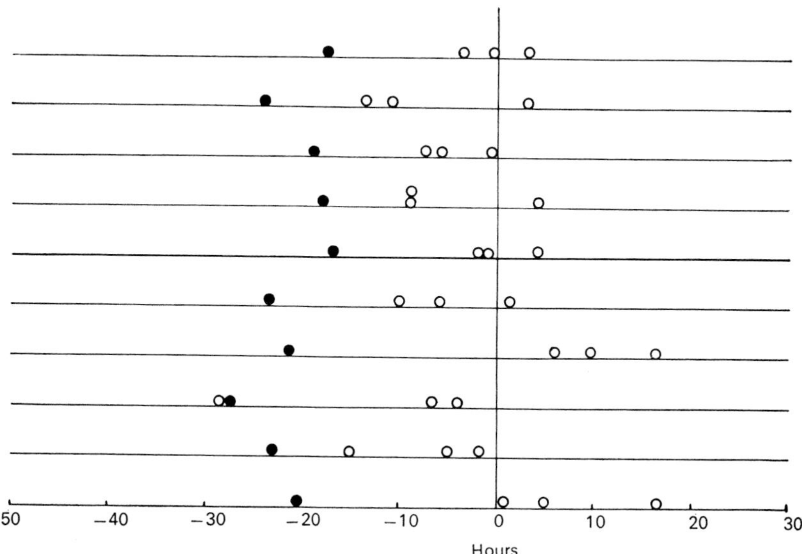

FIG. 15. The domestic fowl. Effects of artificial stimulation of the accelerating type (3 clicks per second) on hatching time; 10 clutches. ● Indicates the stimulated egg and ○ its unstimulated siblings. Zero hours indicates the normally expected time of hatching. From Vince *et al.* (1970), by courtesy of *British Poultry Science*, Edinburgh, Scotland.

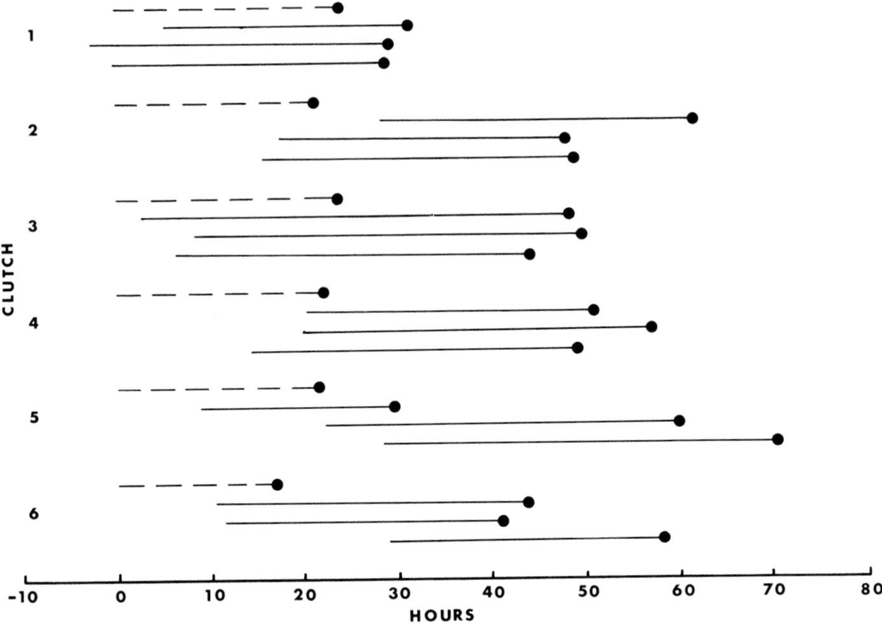

FIG. 16. The domestic fowl. The initiation and duration of lung ventilation in one embryo given accelerating stimulation compared with three unstimulated controls recorded at the same time; six clutches. For each clutch, 0 hours represents the time when the stimulated embryo began to breathe. - - - Accelerated; —— control; ● hatched. From Vince (1973a), by courtesy of *British Poultry Science*, Edinburgh, Scotland.

Effects obtained from the domestic goose (Embden × Toulouse) were similar with respect to earlier hatching of the stimulated eggs, but in their case lung ventilation was advanced less consistently as a result of stimulation (Fig. 18). In this respect the goose is more like the bobwhite quail. The stimulated geese hatched earlier, often breathing for a shorter time than unstimulated controls.

F. The Question of How Early Stimulation from Outside the Egg Can Affect Metabolic and Postural Development

We do not yet know how early in the incubation period embryos can respond to stimulation from outside the shell in such a way that their rate of development and achievement of postures leading to the hatching position can be accelerated or retarded. It is clear, however, that the onset of lung ventilation can be advanced by artificial stimulation from outside the egg (see Figs. 16, 17, and 18). It is clear also that this means that postural changes leading to the hatching position have also been advanced: in Japanese quail

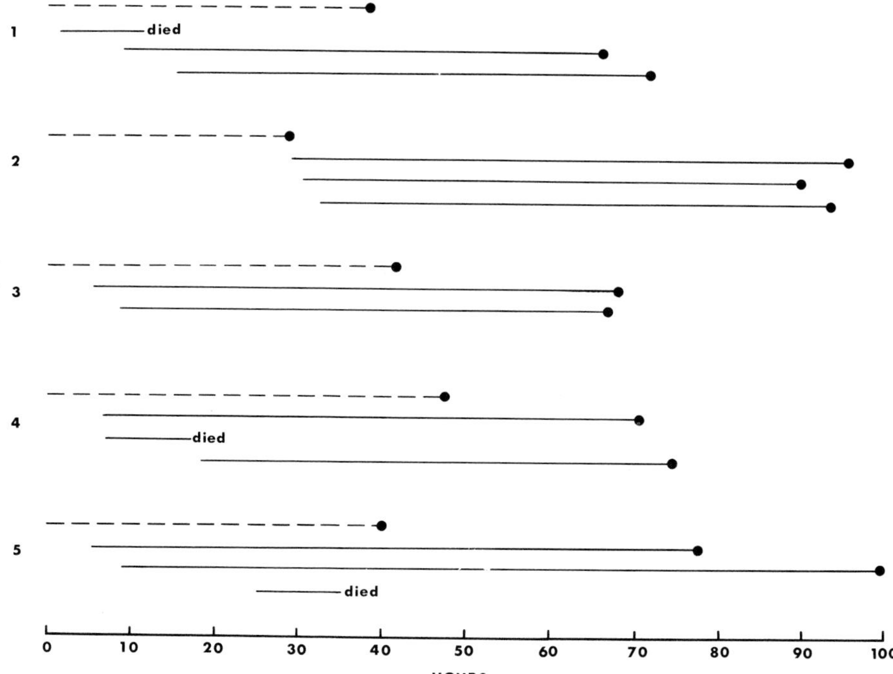

FIG. 17. The duck embryo. The initiation and duration of lung ventilation in one embryo given accelerating stimulation compared with two or three unstimulated controls recorded at the same time; five clutches. For each clutch 0 hours represents the time when the stimulated embryo began to breathe. ---- Accelerated; —— control; ● hatched. From Vince (1973a), by courtesy of *British Poultry Science*, Edinburgh, Scotland.

given accelerating stimulation it has been found that embryos which breathed early as a result of stimulation inflated their lungs in the normal way (M. Vince and B. Tolhurst, unpublished observations) and they also pipped the shell and pierced the membranes early in comparison with unstimulated controls (Vince, Green, & Chinn, 1971). More recent work suggests that both postural and metabolic changes can be affected by external stimulation earlier than this; however, the mechanisms involved in these changes are not yet known.

G. *Possible Responses to Stimulation*

We have evidence that stimulation from outside the shell can advance and (at least in certain species) retard the rate of development and time of hatching, but we do not yet know how the embryo responds to produce these effects. It is clear, however, that the rate of development involves the whole animal and includes the series of postural stages which end in hatching, and

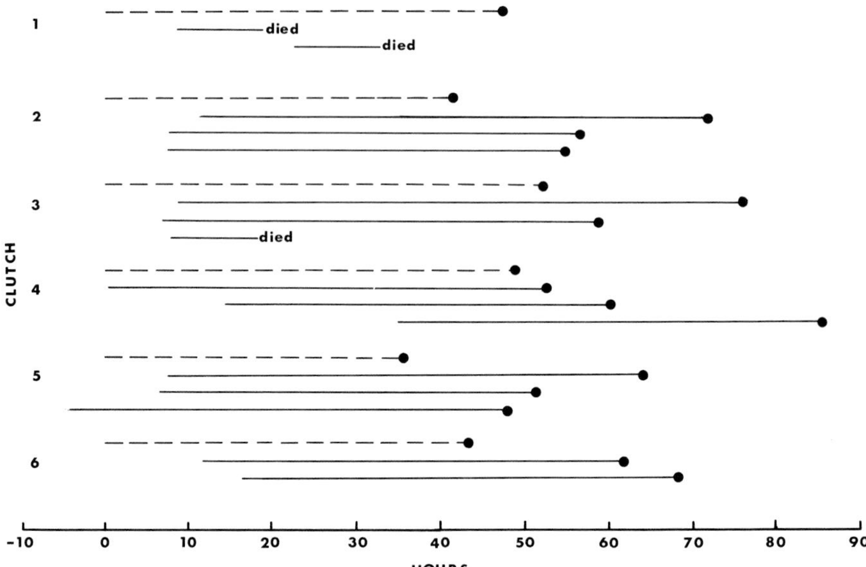

FIG. 18. The domestic goose. The initiation and duration of lung ventilation in one embryo given accelerating stimulation compared with two or three unstimulated controls recorded at the same time; six clutches. For each clutch 0 hours represents the time when the stimulated embryo began to breathe. ---- Accelerated; —— control; ● hatched. From Vince (1973a), by courtesy of *British Poultry Science*, Edinburgh, Scotland.

that these are accompanied by anatomical and physiological changes of equal importance. Development can be advanced in such a way that the embryo emerges from the shell as a viable chick, a state of affairs presumably dependent on regulatory mechanisms in the brain.

However, when development is accelerated, all stages are not affected to the same extent, and it seems that effects of stimulation may vary in different species (for example, we have found no evidence of the advancement of lung ventilation in the bobwhite quail, although we have found this in the Japanese quail and the domestic fowl). The duration of certain events, such as the time taken by the lungs to inflate and the duration of clicking, appears to be unaltered, when development is otherwise accelerated. Also, respiration rates during the preclick and click phases of breathing are very similar when accelerated, retarded, and unstimulated embryos are compared, although the duration of breathing varies among these three groups (Vince, 1973b). Thus, there is no evidence of sudden switching of an overall developmental rate as a result of stimulation. Consistent with this is the finding (Vince *et al.*, 1971) that development can be rather drastically set back, in both species of quail, when stimulation is removed at about the time when the eggs begin regular loud clicking. Under these circumstances previously accelerated embryos breathed silently again for a long period of time and hatched late.

Different types of evidence exist which could suggest ways in which development is accelerated by stimulation. Recent work of Balaban and Hill (1971) supports that of Freeman (1964) in showing an accelerating effect of thyroxine on the time of hatching in the domestic fowl. The administration of thiourea had retarding effects. Balaban and Hill also show an effect of this kind on respiration rates and yolk assimilation. Unfortunately, they have not so far manipulated thyroxine levels in their experimental embryos before day 17. However, their results would be consistent with the hypothesis concerning effects of stimulation that external events act via sensory information on central processes controlling the metabolism.

Alternatively, or rather, as a complementary hypothesis, it seems possible that stimulation could act on embryonic activities which result in the sequence of postural changes leading to hatching. Something of this kind is, in fact, suggested by the work of Hamburger and Oppenheim (1967) on hatching behavior. In their description of activities which lead to hatching, they point out that the successful achievement of each one of these postures occurs when, after many abortive movements, there is a movement of suddenly increased intensity; for example, in tucking, abortive movements occur particularly when "movements are less vigorous than usual [p. 179]" in membrane penetration, this "is accomplished . . . by a particularly vigorous head movement. . . . [p. 183]" and so on. Although these coordinated patterns of behavior are not considered to occur before day 17 in the domestic fowl, it would seem reasonable to suppose that the amplitude of these movements (and/or possibly their rate of occurrence) could be affected by stimulation, either of the accelerating or retarding kind. Another possibility arises from the review of Berry (Volume 2 of this serial publication); from this it would appear that stimulation could affect the rate of development of brain cells. Certainly, rather detailed work is needed here, both with accelerating and retarding stimulation.

V. Passive Effects Determined by Embryonic Anatomy, Physiology, and the Immediate Environment, without Necessarily Involving Sensory Systems

It is clear that the avian embryo responds actively to and interacts with its environment outside the egg toward the end of the incubation period. In addition to this, Kuo (1932c) claims that the direction and extent of an embryo's movement are closely dependent on its structure and the environment within the shell. In this way postural changes can occur partly in a passive manner, which is not directly dependent on the embryonic nervous system. For example, in descriptive work, part of which needs experimental verification, Kuo attributes the change of head movements from lifting and bending at 3–4 days to side-turning from 4–5 days to a marked increase in weight of the head at that time, together with a considerable growth in the length of the neck, which remains slender. It appears to be not strong enough to lift the head at the latter stage. Similarly the "swinging" movements of the embryo are of this passive kind, as is the folded position of the legs beneath the yolk sac. Viewed in this way, the embryo can to some extent be considered as literally shaped by its environment. Its random activity and its own developing structure interact with and are controlled by the developing and changing forces within its very restricting environment.

Kuo's work has been sharply criticized, but his account of forces which cause the embryo to turn within the egg between days 10 and 16 appears to have been accepted (e.g., see Hamburger & Oppenheim, 1967). According to Kuo, this change of position depends on embryonic activity, interacting with the albumen, the yolk sac, and the shell.

There have been fewer suggestions of this kind during the period leading up to hatching (from about 17 days in the fowl), although at this stage the embryo is closely confined by the shell so that its movements are narrowly limited and must indeed be closely guided by its own structure and the shape of the egg. At this stage, embryonic activities leading to hatching have been largely attributed to the maturation of coordinations within the nervous system. It is not difficult, from close observation of development at this stage, to make suggestions about ways in which these coordinated patterns of behavior could be guided in various directions. For example, the success of the prehatching stage is known to depend on the lifting of the head out of the yolk sac and finally turning it to the right (although turns to the left do occur) and tucking it under the right wing with the beak pointing upward toward the airspace membranes. In the rare cases where the beak is finally tucked under the left wing (malposition III, according to Landauer, 1961), the possibility of successful hatching is greatly reduced. The right-turning tendency has been confirmed in the duck by Gottlieb and Kuo (1965) and by Oppenheim (1970). In this paper Oppenheim also reports that the right-turning tendency is lost after draping.

According to the Kuo (1932c), lateral head movements occur predominantly in the right-hand direction throughout incubation, largely owing to the relation between the embryo and the yolk sac. As a result of lying with the left side on the yolk sac or, later, with the left side of the head and body in contact with the yolk, lateral head movements are more extensive when they are directed toward the right and they also occur more frequently in this direction. Such a tendency could possibly be linked with the work of Hamburger and Narayanan (1969), who showed that stimulation of the head and beak produces withdrawal movements of the head from 10 days onward, the movements becoming stronger as incubation progresses. Another factor is that the embryos do not actually take up a position lengthwise of the egg until shortly before hatching. During the pretucking and tucking phases, the embryo is still in a position diagonally across the shell (see Fig. 2 for quail) and about 20° or 30° to the vertical in the fowl, and it is still in the process of turning. At this stage the yolk covers the ventral surface and the embryo's left side, and the neck is bent sideways toward the left at shoulder level and at an angle of about 45°. Thus, the left shoulder and the long curve of the neck are close to the shell with the beak pointing downward and toward the right. The lefts ide of the body appears to be cramped, with the sternum apparently rotated toward the left (in the Japanese quail the right lung usually weighs more than the left—a small difference but found with 30 embryos to be significant at the 5% level). The left tarsal joint is much lower than the right. In addition to the fact that the beak points initially toward the right, the center of the body curves to the left and would appear to provide more space for head movements toward the embryo's right side.

Consideration of the structures within the shell and their relative positions at the stage leading up to tucking could in this way suggest that the right-turning tendency is at least partly dictated by these. However, it is also true that, within the closely packed shell, relationships are unusually obvious, and it can be fatally easy to attribute causation, or a functional relationship, where there is nothing but contiguity. Additionally, in this situation experiment directed toward the elucidation of this kind of relationship can be particularly difficult because of the close physiological connection between the embryo and other structures within the egg.

At this point it is relevant to mention an effect on chick posture which can be related to embryonic development and also to effects of external stimulation on the time of hatching. In the bobwhite quail, which has relatively long toes, a tendency for chicks to have irreversibly curled feet can be associated with late hatching. Extreme cases occur in a small percentage of birds given artificial retarding stimulation at the embryonic stage; but there is also an effect in unstimulated chicks, where the time taken to uncurl the toes after hatching becomes longer when hatching is delayed even a few hours after the normally expected time (Fig. 19). This effect appears to depend on the embryo's posture and its relationship with the yolk sac and the shell. As the yolk becomes smaller and is withdrawn into the body cavity and as the embryo grows larger, the toes, which earlier have been more or less outstretched against the sides of the head [see Fig. 2A(3) and B(3)], curl downward, beneath the head and over the top of the yolk sac, their plantar surfaces toward the shell and the toes bent sideways, toward each other [see Fig. 2C(3) and D(3)]. They would appear to be held firmly in this position by the mental history, (d) by a stimulus or stimulating objects, and (e) the external environmental context. In the foregoing pages I have dealt briefly with the first four of Kuo's factors, and it is relevant now to give some thought to the fifth.

Throughout incubation, under both natural and artificial conditions, embryos are subjected to a considerable amount of stimulation. Although intense stimulation, such as sound, centrifuging, or shaking, can reduce hatchability (see review by Landauer, 1961), a low level of noise and vibration, such as that produced in a forced draught incubator by the fan and the egg-turning mechanisms, is consistent with high hatchability. Also, it has long been recognized that the natural conditions of incubation are far from uniform. For example, in nests of many small passerines there are considerable temperature fluctuations arising from the parent's "on" and "off" periods during the hours of daylight (e.g., see Baldwin & Kendeigh, 1932; Haftorn, 1966). In addition, under natural conditions eggs are turned rather frequently [at intervals of between 10 and 50 minutes in the domestic fowl (Chattock, 1925) and "several times in the course of the day" in the bobwhite quail (Stoddard, 1931)], and in certain species at least, the incubating bird frequently stands, thrusts its beak into the nest lining, and works it about amongst the eggs (Haftorn, 1966). Investigations made using a microphone placed among the eggs in the nest of a blue titmouse (*Parus caeruleus*) suggested that this type of activity is repeated by day and by night at intervals of 5 minutes or less. At the same time parental heart beats could be heard, varying in rate in response to outside sounds. Also parental breathing was prominent, noisy, and liable to fluctuations in rate. "Settling" movements after rising from the eggs and returning to the nest were vigorous (Vince, unpublished observations). At the later stages of incubation, in particular, there is a considerable amount of stimulation provided by parental vocalizations and, at least in some species, a vocal interaction between the embryo and the parent (e.g., see Gottlieb, 1963, 1968; Impekoven, chapter in this volume; Tschanz, 1968). For stimulation acting between siblings see above, Section IV. Thus, it is known that embryos respond to certain types of stimulation during the later stages of incubation, but not yet whether there is any response to potential stimulation, such as egg turning, at earlier stages (a point raised by Gottlieb, 1968). Finally, a word should be said about possible interspecific variations in responsiveness because, although all or almost all species appear to go through the same anatomical and physiological stages of embryonic development (see Oppenheim, this volume), it has not been shown that all interact behaviorally with their environment in the same way; indeed, this is hardly to be expected. For example, the lively interaction resulting in parent/embryo and parent/chick bonds in the guillemot is of obvious relevance to the species' highly specialized cliff-edge environment. Similarly, the establishment of ties between parent and embryo of the duck (Gottlieb, 1963, 1971) and the gull (Impekoven, this volume) could have survival value. In the bobwhite quail, on the other hand, the interaction between siblings could be the more important, and recordings (Vince, unpublished observations) made from several nests of this species suggest that maternal calls on the nest were relatively very quiet during the hatching period. This could be consistent with the views of Stoddard (1931), who stresses the need for concealment in these small ground-nesting birds. Under the prairie-type conditions to which they are adapted and owing to their small size, the most important thing for this species could be to hatch at the same time so that the chicks can leave the nest early and feed early [they are, in any case, more precocious than some other chicks (cf. Vince, 1971, 1972a)]. In this species the parent/embryo tie appears to be less well established at hatching, and the nest exodus occurs rather differently from that reported for ducks by Gottlieb (1971). In quail nests monitored visually and by a microphone placed amongst the eggs, it has been observed that the chicks begin to make sorties from the nest about 3 hours from the time of hatching; at that stage the parents become restless and, finally, about 5 hours from hatching, also begin to leave the nest, taking the remaining chicks (Vince, unpublished observations). A similar situation has been observed in much more detail in the pheasant (*Phasianus colchius*) by P. O. Hopkins (personal communication). Hopkins considers that this pattern—emergence of at least some chicks—leads to termination of incubation and subsequently of brooding periods. From the time of hatching, chick and hen vocalizations play a crucial role in this interaction. However, more work on parent/young interaction is needed both before and after hatching in species adapted to different environments. For example, Bjärvall (1968) reported that mallard ducklings also make sorties from the nest before exodus, and although the initiative for exodus is taken by the female, she does this in response to stimulation from the ducklings, as was also observed by Gottlieb (1963) in hole-nesting ducklings.

VII. Summary and Conclusions

It is clear that there are different types of embryonic activity which also vary with developmental age. Similarly, environmental effects arise in different ways, affect certain aspects of behavior or development more than others, vary with developmental age, and can take a slightly different course in different species. In different species also conditions within and outside the nest vary greatly. Environmental effects can arise from conditions within, or outside, the shell and are, of course, usually but not always mediated directly or indirectly by the nervous system. Thus, the situation within the egg is a constantly changing and very complex one. On the whole, however, the picture outlined in the discussion above is a coherent one, so that different aspects described can clarify each other or pinpoint areas where research is most needed.

The simplest type of interaction considered is that described by Kuo (1932a, 1932b, 1932c) where movement or positioning of the embryo occurs passively, as a result of the movement and/or shape of mainly extraembryonic structures, such as the amnion, yolk sac, or albumen. Although there are difficulties here because of the close physiological interconnection of all structures within the shell, this point of view has obvious validity, and such effects need detailed quantitative analysis.

The work of Hamburger and his associates has shown that embryonic activity of a random, spontaneous kind occurs largely in the absence of sensory input, so that its rate of occurrence is unchanged by deafferentation procedures. In this work, a clear distinction has been made between the random "Type I" activity which continues, although not always at the same rate, throughout the incubation period, and coordinated "Type III' activities, which first become perceptible from about day 17 in the domestic fowl, which may be amenable to stimulation, and which result in the attainment of the hatching position and the emergence of the chick. In addition to this work, Hamburger and his colleagues have shown that reflex activities can be elicited from the intact embryo, although Decker (1970) considers that reflex behavior may be latent or masked until the time of hatching. It has, meanwhile, been pointed out (Gottlieb, 1968; Oppenheim, 1970) that although the rate of occurrence of the random type of motility appears to be unchanged by sensory input, the direction and extent of these movements could be affected in this way. Or, indeed, they could be affected passively by forces of the type considered by Kuo. Thus, it could be profitable to give more detailed consideration to effects of the Kuo type of interaction on the nature, direction, and extent of the Type I activities and also, of course, on the Type III activities.

Work on effects of accelerating or retarding stimulation from outside the egg on the rate of development and hatching time of the embryo raises other problems, for example, the question of how the embryo responds to this stimulation. Experiments are now needed to distinguish between possible developmental and behavioral effects and also to explore the possibility of stimulation having an effect on the development of the brain itself (see Berry, Volume 2 of this series). In addition to this, the advancement of postural changes, as well as, e.g., albumen and yolk sac movements and absorption as a result of external stimulation, at the time when the earliest coordinated

"Type III" movements appear, suggest an extension of the Hamburger type of experiment: whereas no differences were found in the rate of occurrence of activity when this was compared in deafferented embryos and "normals" kept under well-controlled conditions of isolation, it is still possible that a difference might be found if the activity of deafferented (or indeed normal) embryos were compared with the activity of embryos provided with stimulation from outside the shell, during the period between 15 and 17 days of incubation. In fact, the possibility remains that embryos may be more active when they are subjected to the intermittent cooling, vibration, sound, and light changes which appear to be normal accompaniments of natural incubation procedures.

Behavioral differences observed, such as the amount of vocal interaction between parents and embryos, makes obvious the relationship between species-specific patterns and the ecology of the species. It would appear that the amount of parent/embryo interaction could vary even between precocial species, such as hole-nesting ducks, where the parent has to call ducklings from the nest, and small ground-nesting species, like the bobwhite quail, which, as Stoddard (1931) has suggested, may have an overriding need for concealment. Although there is a growing volume of work in this area, interspecific differences throughout the period beginning with lung ventilation and ending with exodus from the nest still need to be examined.

Finally, it would begin to appear that the stage of development between the equivalent of days 13–17 in the domestic fowl, when the random motility rises to a peak, dips, and rises again in rate of occurrence, and before coordinated activities become obvious, is one of the more important and possibly least well understood in the ontogeny of behavior in the avian embryo. By the end of this stage the embryo appears to respond to stimulation from outside the egg in ways which need clarifying. It could be that detailed observation of stimulated and unstimulated embryos during this stage could help us to understand how the coordinated patterns of behavior are set up. This is, in addition, the stage when the Kuo type of interaction could be very important, owing to the rapid growth of the embryo and an increasing rate of absorption of the other shell contents. It is possible that a several-sided attack on this stage of development could be a profitable one.

Acknowledgments

This research was carried out with the support of the Medical Research Council of Great Britain, and of the Department of Psychology, University of Cambridge, England.

References

Balaban, M. & Hill, J. Effects of thyroxine level and temperature manipulations upon the hatching of chick embryos (*Gallus domesticus*). *Developmental Psychobiology*, 1971, **4**, 17–35.

Baldwin, S. P., & Kendeigh, S. C. Physiology of the temperature of birds. *Scientific Publications of the Cleveland Museum of Natural History*, 1932, **3**, 1–196.

Bjärvall, A. The critical period and the interval between hatching and exodus in mallard ducklings. *Behaviour*, 1967, **28**, 141–148.

Bjärvall, A. The hatching and nest-exodus behaviour of mallard. *Wildfowl*, 1968, **19**, 70–80.

Chattock, A. P. On the physics of incubation. *Philosophical Transactions of the Royal Society of London, Series B*, 1925, **213**, 397–450.

Coghill, G. E. *Anatomy and the problem of behaviour*. London: Cambridge University Press, 1929.

Dawes, G. S., & Simkiss, K. The acid-base status of the blood of the developing chick embryo. *Journal of Experimental Biology*, 1969, **50**, 79–86.

Decker, J. D. The influence of early extirpation of the otocysts on development of behavior in in the chick. *Journal of Experimental Zoology*, 1970, **174**, 349–363.

Drent, R. H. *Functional aspects of incubation in the herring gull* (*Larus argentatus pont*). Leiden: Brill, 1967.

Driver, P. M. "Clicking" in the egg-young of nidifugous birds. *Nature* (*London*) 1965, **206**, 315.

Fisher, H. I. The "hatching muscle" in the chick. *Auk*, 1958, **75**, 391–399.

Freeman, B. M. Studies on the oxygen requirements and hatching mechanisms of the domestic fowl. Unpublished doctoral dissertation, University of Leicester, 1964.

Freeman, B. M., & Misson, B. H. pH, pO_2 and pCO_2 of blood from the foetus and neonate of *Gallus domesticus*. *Comparative Biochemistry and Physiology*, **33**, 1970, 763–772.

Gottlieb, G. A naturalistic study of imprinting in wood ducklings (*Aix sponsa*). *Journal of Comparative and Physiological Psychology*, 1963, **56**, 86–91.

Gottlieb, G. Prenatal behavior of birds. *Quarterly Review of Biology*, 1968, **43**, 148–174.

Gottlieb, G. *Development of species identification in birds: An inquiry into the prenatal determinants of perception*. Chicago: University of Chicago Press, 1971.

Gottlieb, G. & Kuo, Z-Y. Development of behavior in the duck embryo. *Journal of Comparative and Physiological Psychology*, 1965, **59**, 183–188.

Green, J. & Vince, M. Foetal development in quail during the final stages of incubation. *British Poultry Science*, 1973, **14**, 185–192.

Haftorn, S. Egglegging og ruging hos meiser basiert på temperaturmålinger og direkte iakttagelser. *Sterna*, 1966, **7**, 49–102.

Hamburger, V. The beginnings of co-ordinated movements in the chick embryo. In G. E. W. Wolstenholme & M. O'Connor (Eds.), *Ciba Foundation symposium on growth of the nervous system*. London: Churchill, 1968.

Hamburger, V. & Balaban, M. Observations and experiments on spontaneous rhythmical behavior in the chick embryo. *Developmental Biology*, 1963, **7**, 533–545.

Hamburger, V., Balaban, M., Oppenheim, R., & Wenger, E. Periodic motility of normal and spinal chick embryos between 8 and 17 days of incubation. *Journal of Experimental Zoology*, 1965, **159**, 1–14.

Hamburger, V., & Narayanan, C. H. Effects of the deafferentation of the trigeminal area on the motility of the chick. *Journal of Experimental Zoology*, 1969, **170**, 411–426.

Hamburger, V., & Oppenheim, R. Prehatching motility and hatching behaviour in the chick. *Journal of Experimental Zoology*, 1967, **166**, 171–204.

Hamburger, V., Wenger, E. & Oppenheim, R. Motility in the chick embryo in the absence of sensory input. *Journal of Experimental Zoology*, 1966, **162**, 133–160.

Horner, F. R. Liberation of the chick from the shell. In W. Wingfield & G. W. Johnson (Eds.), *The poultry book*. London: Wm. S. Orr & Co., 1853.

Kuo, Z.-Y. Ontogeny of embryonic behavior in Aves. I. The chronology and general nature of the behaviour of the chick embryo. *Journal of Experimental Zoology*, 1932, **61**, 395–430. (a)

Kuo, Z.-Y. Ontogeny of embryonic behavior in Aves. II. The mechanical factors in the various stages leading to hatching. *Journal of Experimental Zoology*, 1932, **62**, 453–483.(b).

Kuo, Z.-Y. Ontogeny of embryonic behavior in Aves. III. The structural and environmental factors in embryonic behavior. *Journal of Comparative Psychology*, 1932, **13**, 245–271.(c)

Kuo, Z.-Y. *The dynamics of behavior development: An epigenetic view*. New York: Random House, 1967.

Landauer, W. The hatchability of chicken eggs as influenced by environment and heredity. *Connecticut, Storrs Agricultural Experiment Station*, 1961, No. 1.

Lehrman, D. S. A critique of Konrad Lorenz's theory of instinctive behavior. *Quarterly Review of Biology*, 1953, **28**, 337–363.

Lundy, H. A review of the effects of temperature, humidity, turning and gaseous environment in the incubator on the hatchability of the hen's egg. In T. C. Carter and B. M. Freeman (Eds.), *British Egg Marketing Board symposium*. No. 5, Edinburgh: Oliver & Boyd, 1969.

McCoshen, J. A., & Thompson, R. P. A study of clicking and its source in some avian species. *Canadian Journal of Zoology*, 1968, **46**, 169–172.(a)

McCoshen, J. A., & Thompson, R. P. A study of the effects of egg separation on hatching time and the source of clicking sounds in the embryo of the domestic chicken. *Canadian Journal of Zoology*, 1968, **46**, 243.(b)

Oppenheim, R. W. Some aspects of embryonic behaviour in the duck (*Anas platyrhynchos*). *Animal Behaviour*, 1970, **18**, 335–352.

Pani, P. K. Coleman, T. H., Georgis, H. D., Kulenkamp, A. W., & Kulenkamp, C. M. Influence of contact with eggs containing more advanced embryos on yolk-sac movement into the body cavity of embryonating bobwhite quail. *Poultry Science*, 1969, **48**, 1956.

Ramachandran, S., Klicka, J., & Ungar, F. Biochemical changes in the musculus complexus of the chick (*Gallus domesticus*). *Comparative Biochemistry and Physiology*, 1969, **30**, 631–640.

Romijn, C. Respiratory movements in the chicken during the parafeotal period. *Physiologia Comparata et Oecologia*, 1948, **1**, 24–28.

Stoddard, H. L. *The bobwhite quail*. New York: Scribner's, 1931.

Tschanz, B. Trottellummen: Die Entstehung der persönlichen Beziehungen zwischen Jungvogel und Eltern. *Zeitschrift für Tierpsychologie*, 1968, **4**, 1–103.

Vince, M. A. Social facilitation of hatching in the bobwhite quail. *Animal Behaviour*, 1964, **12**, 531–534.

Vince, M. A. Artificial acceleration of hatching in quail embryos. *Animal Behaviour*, 1966, **14**, 389–394.

Vince, M. A. Effect of rate of stimulation on hatching time in Japanese quail. *British Poultry Science*, 1968, **9**, 87–91.(a)

Vince, M. A. Retardation as a factor in the synchronization of hatching. *Animal Behaviour*, 1968, **18**, 332–335.(b)

Vince, M. A. Embryonic communication, respiration and the synchronisation of hatching. In R. A. Hinde (Ed.), *Bird Vocalisations*. London: Cambridge University Press, 1969.

Vince, M. A. Sone aspects of hatching behaviour. In B. M. Freeman and R. F. Gorden (Eds.), *British Egg Marketing Board symposium*. No. 6. Edinburgh: Oliver & Boyd, 1970.

Vince, M. A. Effect of accelerated hatching on the Japanese quail chick's capacity to get up on to its feet. *Animal Behaviour*, 1971, **19**, 62–66.

Vince, M. A. Effect of external conditions on the capacity of the domestic chick to stand and walk. *British Journal of Psychology*, 1972, **63**, 89–99.

Vince, M. A. Effects of external stimulation on the onset of lung ventilation and the time of hatching in the fowl, duck and goose. *British Poultry Science*, 1973 (in press).(a)

Vince, M. A. Foetal respiration rates and the development of a stable respiratory metabolism in quail. *Comparative Biochemistry and Physiology*, 1973, 44A, 341–354.(b)

Vince, M. A., & Cheng, R. C. H. Effects of stimulation on the duration of lung ventilation in quail fetuses. *Journal of Experimental Zoology*, 1970, **175**, 477–486.(a)

Vince, M. A., & Cheng, R. C. H. The retardation of hatching in Japanese quail. *Animal Behaviour*, 1970, **18**, 210–214.(b)

Vince, M. A., Green, J., & Chinn, S. Acceleration of hatching in the domestic fowl. *British Poultry Science*, 1970, **11**, 483–488.

Vince, M. A., Green, J., & Chinn, S. Changes in the timing of lung ventilation and late foetal development in quail. *Comparative Biochemistry and Physiology*, 1971, **39A**, 769–783.

Vince, M. A., & Salter, S. H. A. Respiration and clicking in quail embryos. *Nature* (*London*), 1967, **216**, 582–583.

Visschedijk, A. H. J. The air space and embryonic respiration. (3) The balance between oxygen and carbon dioxide in the air space of the incubating chicken egg and its role in stimulating pipping. *British Poultry Science*, 1968, **9**, 197–210.

PRENATAL ORIGINS OF PARENT–YOUNG INTERACTIONS IN BIRDS: A NATURALISTIC APPROACH

*MONICA IMPEKOVEN AND PETER S. GOLD**

Institute of Animal Behavior
Rutgers University
Newark, New Jersey, and
Department of Biology
State University of New York at Buffalo
Buffalo, New York

*The experiments presented in Part I of this study were designed by both authors. The experiments presented in Section II by the senior author. The senior author prepared the manuscript.

General Introduction

This paper deals with the behavioral effects of sensory stimulation naturally encountered during embryonic development in gulls. Emphasis will be placed on the auditory modality, but other types of stimulation, notably the ones originating from sources outside the egg, will be briefly discussed also. Section I deals mainly with embryonic responsiveness to stimulation up to prehatching stages; Section II deals with long-term effects that exposure to stimulation prior or around prehatching may have on later prenatal and early postnatal behavior.

I. Sensory Stimulation and Embryonic Motility

A. Review of Some Previous Studies

The determinants of embryonic behavior in birds have been the issue of great concern to a variety of scientists.

Essentially there have been two approaches which are reminiscent of the nature–nurture dichotomy. One school of thought places major emphasis upon stimulation as a prime organizing factor in embryogeny; the other one places emphasis on the autogeny of embryonic behavior (reviewed by Gottlieb, 1970).

Accordingly, some investigators have focused on the role of external organizing factors (e.g., Kuo, 1967), self-stimulative processes (Gottlieb & Kuo, 1965) and the progressive incorporation of the effects of experience (Schneirla, 1965); others have stressed the spontaneity of embryonic motility and the absence of reflexogenic movement in the normal undisturbed embryo (Hamburger, 1963, 1968, 1970; Hamburger, Wenger, & Oppenheim, 1966).

Critical experiments concerning the control of embryonic movement up to prehatching stages have been performed mainly by Hamburger and his colleagues. The possible role of sensory stimulation in embryonic motility has been studied experimentally in the intact embryo with respect to, e.g., proprioceptive and tactile stimuli caused by the embryo touching itself and the swinging of the amnion (Hamburger, 1963; Oppenheim, 1966, 1972a), thermal stimulation (Oppenheim, 1972c), and chemical changes in the environment (Hamburger, 1963). Other experiments on the role of sensory input were performed by means of surgical interference with the embryo (e.g., Decker, 1970; Hamburger & Narayanan, 1969; Hamburger *et al.*, 1966). Most of these studies showed that certain aspects of motility were not affected by such manipulations (reviewed by Oppenheim, 1972c). These results and their interpretations, however, raise various arguments:

1. In some instances the lack of finding immediate effects of stimulation may be due to the fact that only the periodic aspects of motility were quantified. Oppenheim (1972c) points out that the results may have differed if frequency or amplitude of movement or their coordination had been considered.

2. While one has to accept that the temporal aspects, and in some instances also the frequency aspects of embryonic movement are independent of certain types of sensory input, this conclusion has often been identified with the statement that spontaneous motility patterns are uninfluenced by sensory input in general (e.g., Hamburger, 1963; but see Hamburger, this volume).

First, there is a difference between the question as to whether the periodic aspects of embryonic motility are dependent or independent of sensory input and whether such input (naturally) can influence embryonic behavior. The finding that certain types of stimulation affects embryonic motility would not in the least negate that there is an autogenous basis to such movement. As stated by Gottlieb (1970), embryonic motility, while primarily self-generated, is also susceptible to inhibitory and/or excitatory influences of sensory stimulation. According to Oppenheim (1972a), such influences may be transient, or they may be brief and momentary in the sense that they do not alter the frequency of movement over longer time periods. This author also demonstrates that sensory stimulation may elicit a response the first few times it is applied but that responsiveness may wane rapidly with repeated stimulation. It is concluded that while the occurrence and the temporal distribution of embryonic movements are not in any significant way dependent on stimulation, such stimulation can nevertheless evoke responses, yet without altering the basic periodicity pattern.

Second, it seems premature to generalize about the independence of embryonic movement from sensory input since (a) the relevance of the deafferentation experiments to the normal intact embryo has been questioned (Gottlieb, 1968; Hinde, 1970), and (b) only certain types and patterns of stimulation have been examined in the intact embryo (reviewed by Oppenheim, 1972c).

One reason for rarely finding effects of sensory stimulation may be that most investigations have been carried out in the domestic chick raised under conditions of artificial incubation, and that this type of incubation does not reveal all possible types of naturally occurring stimulation with particular reference to extraovular sources. As pointed out by Olsen (1930), artificial incubation bears little resemblance to the natural incubation pattern of the domestic chicken. Thus, an important starting point for further research would be the observation of stimulation to which the embryo is exposed, naturally, from its external environment.

3. It has been argued that even though the embryo can be affected by certain kinds of stimulation, this is primarily so when this stimulation is "unnaturally" intense, which questions the biological significance of such effects (Oppenheim, 1972c). This is true for studies by two Russian investigators, Sviderskaya (1967) using a 2-kHz sound at 95 dB, and Bursian (1964) using a 500-W light. It has not been considered, however, that some of the stimuli used in such experiments may have been unnatural per se in the sense that stimulation bore no resemblance to the kinds of stimuli and the way they may act under natural conditions of incubation. This argument applies to the above Russian authors but also to some extent to studies exploring the effect of tactile stimulation (e.g., Hamburger, 1963; Oppenheim, 1972a), derived from the embryo touching itself. It is, of course, much more difficult to experimentally approximate intraovo stimulation to the natural situation than sources of stimulation originating from outside the egg.

4. Hamburger (1970) specifies that embryonic motility in the chicken is independent of sensory stimulation till at least around day 15–17, that is, until the stage when preparatory movements to hatching are first observed. The change from seemingly uncoordinated motility to integrated movements is not abrupt, but a more gradual transition. Uncoordinated movements persist during intervals in the episodes of integrated prehatching and hatching movements until close to climax (Hamburger & Oppenheim, 1967). Hamburger (1968, 1970) and his collaborators have not experimentally explored whether random-type motility is independent of sensory stimulation until around day 17, and subsequently is dependent on it, or whether possible effects of stimulation are restricted to the newly appearing coordinated movements. For prehatching stages there is clear evidence in the domestic chicken and the Peking duck that motility is affected (whether dependent or not) by naturally occurring stimulation, as measured by the rate of oral activity (Gottlieb, 1965).

5. The apparent lack of effects of sensory stimulation on embryonic motility might lead to the conclusion that, therefore, the embryo is impervious to such stimulation as a whole. However, Oppenheim (1972a and 1972c), stresses that while many aspects of embryonic behavior are independent of sensory input, this should not be interpreted to mean that, therefore, stimulation plays no role whatsoever in the ontogeny of ongoing or later behavior. Even in the absence of any sign of immediate overt response, the question of long-term changes remains open. As suggested by Gottlieb (1968, 1971), the functional status of a sensory system can be demonstrated in absence of an immediate motor response when later behavior can be shown to be the consequence of earlier sensory stimulation. With this statement, Gottlieb refers to developmental stages prior to the formation of effective connections with the motor system. However, Gottlieb

(1971) himself has demonstrated that even after the formation of such connections, exposure to certain types of auditory stimulation to which duck embryos are behaviorally unresponsive affects their subsequent responsiveness to a different type of sound.

Further experimental study is required to explore the effects that sensory stimulation perceived *prior* to prehatching may have on the embryo's responsiveness in later developmental stages.

Tinbergen (1968) states that the ethologist no longer must relegate the study of embryology to the experimental embryologist but must himself embark on a program of the study of embryos, in order to discover to what extent development is influenced by the environment. Using the ethologist's approach, a first step has been made in this direction.

The present study has been initiated by observations of the behavior in wild birds during incubation in order to discover the naturally occurring kinds of (extraovo) stimulation that embryos are exposed to which might affect their behavior and subsequent development. Subsequent experiments are derived from questions that arose from such observations.

B. *The Behavior during Incubation in Wild Birds, with Special Reference to Gulls (Laridae)*

As can be seen from a hide (or blind) amidst a laughing gull colony (or herring gull colony: Baerends, 1970; Drent, 1970; or black-headed gull colony: Beer, 1961) during incubation, an incubating gull responds to its eggs in several ways. Beer (1961) described the following for the black-headed gull (*Larus ridibundus*):

> The incubating bird rises from time to time and assumes a stooping posture which is often accompanied by up-and down movements of the feet. Frequently the feet tend to move a little in towards the center of the nest and so trundle the eggs in front of them. The gull then may bend its head even further down to bring the underside of its lower mandible into contact with the surface of the eggs. The bill and feet tend to move towards each other with the eggs between them. The feet are often trampled rapidly and the movements of the bill and feet serve to move the eggs about in the nest. As a result the eggs become arranged or rearranged two in front and one behind—the pattern that matches the arrangement of the brood patches .... Resettling on the nest consists of the following movements: (a) Chest-dropping: The bird leans over the eggs until it tips forward onto its breast. (b) Waggling: Immediately after dropping the gull moves its legs up and down and its tail from side to side. (c) Quivering: The final resettling movement can be recognized as a trembling or shaking of the surface layers of the body .... Sometimes a gull will quiver without having lowered onto the eggs immediately beforehand .... It seems likely that quivering serves to increase the intimate contact between the surfaces of the eggs and the brood patches [Pp. 66–71].

In the laughing gull, the species investigated in the present study, these motor patterns look much the same. However, in addition it was noted that the incubating bird often utters a certain kind of call when it rises from the eggs, either before or after shifting them, and most frequently during the act of resettling, i.e., during the time that close contact is reestablished. Onomatopoetically these calls can be described as ranging from a low intensity brief "uh" to a more drawn out "uhr" or disyllabic "u-ruh" (Fig. 1).

Both mates of a pair take turns in incubation. During nest-reliefs they often interact vocally. The oncoming mate may, at any stage of incubation, utter a series of calls referred to as "crooning" as it relieves the sitting bird for incubation (Fig. 2a). Various intermediates between "crooning" and "uhr" calls can be heard, in particular by the sitting bird in response to activities of the "off-duty" mate near the nest (Fig. 2b, c). "Crooning" is a smooth, long-drawn and low-pitched call.

When the gullery is being disturbed during incubation, the birds rise from their eggs in the air and utter long series of "kow" calls. Even a single gull on the nest may suddenly sleek its feathers and "kow"-call, for example, in response to a harmless clapper-rail walking by. (Clapper-rails nest within the gull colony.) "Kow" calls differ from "crooning" in their sudden onset, shorter duration, higher pitch, and harshness (Fig. 3).

By means of the "long" call, a sitting bird may respond to the oncoming mate or a neighboring gull. While not very frequent in mid-incubation, this call becomes more common around hatching (described further in Section II). The "long" call is often introduced by a disyllabic note referred to as "ke-hah," and is composed by a string of shorter notes followed by a string of longer notes, all of a harsh quality (Fig. 4).

The sonagram of vocalizations (made on a Kay Electric Co. Sonagraph with a narrow band filter, Figs. 1–4) enables one to easily distinguish between different calls in the repertoire of a bird. While conveying a fairly good impression about pitch (kHz), frequency modulation, harmonics, and temporal characteristics, amplitude modulation is not clearly depicted in this way. There are several other types of calls that laughing gulls utter during incubation, but they have not been considered for the present study.

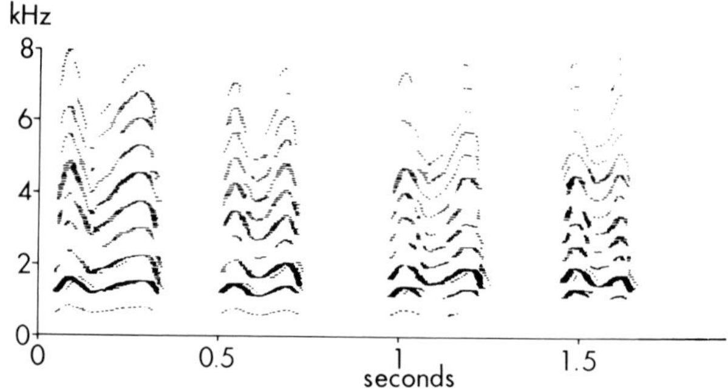

FIG. 1. Sonagram of "uhr" calls uttered during resettling.

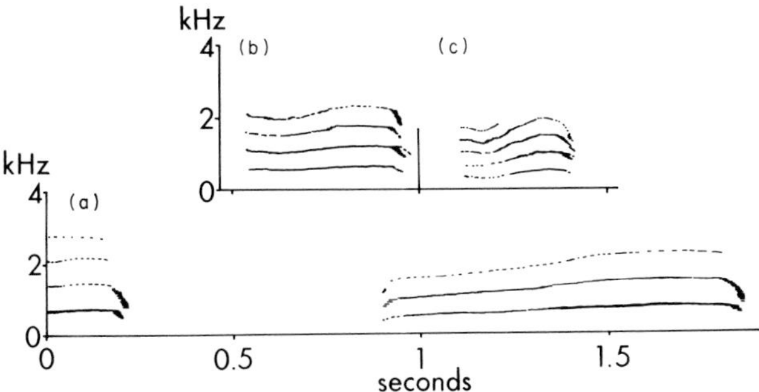

FIG. 2. Sonagram of "crooning" (a) and of two intermediates between "crooning" and "uhr" calls (b) and (c) as heard during incubation.

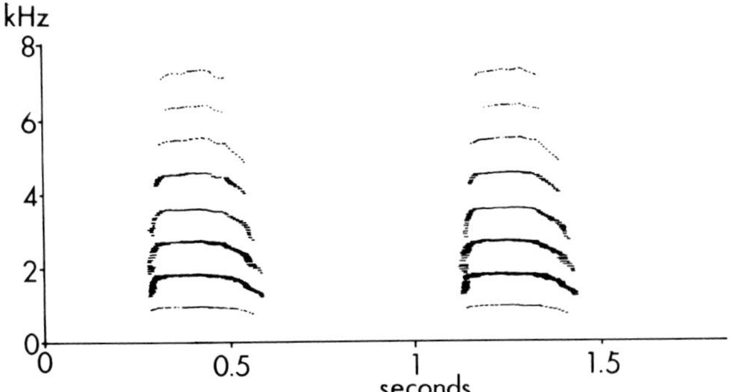

FIG. 3. Sonagram of "kow" calls.

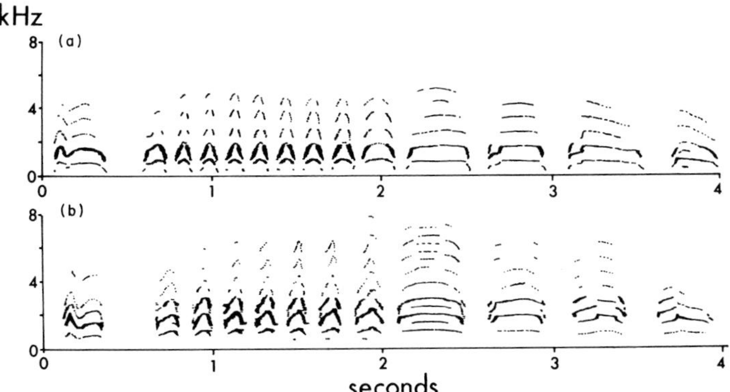

FIG. 4. Sonagram of "long" calls of two individuals (a and b used in experiments on personal recognition—see Section II).

In summary, the eggs are being exposed from time to time to light, a change in temperature, mechanical stimulation due to shifting, and various kinds of sound. Drent, Postuma, and Joustra (1970) have shown experimentally for the herring gull, by means of artificial eggs, containing thermistors, placed in nests, how the egg temperature can change in response to the incubating bird rising from its nest. Even during minor signs of alarm, during which the bird does not rise, the temperature can drop somewhat, probably due to the loss of close contact between the brood-patch and the eggs. As to sound, possibly not only auditory aspects of such vocalizations, but also vibrational properties penetrate the eggshell by means of direct contact with the parent's body. Table I illustrates some quantitative aspects of such an analysis in the laughing gull in mid-incubation. The table does not convey a true picture of the total amount of calling, since for purposes not related to this issue, 20–minute periods were selected when the mate of the sitting bird was not in the vicinity (see Impekoven, 1973a). The picture remains much the same until around 20–21 days of incubation. During the last 2 or 3 days before hatching (i.e., day 21–23) when the eggs are pipping and vocalizing, rising, calling, and reliefs for incubation become more frequent (see Section II) (Impekoven, 1973a).

While the general incubation pattern is probably similar for a wide variety of bird species, vocalizations seem to be particularly prevalent in gulls and other birds breeding in open (and dense) colonies. However, it has been found that even broody chickens vocalize in conjunction with shifting their eggs and resettling (Guyomarc'h, 1972).

C. *Questions Raised by the Naturalist*

According to Tinbergen (1963), ethology is characterized by the biological approach, which essentially entails four problems. One concerns the immediate causes of behavior, another its developmental history in the individual, a third its biological function, and a last its evolution in the species. Questions about causation and biological function both start from the observable aspect of a life process, in this case, behavior. However, in the search for causes one looks, as it were, "back in time," while in the study of function one looks "forward" in an attempt to discover the consequences of the observed behavior.

For the naturalist questions about function are particularly prevalent. Experiments can be designed to find out about the effect or functional significance of behavior, and this with the idea in mind that animals have evolved in a way that their behavior and responsiveness is adapted to their environment.

Some of the writings in classical ethology reveal a tendency to derive

TABLE I
FREQUENCY OF CERTAIN BEHAVIOR PATTERNS OF INCUBATING LAUGHING GULLS BETWEEN DAYS 8–15 OF INCUBATION, DURING 20-MINUTE OBSERVATION PERIODS

Behavior patterns	Median no. per 20 minutes (range)	No. birds
Rising–resettling	.8 (0–5)	25
% Rises accompanied by egg shifting	49.7 (0–100)	17 (Rising birds only)
"Uhr" calls during rising–resettling	2 (0–19)	17 (Rising birds only)
Quivering while sitting	.5 (0–8)	25
Other types of calls (directed at other gulls)	.1 (0–2)	25

questions about the causation and the development of behavior from questions about its biological function (see also Beer, 1963, 1964, 1968). In this way, some important facts about the causation of behavior were discovered. For instance, Tinbergen and Perdeck (1950) noted a red spot on the lower mandible of the herring gull's beak. The question arose, what is it good for and thus what is its functional significance? It was suspected that one of the effects may be to elicit and direct the food-begging response of the newly hatched gull chick. Experiments demonstrated that this assumption was correct. At the same time they revealed causal factors for the elicitation of the pecking response in the gull chick, but without leading to a profound understanding on how the red spot produced its effect. Recently, however, more detailed analyses of the stimulus parameters eliciting pecking in gull chicks have been reported (e.g., Hailman, 1967).

The evolutionary functional approach led to the concept of the Innate Releasing Mechanism (IRM) (Tinbergen, 1951), a mechanism by which an animal has become phylogenetically adapted to responding preferentially to certain types of naturally occurring stimuli in its environment, including other members of its species (e.g., Schleidt, 1963). While, as has been discussed (Quine & Cullen, 1964), this concept has many disadvantages which can make it misleading, the basic idea, i.e., that of adaptiveness, retains its value in that it calls our attention to the existence of natural stimulus preferences which may be overlooked or ignored by the investigator who lacks a naturalistic approach. The existence of such preferences does not preclude that they have arisen through developmental processes in interaction with the environment and that they can be modified by further experience. The classical ethologists have been criticized for their lack of interest in the origin of behavior before it begins to function in adaptation to the postnatal environment (Lehrman, 1970). However, adaptive preferences may arise prenatally, as shown, for instance, with respect to the species specificity of responsiveness to maternal calls in chick and duck embryos (Gottlieb, 1965, 1971).

Let us return to the observations of the incubation behavior in gulls. Seeing laughing gulls shift and chill their eggs and hearing them call, partly in relation to activities with the eggs, leads to the suspicion that some of this behavior, apart from possibly having an effect on the parent and on species mates, also affects the embryos' perception and behavior. Consequently, ideas about the causation of embryonic behavior can be derived from questions about suspected effects of the adults' behavior. The first step is to see whether this kind of stimulation affects the embryo, and if so, how. As a next step the parameters of the stimulus situation will have to be analyzed in greater detail. Lastly, the long-term effects resulting from such stimulation need to be explored. In the laughing gull, the first question has been tackled, with respect to some of the parental calls described above.

D. *Parental Calls and Embryonic Motility in the Laughing Gull (*Larus atricilla*)*

For these experiments, the parental calls were selected that not only may be important to the embryo but also to the young after hatching (see Section II): "crooning," as heard during nest-reliefs, "kow," as heard in alarm situations, and "long" calls, not very frequently heard during incubation in response to the mate or a neighboring gull. (Figures 2, 3, and 4 represent the calls used in these experiments.)

1. Methods

a. Subjects. The eggs used for this study were collected in the gullery on the day they were laid. They were incubated in forced-draft incubators at 37–38°C and 70% relative humidity in the absence of more advanced vocalizing embryos, in a room out of earshot of any extraneous noises. In other words, prior to testing, these embryos were not exposed to sound other than the low-frequency hum of the incubator fan.

b. Apparatus. A Uher 4200-L tape recorder and a Telmar loudspeaker, Model VS-300, were used. A test tape was made containing 30-second periods of sound, interspersed by 3 minutes of silence between successive sound periods. In the case of "crooning," the 30-second periods contained three sequences of five notes each. The intervals between successive notes measured .7 seconds, the interval between the sequences approximately 2 seconds. The duration of individual "croons" was just under 1.0 second

(Fig. 2a). With respect to the "kow" call, the 30-second period contained a natural sequence of 36 such calls. "Kows" are very regularly spaced in time at .5 seconds between successive notes. The duration of "kow" calls was .4 seconds (Fig. 3). As for the "long" call, four such vocalizations were recorded at equal intervals of 4.2 seconds over the 30-second period. The duration of a "long" call (including intervals between notes) was just under 4 seconds (Fig. 4). In general, an attempt was made to arrange intercall intervals in such a way that the total sound period was approximately 15 seconds within the 30-second period for all three calls examined.

Sound level measurements were made on a General Radio sound level meter, scale B, slow. A keyboard-operated Rustrak event recorder was used for recording motility. The embryos were tested in a sound-attenuated box with a transparent top, at approximately incubation temperature.

c. Procedure. Observation windows were made over the airspace, and the underlying membrane was made transparent with petroleum jelly (Kuo, 1932). Each embryo was adapted to the test situation for a period of 8–10 minutes. A sequence of "crooning," "kow" calls, and "long" calls was presented twice to each embryo, and the order of the calls was alternated with subsequent embryos. The first sequence will subsequently be referred to as the first subtest, the second sequence as the second subtest. The calls were played at a sound level of 80–85 db, corresponding roughly to the level of "kow" and "long" calls previously measured at a gull's nest. "Crooning" reaches a level of only 70–75 db. Each embryo was tested at one of the following developmental stages: approximately day 15, 17, 19, and 21 (see Table II).

The right leg or parts thereof and/or the right wing were visible nearly at all times and in all developmental stages tested, while other parts of the embryos, e.g., the head and the body, were visible only at times and more in some stages than in others. Consequently, activity measures were based on the movements of the right leg and/or wing. Most of these movements consisted of apparently random jerks, i.e., type I movements according to Hamburger's and Oppenheim's (1967) nomenclature.

The 30-second sound periods were compared with a 1-minute period preceding and a 1-minute period following such a test period. The frequency values of the two 30-second periods of each call were combined for analysis, and the values of the pre-and postperiods were combined and averaged. Accordingly, the frequency of motor patterns has been presented as per minute of exposure to either sound or no-sound conditions. In a more detailed analysis, the frequency values of the two 30-second periods of the calls (first and second subtest) were examined separately. For comparison between the different test conditions, the Wilcoxon matched-pair signed ranks test was used (two-tailed p-values). [For day 21, the frequencies measured during the silent periods were higher than the ones given by Oppenheim (1972b). It is unknown to what extent this difference is due to the recording technique and to what extent to differences in development stage. In the present study only 6 out of 10 birds were tucked or draped on day 21, whereas all of Oppenheim's birds were tucked or draped at this stage.]

2. Results

As shown in Table II, "crooning" clearly increased motility while "kow" calls decreased it, although the latter is true only for day 15. "Long" calls had no effect.

A more detailed analysis reveals that the effects of "crooning" were at all stages greater in the first subtest than in the second subtest. For days 17, 19, and 21, the differences are statistically significant in the first subtest, and for day 15 the difference is below the .1 level. In the second subtest, effects, while not statistically significant, point in the same direction and, thus, add to the differences when the two subtests are combined for analysis. With respect to "kows," the subtests do not show statistically significant effects.

The frequency of movement during no-sound periods preceding and following a sound period did not differ. A quantitative analysis of the temporal aspects of motility is still pending.

In a pilot experiment carried out in an earlier season using a somewhat different time schedule for the presentation of calls, with embryos just prior to or after menbrane penetration and hearing other embryos vocalize, "crooning" similarly increased the frequency of leg movements and bill-clapping, while the other calls had no such effects.

3. Discussion

The result that "crooning" increased activity more in the first than in the second 30-second exposure indicated that responsiveness wanes with repeated stimulation, a finding that agrees with earlier ones using sound (Gottlieb, personal communication), as well as tactile and photic stimulation (Oppenheim, 1968, 1972a). The fact that the frequency of movement during no-sound periods following and preceding sound periods did not significantly differ implies that the effects of the calls did not exceed the period of stimulation.

The activity-increasing effect of "crooning" stands in contrast to findings with the mallard maternal call (Gottlieb, 1971). First, no reliable behavioral effects were demonstrated before 81% of the incubation period was completed, while in the present experiment effects were discovered after only 62.5–65% of the incubation period. The domestic mallard (Peking) hatches after 27 days of incubation, whereas the laughing gull hatches after

TABLE II

EFFECT OF PARENTAL CALLS ON EMBRYONIC MOTILITY IN LAUGHING GULLS[a]

Stage[b]	15 days + 1 hour		16 days + 22 hours		18 days + 20 hours		20 days + 19 hours	
n	12		12		10		10	
Test condition	ns	s	ns	s	ns	s	ns	s
Croon	34.2 (7.9)	47.6 (16.5)	27.7 (5.6)	49.4 (20.5)	26.5 (6.3)	49.2 (17.8)	20.8 (7.2)	30.8 (11.9)
p (no sound vs. sound)	< .05		< .02		< .01		< .02	
Kow	36.5 (9.1)	29.2 (12.8)	30.9 (6.3)	32.2 (11.7)	29.1 (7.9)	28.4 (10.6)	19.9 (9.1)	15.7 (9.2)
p (no sound *vs.* sound)	< .05		NS		NS		NS	
Long	33.5 (11.7)	34.4 (13.5)	24.6 (3.9)	33.8 (13.0)	28.7 (8.7)	31.3 (7.6)	23.2 (7.9)	19.7 (8.4)
p (no sound *vs.* sound)	NS		NS		NS		NS	

[a]Mean number of movements per minute (and SD) during no sound (ns) and sound (s) conditions at different stages of incubation.

[b]Standard deviations are 3.5 hours, 2.9 hours, 1.5 hours, and 2.1 hours.

23–24 days. The first stage at which the laughing gull was found to respond to parental calls (day 15) corresponds roughly to day 13 in chicken embryos, at which stage only cochlear microphonic effects of sound stimulation have thus far been reported (Vanzulli & Garcia-Austt, 1963). The question arises to what extent apparent differences in the developmental onset of responsiveness to sound are due to species differences and to what extent to the different methods of study. In the case of the domestic mallard and chick, the head was pulled out of the shell for recordings of bill-clapping, while in the present investigation the embryo was left in its natural position within the shell for observations of overall motility.

Second, in ducks, inhibitory effects were found at first, and only the hearing of self-vocalizations or the vocalizations of brood-mates led to a change from inhibition to activation. The number of embryos used in this study is insufficient to test the validity of the results with the "kow" call, and replication is required. Heaton (1971b), analyzing the maternal call of the mallard with respect to its critical elements, found bidirectionality of responsiveness depending on repetition rate. Three days before hatching, the rate (4/second) approximating that of the exodus call produced excitation, while a slower rate (1/second), resembling that of calls uttered in alarm situations, produced inhibition.

This very early onset of behavioral responsiveness to acoustical stimulation is unprecedented. In this connection it is important to note that a Ballisto-cardiograph (Rogallo, 1964), which had been used to record the activity of some of the embryos, picked up vibrations when "crooning" was played through the loudspeaker, but not when "kow" calls of "long" calls were played. This response occurred even when an egg containing a dead embryo was placed on this apparatus. This finding suggests that the effect of "crooning" on the embryo may not be due solely, or even primarily, to auditory stimulation (e.g., predominance of low auditory frequencies to which the developing auditory system first responds) but to very low-frequency vibrations which might act as a source of proprioceptive and/or tactile stimulation. Such a finding would be interesting in view of the fact that the tactile and the proprioceptive system begin to function before audition, as shown for the domestic chick and duck (reviewed by Gottlieb, 1968).

Future investigations will have to be directed at analyzing the "crooning" call in order to determine what its effects are due to. The same question applies to the possible inhibitory effect of "kow" calls on day 15 which in contrast to "crooning" are characterized by sudden onset, higher pulse rate and pitch, and harshness.

Calls uttered during resettling ("uhr" calls, see Figs. 1 and 2 a, b) and in response to the mate near the nest, share certain characteristics with "crooning," and thus they could also stimulate embryonic activity, although this remains to be demonstrated. A further step in this study must be to see how far back in embryonic life behavioral effects of parental calls can be traced.

E. Other Parental Stimuli: Suggestions for Future Research

Other sources of low-frequency sound in nature are the knocking together of the eggs in the nest in response to being shifted by the incubating bird (Gottlieb, 1968). The effect of egg shifting (acting as tactile, vestibular, and proprioceptive stimulation) needs to be investigated also. Olsen (1930) and Drent (1970) discuss the function of egg shifting, but exclusively with respect to the physical development of the embryo. Instead of turning the eggs around their two axes the way the incubating bird does, studies have been performed rotating eggs on a disc, Oppenheim and Gottlieb (personal communication), in preliminary findings, have been unable to replicate a result by Visintini and Levi-Montalcini (1939) according to which 8-day-old chicken embryos are responsive to such rotation, nor have they so far found any evidence of vestibular sensitivity up to at least day 15 with this method. The reports of Decker (1970) and Kovach (1968, 1970) agree with this finding.

Behavioral effects of natural lighting will have to be explored more closely, since up to the present, studies have used strong artificial lights and have used temporal patterning not resembling the one occurring during rising and resettling of an incubating bird (e.g., Bursian, 1964). A recent investigation has shown that sunlight penetrates even pigmented eggshells of Japanese quail and thus affects the pupillary response of the embryo (Heaton, 1971a). Similarly, experiments on possible effects of temperature have not been derived from the kinds of changes observed during natural incubation (Oppenheim, 1972c). The approximation of the experimental stimulus to the natural situation is important for the following reasons: if by using unnatural types of stimulation, an effect on embryonic behavior is found, its biological significance can be questioned; if no effect is shown, it can be argued that the embryo may not be equipped to respond to the stimulus.

For the demonstration of certain prenatal stimulation effects, it may be important that different stimulus modalities be temporally correlated in the way this happens in nature; e.g., egg shifting goes hand-in-hand with exposure to light and lowering of the temperature, and calling may either precede or more frequently follow shifting. Possibly there is some sort of summation of effectiveness of different stimulus modalities (Tinbergen, 1951), as has been shown by ethologists for the responsiveness of adult animals in various behavioral contexts (reviewed by Hinde, 1970).

Alternatively, or in addition, a pattern of "conditioning" has been suggested (Schneirla, 1965) whereby stimuli of one sensory modality maturing at a later developmental stage become effective, or more effective, by temporal association with stimuli pertaining to a sensory system functioning at an earlier stage. Schneirla cites as an example Lehrman's observation on the redwinged blackbird, for which he postulates that the parent's form may become effective in eliciting feeding responses by the young through association with vibratory stimuli in the late nonvisual phase. Some experimental evidence has been obtained for kittens that associations are formed between olfactory and visual cues during a developmental period in which orientation to the "home" is determined primarily by smell (Rosenblatt, Turkewitz, & Schneirla, 1969).

Similarly, e.g., embryonic responsiveness to calls, emitted in conjunction with egg shifting, providing tactile, vestibular, and proprioceptive stimulation, may be facilitated by such associations. Whether such facilitation would follow the pattern of classical conditioning is another question. Gos (1935), Hunt (1949), and Sedláček (1964) have demonstrated that classical conditioning to sound is possible during late embryonic stages, but it is difficult to see how these experiments can be interpreted within the more naturalistic context which this problem entails.

Finally, studies will have to be performed to show whether and how such prenatal stimulation influences later prenatal and early postnatal behavior.

II. Sensory Stimulation and Later Behavior

A. *Review of Some Previous Studies*

There is experimental evidence on the fact that during prehatching stages sensory stimulation influences the embryos' behavior and perception and that some of these effects express themselves in later prenatal and early postnatal behavior. The precise stages of the development during which such long-term effects are established have not always been examined. In the articles cited below references are included in which prenatal treatments were carried out during the last 7–9 days of incubation, yet it is uncertain whether effects on later behavior resulted from this whole period. As for earlier embryonic stages most experimental investigations were not related to, nor directly based on, the kinds of stimulation provided by an incubating bird.

Gold (1971) found effects of egg turning on hatching times and some aspects of social attachment. Dimond (1970), Adam and Dimond (1971), and Gold (1971) demonstrated effects of egg lighting on behavior in an imprinting situation. Vince (1969 and in this volume) showed most elegantly how "clicking" in quail embryos functions to synchronize the hatching of the brood.

For studies on the effects of prenatal sound stimulation on later prenatal and early postnatal behavior, a clear distinction should be made between the differential effects that such stimulation may have. While the ability to discriminate between two types of calls and the strength or frequency of response to one type of call may go hand in hand, the former is reminiscent of the directing or orienting properties of a stimulus, while the latter of its eliciting properties, a distinction made in ethology (Franck, 1966; Tinbergen, 1951).

Gottlieb (1971) presented evidence on the selective responsiveness to conspecific maternal calls in duck embryos and the role of hearing self-vocalizations or the calls of brood-mates for the development of this selective response in ducklings, as displayed in simultaneous choice-discrimination tests after hatching. Grier, Counter, and Shearer (1967) exposed chicken embryos to artificial low-frequency sounds. This experience increased the postnatal attractiveness of the familiar sound over a novel sound presented successively.

On the other hand, Gottlieb (1971) found that, in the peking duck embryo, exposure to the vocalizations of brood-mates or self affected the eliciting properties of the conspecific maternal call later prenatally. Prior to the exposure to such vocalizations, the maternal call exerts an inhibitory influence on the rate of bill-clapping; subsequent to exposure, the maternal call exerts an activating influence. Research by the senior author (Impekoven, 1973b) shows that prenatal exposure to conspecific maternal calls significantly increases the strength of responsiveness to such calls after hatching, as measured by the distance traveled toward the sound source. With respect to the laughing gull, a study will be reviewed here on the effect of prenatal experience of parental vocalizations in affecting the frequency of pecking in neonatal chicks (Impekoven, 1971).

Furthermore, while Gottlieb directed his attention to questions of species identification, Tschanz (1968) has been concerned with the prenatal origins of individual recognition. Concerning gulls, species recognition has not been examined, but individual recognition has been established postnatally in the black-billed, the ring-billed and the laughing gull (Evans, 1970a and 1970b; Beer, 1970a, 1970b). An investigation on possible prenatal origins of such recognition will be presented in this paper.

B. Prenatal Experience of Parental Calls and Pecking in the Laughing Gull

Experiments by C. G. Beer (personal communication) show that laughing gull chicks, raised communally in incubators till 24 hours after hatching, in contrast to chicks raised by their own parents, do not approach "crooning," the call with which the parents lure their young, but crouch. This finding supports the idea that prenatal (and/or early postnatal) experience of parental calls is a necessary prerequisite for their later effectiveness in enhancing filial responses.

"Crooning" and intermediates between "uhr" calls and "crooning" (Fig. 2a, b,c), are uttered with increasing frequency prior to hatching (see Section I). Shortly after hatching adult gulls "croon" in the context of feeding their young. On the first day after hatching the chicks usually do not locomote but snuggle or sit nestled against the parent's plumage. Before feeding, the parent adopts a stooping posture so that its beak is presented directly in front of the chicks' eyes. According to Beer (1970a, 1970b) and my own observations (Impekoven, 1971), the parents then croon when the chicks do not peck immediately. This call appears to enhance the pecking at the parent's bill-tip (or the food the parent presents to them). Chicks of a clutch of eggs may hatch on successive days; thus late-hatchers are exposed to "crooning" uttered during such feedings while still in the egg. An experiment was designed to discover whether "crooning" perceived prior to hatching would have any bearing on its effectiveness in eliciting pecking after hatching (Impekoven, 1971). A control experiment was carried out using "kow" calls in order to see whether the effect of "crooning" would be a discriminative one.

Eggs were incubated communally. One group of embryos was exposed repeatedly to "crooning" during the later stages of pipping, another group to "kow" calls (prenatal treatment:sound); a third group was exposed to no calls at all (no sound) (Table III). Postnatally, all groups were confronted with a pecking situation with a red stimulus rod (in an attempt to imitate the parent's red beak). Each chick was tested first for $\frac{1}{2}$ minute with sound, and then for $\frac{1}{2}$ minute without sound, or vice versa. Chicks with prenatal "crooning" experience pecked more often when hearing this call than when not, while the reverse was true for chicks without this experience. However, the total pecking scores (during sound and no-sound conditions combined) were not different under the two treatments. Chicks pecked less frequently when hearing "kow" calls as compared to silent periods, regardless of prenatal experience with this call (Table III).

The results show that late prenatal experience of "crooning" affects the postnatal responsiveness to this call. As to the question whether "crooning" heard prenatally enhances the pecking response, the results do not provide a clear answer. In this study the eggs were collected at widely different stages of incubation; thus, they may have had differential auditory experience prior to the controlled experimental exposures. In connection with the experiments presented in part 1, it would be interesting to explore how long-term exposures to "crooning" and other calls beginning much earlier in incubation and lasting through to hatching would affect later pecking and approach behavior.

C. *Prenatal Origins of Personal Recognition, a Comparison between the Laughing Gull and the Guillemot* (Uria aalge)

"Crooning" does not clearly differ between individual laughing gulls according to spectrographic analysis, and Beer (in press) found that chicks raised in the gullery up to 12–24 hours posthatching show no discri-

TABLE III
RESPONSIVENESS OF LAUGHING GULL CHICKS TO "CROONING" AND "KOW" CALLS IN A PECKING TEST[a]

Call	Condition	n	Sound	No-sound	p
"Crooning"					
Prenatal	Sound[b]	28	9.25 (5.60)	5.39 (4.80)	<.001
Treatment	No-sound	21	6.19 (6.36)	9.70 (6.13)	<.02
p (Sound *vs.* no-sound)			<.01		
"Kow"					
Prenatal	Sound[b]	12	3.58 (4.85)	8.91 (6.00)	<.01
Treatment	No-sound[c]	12	4.67 (3.49)	11.58 (5.36)	<.001

[a]Mean number of pecks (and SD) toward a red stimulus rod. Statistical tests: Analysis of variance on square-root transformed scores for comparison between different test conditions, Mann-Whitney U for comparison between groups with different prenatal experience (p values, two-tailed).

[b]Difference between prenatally "crooning" and "kow" experienced chicks, $p < .002$.

[c]These chicks had been tested with "crooning" before, which may have affected their performance in the subsequent test with "kow" calls.

mination between the "croon" of a parent and a neighbor. In some other gull species (Evans, 1970a and b), "crooning" differs quite markedly between individuals, and personal recognition on the basis of this call has been well established postnatally. In guillemots (*Uria aalge*), a species living in dense colonies on bare cliff ledges without nests, Tschanz (1968) showed beautifully that embryos learn during pipping the characteristics of their parents' luring calls (which after hatching function like "crooning" in the laughing gull). This call consists of a string of pulses of sound. The duration of the pulses and the duration of the intervals between them vary within a call, in a pattern that is characteristic for an individual. Furthermore, the calls differ in their pitch and timbre.

Tschanz first established that neonatal chicks raised in the wild vocalize to and approach discriminately a loudspeaker emitting the parental luring call. Subsequent experiments were designed to discover whether embryos raised in incubators could be "taught" to respond selectively to one (or two) parental calls to which they were repeatedly exposed during pipping.

In a first experiment, embryos were exposed to a series of five luring calls of an adult bird, once every hour during daytime, from the time they began to vocalize in the unpipped egg until a time when the pipping hole was enlarged, i.e., over a period of about $2\frac{1}{2}$ days. Each embryo was exposed individually. It was taken from the incubator and placed in front of the speaker. Embryos were treated in this way between 13–41 times, on the average 24 times. During late pipping the embryos were presented with successive discrimination tests between the familiar and an unknown luring call. The test tape consisted of a series of five calls of each kind with pauses of 5 seconds between successive calls. With this method, it was found that the embryos vocalized more frequently in response to the familiar call than to a novel call, and that a larger number of familiar calls were responded to. In another experiment it was shown that embryos that were not exposed to sound during the stages in which the pipping hole was formed and enlarged, subsequently were less responsive to sounds played to them than embryos with such experience.

Postnatally some chicks were tested in a simultaneous choice-discrimination test. They were tested between 0–23 hours after hatching, so that about 3–42 hours had elapsed between the last prenatal exposure and the postnatal test. It was noted that the chicks called, turned toward the speaker, approached it, snuggled up against it, and pecked at it in response to the familiar call. The only quantitative measure taken was which call they approached, and it was found that the young very clearly preferred the call they had been exposed to prenatally. By means of successive discrimination tests, it was shown that prenatal experience with a particular luring call did not increase responsiveness to such calls in general, but only to the familiar call.

Without presenting individual data on this, it is stated that in exceptional cases three exposures (to five calls each) in 3 hours were sufficient to establish a preference and that once the preferences were formed, they were retained over a long time. This finding is exemplified by chicks that 42 hours later preferred the call learned in the egg to an unknown one, even if in the meantime they were exposed to a second call which they learned in addition. Birds that prenatally were not exposed to a luring call were less likely to approach the calls after hatching, and they showed no preferences between luring calls of different individuals.

According to Tschanz, guillemot chicks do not develop selective responsiveness to any of the other parental calls until just before they leave their rearing site, i.e., leap from the cliff ledge a few hundred yards down into the sea where they rejoin their parents. At this stage they learn the individual characteristics of a call referred to as "crowing," which initiates and accompanies this spectacular exodus. [For a more extensive review of Tschanz's findings, see Beer (1972) and Vince, this volume.]

In the laughing gull, the call which differs most clearly between individuals is the "long" call (see Section I). This call, frequently heralded by a note referred to as "ke-hah," is composed of a string of short notes followed by a string of long notes and is often terminated by one or more "head-toss" notes. The number of short notes, their frequency modulation, and the duration of the intervals between them vary most clearly between individuals (Beer, 1970b) (Fig. 4). Beer (in press) found that 24 hours after hatching already chicks are clearly able to discriminate parents from neighbors on the basis of this call. The ability to discriminate is even better developed after 1–3 weeks (Beer, 1970a, 1970b). Like "crooning," this call is uttered more frequently during the pipping of the eggs than earlier in incubation; thus embryos in the field are exposed to such calls. Shortly after hatching, the "long" call does not seem to serve any obvious function in parent—young interactions, but later, at feeding times, when the parents return to their offspring, hidden in the vegetation, this call, emitted when landing, stimulates the chicks to vocalize in return and thus to inform their parents of their exact location. Experiments were designed, approximating Tschanz's methods, to discover whether gull embryos can learn the personal characteristics of a "long" call.

1. Methods

a. Subjects. The embryos were incubated in force-draft incubators at 37–38°C and 70% relative humidity in a room out of earshot of other extraneous noises. Some of these embryos had been used earlier for experiments described in Section I, but it was found that their performance during the tests did not differ in any significant way from embryos not previously tested. After hatching, the chicks were isolated in small cardboard boxes. Nineteen (out of 34) chicks were kept in a room out of earshot of any noises other than the vocalizations of their brood-mates. Fifteen chicks were kept in a room from which human voices were audible at times. Inspection of the results does not reveal any significant differences between these two groups.

b. Apparatus. A Uher 4200-L tape recorder with Telmar loudspeakers, Model VS 300, and a Sony Stereo tape recorder with loudspeakers were used. The "training" tape consisted of a series of five "long" calls (either 1 or 2) (Fig. 4) alternating with five "croons." The whole sequence was 2 minutes long. The prenatal test tape consisted of a sequence of two "long" calls 1, two "long" calls 2, and the same over again. After each "long" call, a pause of 16 seconds was inserted. The whole test tape lasted 80 seconds. The postnatal test tape consisted of 10 calls, either "long" call 1, or 2, or both alternating. After each call there was a pause of approximately 8 seconds. The whole tape lasted 2 minutes. Sound level measurements were made on a General Radio Soundlevel meter, scale B, slow. All tapes were played at 80–85 dB, with a background noise level of 50 dB or less. In the prenatal tests, bill-clapping was recorded by means of a Rustrak event recorder. Postnatally the behavior was recorded on specially designed recording sheets. Vocalizations were recorded both pre- and postnatally on tape. Embryos were tested either in a sound-attenuated box with a transparent top, at approximately incubation temperature, or under a heat lamp on a table. Chicks were tested in an "arena" (30 × 160 cm) painted in black (modified after Beer, 1970a, 1970b). Speakers were placed on opposite ends of the runway, and the distance from the center to either of the speakers measured 50 cm. This test was carried out in a room where the temperature ranged between 20–24° C.

c. Procedure. Embryos were exposed singly or in age-matched groups to either of the two training tapes. They were taken from the incubator and carried to another room (out of earshot) where they were placed in front of the speaker. One half of the embryos were exposed to "long" call 1, the other half to "long" call 2. A control group was repeatedly taken from the incubator and exposed to no sound. Exposures were started at the first signs of pipping (corresponding to tucking, see Oppenheim, 1972) and were continued until the beginning of hatching every hour over a 12 hour day. Thus, each embryo was exposed between 5 and 38 times, on the average 27 times. (These figures are similar or somewhat lower than the number of parental "long" calls that embryos would experience in the gullery.) the prenatal discrimination test was carried out during late pipping (i.e., when a large pipping hole had been formed). Each embryo was tested singly starting with 1 minute of no-sound. Then the tape was switched on, and for 80 seconds the described sequence of the two "long" calls was played. This exposure was again followed by a minute of no-sound. During the whole test bill-clapping (as visible through the enlarged pipping hole) and vocalizations were recorded. Postnatally chicks were tested at the age of 5–12 hours, i.e., about 8–24 hours after the last prenatal exposure, either in two successive discrimination tests (9 chicks) or in one simultaneous choice discrimination test (25 chicks).

Each chick was placed singly into the center of the arena at a right angle to the speakers and was first observed during 1 minute of no-sound. Subsequently the tape was switched on for a duration of 2 minutes. The test was concluded after another minute of no-sound. Vocalizations, orientation with respect to the sound source, and position in the arena were recorded during no-sound periods as well as during and after each "long" call. Qualitative notes were taken on pecking, snuggling, and crouching.

Behavior patterns were compared between call periods and no-sound periods preceding and following such call periods. The frequency of bill-clapping and vocalizations was subsequently calculated as per minute of exposure of either sound or no-sound condition. The statistical tests used in the data analysis were the Wilcoxon matched pair, signed ranks for comparison between different test conditions, and the Mann-Whitney U for comparison between groups with different prenatal experience (two-tailed *p*-values).

B. Results

1. Prenatal Discrimination Tests (Table IV)

The rate of bill-clapping of the no-sound group tended to be lower during "long" call periods than during the no-sound pre- and posttest. This group did not differentiate between "long" call 1 and "long" call 2 (not shown on table). Embryos with prenatal exposures to either "long" call clapped more during the familiar call than during silent periods and more in response to the familiar than the unfamiliar call.

In contrast to no-sound experienced embryos, the frequency of bill-clapping was higher in the presence of either "long" call in sound-experienced embryos. Even though the baseline values during no-sound conditions were also somewhat higher, the differences are far from reaching statistical significance.

Embryos with no-sound experience vocalized as frequently during the sound as during the no-sound condition and with equal frequency during "long" call 1 and "long" call 2 exposures. Embryos with "long"call experience showed a tendency to vocalize more frequently during the familiar call than during the no-sound period and during the unfamiliar call. However, as compared to the control value in the no-sound experienced embryos, the rate of calling was not significantly raised to either "long" call.

2. Postnatal Discrimination Tests (Table V)

The rate of vocalization of no-sound experienced chicks was depressed in the presence of "long" calls, as compared with no-sound periods. The rate did not differ between "long" call 1 and "long" call 2 (not shown on Table V). Embryos with prenatal sound exposure vocalized more frequently in response to the familiar "long" call than during no-sound periods and more frequently during the familiar call than during the unfamiliar one. Compared to no-sound experienced chicks, the rate of vocalization in sound-experienced chicks was significantly higher in the presence of both "long" calls.

Only 8 of 14 no-sound experienced chicks oriented toward the calls, while 19 of 20 sound-experienced chicks turned toward either of the speakers. Sound-experienced chicks did not display a preference for the familiar call with respect to orientation or position. Many chicks did not show any signs

TABLE IV
PRENATAL DISCRIMINATION TESTS WITH "LONG" CALL 1 AND 2[a]

Behavior	Prenatal treatment	Prenatal test condition					
		n	No-sound	*p*	1 or 2 (Familiar)	*p*	1 or 2 (Unfamiliar)
Bill claps	No-sound	20	13.37 (9.15)	$= .1$	11.63 (11.39)		11.63 (11.39)
	Sound (1 or 2)	28	16.73[b] (12.97)	$< .005$	20.84 (15.57)	$< .01$	15.54[b] (12.50)
p (No sound *vs.* sound)			NS		$< .001$		$< .05$
Vocalizations	No-sound	20	4.62 (6.83)		4.62 (3.84)		4.62 (3.84)
	Sound (1 or 2)	28	4.41[c] (7.31)	$= .1$	5.11 (6.44)	$= .1$	4.39[c] (4.91)
p (No sound *vs.* sound)			NS		NS		NS

[a]Mean number of bill claps and vocalizations per minute (and SD).
[b]NS.
[c]NS.

TABLE V
POSTNATAL DISCRIMINATION TESTS WITH "LONG" CALL 1 AND 2[a]

Prenatal treatment	N	Postnatal test condition: No-sound	p	1 or 2 (Familiar)	p	1 or 2 (Unfamiliar)
No sound	14	29.17 (16.55)	<.02	22.43 (15.14)		22.43 (15.14)
Sound (1 or 2)	20	30.88[b] (13.17)	<.01	35.82 (14.17)	<.1	33.40[b] (15.45)
(Sound *vs.* no sound)		NS		<.005		<.02

[a]Mean number of vocalizations per minute (and SD).
[b]NS.

of approach. Five of 14 no-sound experienced chicks and 8 of 20 sound-experienced chicks shuffled as far as 5–10 cm toward or away from the sound source.

Some of the chicks that had not been exposed prenatally to "long" calls and "crooning" crouched in response to the calls postnatally. (Out of 14 birds tested, seven crouched in response to either "long" call. They maintained the crouching position, their belly and chin pressed against the ground, for a prolonged period of time.) Although some of the chicks with prenatal sound exposures also crouched, they tended to maintain this position only for a few seconds at the onset of the sound. (Out of 20 birds tested, six crouched in response to the unfamiliar call, and four out of these six crouched also to the familiar call.) Lastly, pecking and snuggling movements did not seem to reveal any difference between groups with different prenatal treatment and in response to the different calls.

C. *Discussion*

1. PRENATAL TESTS

The vocalizations uttered during no-sound conditions by both sound and no-sound treated embryos may have been due to the testing conditions. In some cases the temperature was lower than in the incubator. Lowering the temperature in itself induces increased vocalization (e.g., Lind, 1961). In other cases where the temperature was kept at incubator levels, very few vocalizations were heard, both in the absence and the presence of sound.

The data show that pipping embryos are able to form individual preferences on the basis of repeated exposure to either of two "long" calls (in conjunction with "crooning"), as displayed by the rate of their bill-clapping, and to a far lesser extent on the basis of their vocalizations.

Since Tschanz does not present any quantitative data on bill-clapping, no comparisons can be made with his embryos. With respect to vocalizations, laughing gull embryos discriminated far less well than guillemots between the calls of two individuals.

2. Postnatal Tests

Similar to the prenatal tests, the vocalization frequently emitted during no-sound conditions may have been due to the low temperature (20–24° C), but in addition also possibly to the strangeness of the visual environment.

It is interesting that birds without prenatal experience of "long" calls postnatally showed a clear-cut suppression of vocalization in the presence of calls. This effect was temporary. In subsequent tests, not reported in the present study, the calling rate was the same between no-sound and sound conditions. With respect to other behavior patterns, activity reducing effects are suspected but not clear. Tschanz has not recorded calling in his postnatal tests; however, the fact that fewer of his no-sound experienced chicks approached the sound source suggests that their activity was also depressed. In contrast to the no-sound group, the call rate of sound-experienced chicks was increased in the presence of calls, as compared to silent periods.

Personal recognition is indicated merely on the basis of differential rates of vocalization. This result differs from Beer's (in press) with field-raised chicks 24 hours after hatching who recognize their parents' voices clearly, as shown by their orientation and position but not their vocalizations. The present findings also differ much from Tschanz's, suggesting that laughing gulls are less able than guillemots to form and retain preferences, based on prenatal experience with individual calls.

If one distinguishes again between eliciting and directing properties of the stimulus, in this particular case a "long" call, the present results imply that prenatal experience influences its eliciting rather than its directing aspects.

To no-sound experienced chicks, "long" calls were a novel stimulus and thus may have elicited a "listening" response or "orienting reflex" which is subject to habituation with repeated presentation of the novel stimulus (Sokolov, 1960). Prenatal experience with a "long" call may lead to habituation with respect to long calls in general, shown by the fact that the rate of vocalization was not reduced in response to the unfamiliar call. Exposure to one "long" call leading to habituation with respect to other "long" calls implies stimulus generalization. The extent of such generalization needs to be further explored.

Beyond habituation to "long" calls in general, hearing repeatedly a specific "long" call establishes a slight preference for this call, shown by the fact that the rate of vocalization was increased toward the familiar call as compared with the unfamiliar call and no-sound periods.

Concerning the establishment of individual preferences, the present findings should be complemented by a study in the wild. By taking eggs from the gullery during a late pipping stage and testing them both pre- and early postnatally with the calls of their parents versus calls of neighbors, one would be able to prove or disprove the validity of the present results with respect to the natural situation.

There are several possible conditions under which the first signs of personal recognition may develop prenatally, but experimental work on this issue is still pending.

1. It is conceivable that personal recognition, i.e., familiarization with particular calls, develops on the basis of sheer exposure to certain calls. Such a process appears to be tacitly implied by Grier *et al.'s* (1967) findings on auditory imprinting in the domestic chicken. It is assumed that postnatal imprinting to visual stimuli occurs in this way (reviewed by Bateson, 1966).

2. Another possibility for the development of personal preferences is through association of certain calls with movement, light, and temperature change, as occurring naturally during the shifting of the eggs, and in the experiments by taking the eggs from the incubator and placing them in front of the speaker.

3. Third, preferences for a "long" call might develop on the basis of its association with "crooning," with which it was paired during the training periods. "Crooning," unlike "long" calls, increases responsiveness prenatally without prior exposure to this call (see Section I; also see Impekoven, 1971).

4. As a last possibility, personal preferences might arise on the basis of response contingencies between the parent and the embryo: the present experimental design according to which embryos were exposed to calls out of earshot of other embryos does not resemble the natural situation. In the gullery, various gulls are calling most of the time. If the embryos should begin to learn anything about their parents' calls, it would appear that this is due to the fact that their activity cycle coincides in time with that of their parents. When the embryos are most active, either "spontaneously" or as a consequence of being exposed to light and egg shifting by the parent, this stimulates the parent to respond by vocalizing, while other birds may call also, yet at times not coinciding with activity periods other than those of their own embryos. This fact has been pointed out by Tschanz (1968) for the guillemot, and it holds for gulls and probably many other species as well.

The question about functions of the response contingencies between the incubating parent and its embryos led the senior author to an experimental study in Peking ducks in which maternal calls of the mallard or the chicken were made contingent upon the embryos' motor activity. While this study provides evidence that responsiveness is selectively increased to the conspecific maternal call as a function of such a contingency, as compared with yoked controls (for which the calls were noncontingent), no discrimination tests were performed to explore whether individual characteristics of the test call had been learned (Impekoven, 1973b).

Finally, in connection with the difference found between gulls and guillemots, the differential needs of these two species must be stressed once more: guillemots live almost shoulder-by-shoulder, without a defined nest area. Thus the young might easily get mixed up or lost soon after hatching if no personal ties existed. Laughing gulls live grouped but spaced out one to several meters with no danger of meeting a strange gull until several days after hatching. As a consequence, less selection pressure may have been exerted on this species toward learning quickly and retaining the learned calls over long periods of time. However, the prenatal discrimination tests in particular do suggest that even where such pressure is not as striking as in guillemots, a disposition toward learning specifics about a call at this stage nevertheless exists.

Guillemots belonging to the family Alcidae should really not be compared with the laughing gull, a member of the family Laridae which is only distantly related and thus differs in many ways, but rather with a close relative such as the Razorbill (*Alca torda*). Such a comparison is now under way. The first findings indicate that razorbills, living more spaced out than guillemots, do not clearly learn their parents' individual characteristics during pipping. The conclusions derived from such results concerning behavioral adaptedness to the species-typical environment are in line with the ones presented for the laughing gull (Tschanz, personal communication).

III. Concluding Remarks

It is not meant as an affront to other approaches to point out that they have not dealt with certain kinds of questions. Our aim has not been to criticize experimental embryologists, but only to draw attention to the possible role of the natural (species-typical) environment in structuring the embryo's behavior and perception; or, in other words, to illustrate possible relationship between problems of prenatal development and adaptedness.

IV. Summary

In the past little attention has been given to the natural (species typical) sources of extraovular stimulation, since most studies have been carried out on domesticated species raised in artificial environments. The range and patterns of stimuli of possible relevance to the embryo's immediate ongoing and later behavior becomes more restricted if one does not take into account the kinds of stimuli and incubating hen might provide.

The present study presents qualitative and quantitative data on the incubation behavior of wild birds with special reference to gulls. On the basis of such observations, questions were raised about the possible effect of stimulation provided by an incubating bird. Emphasis is placed on the auditory modality (parental calls), but other types of stimulation occurring in association with rising and resettling on the eggs are also briefly discussed.

An experiment was designed to discover whether and how laughing gull embryos would respond to certain parental calls to which they are naturally exposed during incubation. It was found that one such call ("crooning" in Fig. 2) enhances aspects of embryonic motility, another call ("kow" in Fig. 3) appears to have depressing effects (Table II) as early as day 15 (62.5–65% of the completed incubation period). These results are discussed in the light of the relevant literature and suggestions for future research are made.

Another experiment carried out earlier by the senior author is reviewed which shows that late prenatal experience of parental calls ("crooning") bears on their effectiveness in eliciting feeding responses in neonatal gull chicks (Table III).

A third experiment presents evidence on prenatal origins of personal recognition of certain parental calls ("long" calls, Fig. 4) (Tables IV and V). These findings are compared with the ones of a previous study on this subject carried out with guillemots, and the species differences in the apparent ability to form and retain individual preferences at this stage are discussed in view of their adaptive significance.

Acknowledgments

Part of this work was carried out in the Brigantine Wildlife Refuge. We are grateful to the United States Fish and Wildlife Service for permission to work in the Refuge and to collect eggs, and to the Refuge manager and personnel for their hospitality and cooperation.

We also wish to thank Miss Heidi Hardacker for her assistance in carrying out some of the experiments. The senior author is indebted to Dr. J. S. Rosenblatt for comments on the manuscript.

The research was supported in part by PHS-Grant GM-16727 (to Dr. C. G. Beer), by NIGMS Training Grant GM-1135, by a grant from The Neurosciences Program of the Alfred P. Sloan Foundation, and in part by a Faculty Research Fellowship from the Research Foundation of the State University of New York (to P.S.G.). The present paper constitutes publication No. 143 from the Institute of Animal Behavior.

References

Adam, J., & Dimond, S. J. The effect of visual stimulation at different stages of embryonic development on approach behaviour. *Animal Behaviour*, 1971, **19**, 51–54.

Baerends, G. P. An ethological analysis of incubation behaviour in the herring gull. *Behaviour, Supplement*, 1970, **17**, 135–234.

Bateson, P. P. G. The characteristics and context of imprinting. *Biological Reviews of the Cambridge Philosophical Society*, 1966, **41**, 177–210.

Beer, C. G. Incubation and nest-building behaviour of black-headed gulls. I. Incubation behaviour in the incubation period. *Behaviour*, 1961, **18**, 62–106.

Beer, C. G. Ethology—the zoologist's approach to behaviour. Part 1. *Tuatara*, 1963, **11**, 170–177.

Beer, C. G. Ethology—the zoologist's approach to behaviour. Part 2. *Tuatara*, 1964, **12**, 16–39.

Beer, C. G. Ethology on the couch. *Science and Psychoanalysis*, 1968, **12**, 198–213.

Beer, C. G. On the responses of laughing gull chicks (*Larus atricilla*) to the calls of adults. I. Recognition of the voices of the parents. *Animal Behaviour*, 1970, **18**, 652–660. (a)

Beer, C. G. On the responses of laughing gull chicks (*Larus atricilla*) to the calls of the adults. II. Age changes and responses to different types of call. *Animal Behavior*, 1970, **18**, 661–677. (b)

Beer, C. G. Individual recognition of voice and its development in birds. *Proceedings of the 15th International Ornithological Congress*, 1972. Pp. 339–356.

Beer, C. G. A view of birds. Minnesota symposia on child Psychology. J. P. Hill (ed.). Vol. 7. The University of Minnesota Press. Minneapolis (in press).

Bursian, A. V. The influence of light on the spontaneous movements of chick embryos. *Bulletin of Experimental Biology and Medicine* (*USSR*), 1964, **58**, 7–11.

Decker, J. D. The influence of early extirpation of the otocysts on development of behavior in the chick. *Journal of Experimental Zoology*, 1970, **174**, 349–364.

Dimond, S. J. Visual experience and early social behavior. In J. H. Crook (Ed.), *Social behaviour in birds and mammals*. New York: Academic Press, 1970.

Drent, R. H. Functional aspects of incubation in the herring gull (*Larus argentatus*). *Behaviour Supplement*, 1970, **17**, 1–132.

Drent, R. H., Postuma, K., & Joustra, T. The effect of egg temperature on incubation behaviour in the herring gull. *Behaviour, Supplement*, 1970, **17**, 239–258.

Evans, R. M. Parental recognition of the "mew call" in black-billed gulls (*Larus bulleri*). *Auk*, 1970, **87**, 503–513. (a)

Evans, R. M. Imprinting and the control of mobility in younf ring-billed gulls (*Larus delawarensis*). *Animal Behaviour Monographs*, 1970, **3**, 195–248. (b)

Franck, B. Möglichkeiten zur vergleichenden Analyse auslösender und richtender Reize mit Hilfe des Attrappenversuchs, ein Vergleich der Successiv- und Simultanmethode. *Behaviour*, 1966, **27**, 150–159.

Gold, P. S. The effects of sensory stimulation during embryogeny on growth, hatching and imprinting in the domestic chick. Unpublished doctoral dissertation, New York University, 1971.

Gos, M. Les réflexes conditionnels chez l'embryon d'oiseau. *Bulletin de la Société Royale des Sciences de Liège*, 1935, **4**, 194–199, 246–250. Cited by Gottlieb (1968), p. 156.

Gottlieb, G. Prenatal auditory sensitivity in chickens and ducks. *Science*, 1965, **147**, 1596–1598.

Gottlieb, G. Prenatal behaviour of birds. *Quarterly Review of Biology*, 1968, **43**, 148–174.

Gottlieb, G. Conceptions of prenatal behavior. In L. R. Aronson, E. Tobach, D. S. Lehrman, & J. S. Rosenblatt (Eds.), *Development and evolution of behavior*. San Francisco: Freeman, 1970. Pp. 111–137.

Gottlieb, G. *Development of species identification of birds. An inquiry into the prenatal determinants of perception.* Chicago: University of Chicago Press, 1971.

Gottlieb, G., & Kuo, Z. Y. Development of behavior in the duck embryo. *Journal of Comparative and Physiological Psychology*, 1965, **59**, 183–188.

Grier, J. B., Counter, S. A., & Shearer, W. M. Prenatal auditory imprinting in chickens. *Science*, 1967, **155**, 1692–1693.

Guyomarc'h, J.-C. Les bases ontogénétiques de l'attractivité du gloussement maternel chez la poule domestique. *Revue de Comportement Animal*, 1972, **6**, 79–94.

Hailman, J. P. The ontogeny of an instinct. The pecking response in chicks of the laughing gull and related species. *Behaviour, Supplement*, 1967, **15**, 1–159.

Hamburger, V. Some aspects of the embryology of behaviour. *Quarterly Review of Biology*, 1963, **138**, 342–365.

Hamburger, V. The beginnings of coordinated movements in the chick embryo. In C. E. W. Wolstenholme & M. O'Connor (Eds.), *Ciba Foundation symposium on growth of the nervous system*. London: Churchill, 1968.

Hamburger, V. Embryonic motility in vertebrates. In O. Schmitt (Ed.), *The neurosciences: Second study program*. New York: Rockefeller University Press, 1970. Pp. 141–151.

Hamburger, V. Motility in birds and mammals. In G. Gottlieb (Ed.), *Developmental Studies of Behavior and the Nervous System*, Vol. 1. New York: Academic Press, 1973.

Hamburger, V., & Narayanan, C. H. Effects of the deafferentation of the trigeminal area on the motility of the chick embryo. *Journal of Experimental Zoology*, 1969, **170**, 411–416.

Hamburger, V., & Oppenheim, R. W. Prehatching motility and hatching behavior in the chick. *Journal of Experimental Zoology*, 1967, **166**, 171–204.

Hamburger, V., Wenger, E., & Oppenheim, R. W. Motility in the chick embryo in the absence of sensory input. *Journal of Experimental Zoology*, 1966, **162**, 133–160.

Heaton, M. B. Early visual function in Bobwhite and Japanese quail embryos as reflected by the pupillary reflex. *Developmental Psychobiology*, 1971, **4**, 313–332. (a)

Heaton, M. B. Stimulus coding in the species-specfic perception of Peking ducklings. An analysis of critical elements and their prenatal emergence. Unpublished doctoral dissertation, North Carolina State University, 1971. (b)

Hinde, R. A. *Animal behaviour. A synthesis of ethology and comparative psychology*. (2nd ed.) New York: McGraw-Hill, 1970.

Hunt, E. L. Establishment of conditioned responses in chick embryos. *Journal of Comparative Psychology*, 1949, **42**, 107–117. Cited by Gottlieb (1968), p. 156.

Impekoven, M. Prenatal experience of parental calls and pecking in the laughing gull chick (*Larus atricilla*). *Animal Behaviour*, 1971, **19**, 471–476.

Impekoven, M. The response of incubating laughing gulls (*Larus atricilla*) to calls of hatching chicks. *Behaviour*, in press, 1973. (a)

Impekoven, M. Response-contingent prenatal experience of maternal calls in the Peking duck (*Anas platyrhynchos*). *Animal Behaviour*, 1973, **21**, 164–168.

Kovach, J. K. Spatial orientation of the chick embryo during the last five days of incubation. *Journal of Comparative and Physiological Psychology*, 1968, **66**, 283–288.

Kovach, J. K. Development and mechanisms of behavior in the chick embryo during the last five days of incubation. *Journal of Comparative and Physiological Psychology*, 1970, **73**, 392–406.

Kuo, Z. Y. Ontogeny of embryonic behavior in Aves: I. Chronology and general nature of behavior of chick embryos. *Journal of Experimental Zoology*, 1932, **61**, 395–430.

Kuo, Z.-Y. *The dynamics of behavior development. An epigenetic view*. New York: Random House, 1967.

Lehrman, D. S. Semantic and conceptual issues in the nature-nurture problem. In L. R. Aronson, E. Tobach, D. S. Lehrman, & J. S. Rosenblatt (Eds.), *Development and evolution of behavior*. San Francisco: Freeman, 1970.

Lind, H. Studies on the behaviour of the black-tailed godwit (*Limosa limosa*). *Meddelelse fra Naturfredningsradets Reservatudvalg (Copenhagen)*, 1961, **66**, 1–157.

Olsen, M. Influence of turning and other factors on the hatching power of hen's eggs. Unpublished master's thesis, Iowa State College, 1930.

Oppenheim, R. W. Amniotic contractions and embryonic motility in the chick embryo. *Science*, 1966, **152**, 528–529.

Oppenheim, R. W. Light responsivity in chick and duck embryos just prior to hatching. *Animal Behaviour*, 1968, **16**, 276–280.

Oppenheim, R. W. An experimental investigation of the possible role of tactile and proprioceptive stimulation in certain aspects of embryonic behavior in the chick. *Developmental Psychobiology*, 1972, **5**, 71–91. (a)

Oppenheim, R. W. Prehatching and hatching behaviour in birds: A comparative study of altricial and precocial species. *Animal Behaviour*, 1972, **20**, 644–655. (b)

Oppenheim, R. W. The embryology of behavior in birds: A critical review of the role of sensory stimulation in embryonic movement. *Proceedings of the 15th International Ornithological Congress*, 1972. Pp. 283–302. (c)

Quine, D. A., & Cullen, J. M. The pecking response of young Arctic terns and the adaptiveness of the releasing mechanism. *Ibis*, 1964, **166**, 145–173.

Rogallo, V. L. Measurement of the heartbeat of bird-embryos with micro-meteorite transducer. *NASA Technology Utilization Report*. 1964, **SP-5007**.

Rosenblatt, J. S., Turkewitz, G., & Schneirla, T. C. Development of home orientation in newly born kittens. *Transactions of the New York Academy of Sciences*, 1969, **31**, 231–250.

Schleidt, W. M. Wirkungen aeusserer Faktoren auf das Verhalten. *Fortschritte der Zoologie*, 1963, **16**, 469–499.

Schneirla, T. C. Aspects of stimulation and organization in approach/withdrawal processes underlying vertebrate behavioral development. In D. S. Lehrman, R. A. Hinde, & E. Shaw (Eds.), *Advances in the study of behavior*. Vol. 1. New York: Academic Press, 1965.

Sedláček, J. 1964. Further findings on the conditions of formation of the temporary connection in chick embryos. *Physiologia Bohemoslovenica*, 1964, **13**, 411–420.

Sokolov, E. N. Neuronal models and the orienting reflex. In M. A. B. Brazier (Ed.), *The central nervous system and behavior*. Madison, Wis.: Madison Printing Co., 1960. Pp. 187–276.

Sviderskaya, G. E. Effect of sound on the motor activity of chick embryos. *Bulletin of Experimental Biology and Medicine (USSR)*, 1967, **63**, 24–28.

Tinbergen, N. *The study of instinct*. London: Oxford University Press (Clarendon), 1951.

Tinbergen, N. On aims and methods of ethology. *Zeitschrift für Tierpsychologie*, 1963, **20**, 410–432.

Tinbergen, N. On war and peace in animals and man. *Science*, 1968, **160**, 1411–1418.

Tinbergen, N., & Perdeck, A.C. On the stimulus situation releasing the begging response in the newly hatched herring gull chick (*Larus argentatus argentatus*). *Behaviour*, 1950, **3**, 1–39.

Tschanz, B. Trottellummen. Die Entstehung der persoenlichen Beziehung zwischen Jungvogel und Eltern. *Zeitschrift für Tierpsychologie*, 1968, Beiheft (= Suppl.) **4**, 1–103.

Vanzulli, A., & Garcia-Austt, E. Development of cochlear microphonic potentials in the chick embryo. *Acta Neurologica Latinoamericana*, 1963, **9**, 19–23.

Vince, M. A. Embryonic communication, respiration and the synchronisation of hatching. In R. A. Hinde (Ed.), *Bird vocalizations*. London: Cambridge University Press, 1969.

Vince, M. A. Some environmental effects on the activity and development of the avian embryo. In G. Gottlieb, (Ed.), *Developmental Studies of Behavior and the Nervous System*, Vol. 1. New York: Academic Press, 1973.

Visintini, F., & Levi-Montalcini, R. Relazione tra differenciazione strutturale dei centri e delle vie nervose nell'embryone di pollo. *Schweizer Archiv für Neurologie und Psychiatrie*, 1939, **43**, 1–45. Cited by Gottlieb (1968), p. 154.

AUTHOR INDEX

Numbers in italics refer to the pages on which the complete references are listed

SUBJECT INDEX